Processing and Properties of Advanced Ceramics and Composites IV

Processing and Properties of Advanced Ceramics and Composites IV

Ceramic Transactions, Volume 234

Edited by
J. P. Singh
Narottam P. Bansal
Takashi Goto
Jacques Lamon
Sung R. Choi
Morsi M. Mahmoud
Guido Link

The American Ceramic Society

A John Wiley & Sons, Inc., Publication

Published by John Wiley & Sons, Inc., Hoboken, New Jersey.
Published simultaneously in Canada.

For general information on our other products and services or for technical support, please contact our Customer Care Department within the United States at (800) 762-2974, outside the United States at (317) 572-3993 or fax (317) 572-4002.

Wiley also publishes its books in a variety of electronic formats. Some content that appears in print may not be available in electronic formats. For more information about Wiley products, visit our web site at www.wiley.com.

Library of Congress Cataloging-in-Publication Data is available.

ISBN: 978-1-118-27336-4
ISSN: 1042-1122

Printed in the United States of America.

10 9 8 7 6 5 4 3 2 1

Contents

COMBUSTION SYNTHESIS AND SHS PROCESSING

MICROWAVE AND MILLI-METER PROCESSING AND ITS FIELD EFFECTS

COMPOSITES

FOREIGN OBJECT DAMAGE

TESTING, EVALUATION, AND MICROSTRUCTURE–PROPERTY RELATIONSHIPS

MODELING

Preface

This volume contains papers presented at three international symposia—"Innovative Processing and Synthesis of Ceramics, Glasses and Composites," "Advances in Ceramic Matrix Composites," and "Microwave Processing of Materials" held during the Materials Science & Technology 2011 Conference & Exhibition (MS&T'11), Columbus, OH, October 16–20, 2011. These conference symposia provided a forum for scientists, engineers, and technologists to discuss and exchange state-of-the-art ideas, information, and technology on advanced methods and approaches for processing, synthesis and characterization of ceramics, glasses, and composites. Over 100 presentations, including invited and contributed talks as well as posters were made by participants from 16 countries. The speakers represented universities, industries, and government research laboratories.

Thirty two papers comprising aspects of synthesis, processing, properties, and testing of ceramics, glasses, and composites that were discussed at the symposia are included in this proceeding volume. Each manuscript was peer-reviewed using The American Ceramic Society's review process.

The editors wish to extend their gratitude and appreciation to all the authors for their timely submissions and revisions of manuscripts, to all the participants and session chairs for their time and effort, and to all the reviewers for their valuable comments and suggestions. Financial support from The American Ceramic Society (ACerS) is greatly acknowledged. Thanks are due to the staff of the meetings and publications department of ACerS for their invaluable assistance.

We hope that this volume will serve as a useful reference for the professionals working in the field of synthesis and processing of ceramics and composites as well as their properties.

J. P. SINGH
NAROTTAM P. BANSAL
TAKASHI GOTO
JACQUES LAMON
SUNG R. CHOI
MORSI M. MAHMOUD
GUIDO LINK

Synthesis and Processing

EFFECT OF PARTICLE SIZE AND TEMPERATURE ON THE SINTERING BEHAVIOUR OF GLASS COMPACTS

Adele Dzikwi Garkida[1], Jiann-Yang Hwang[2], Xiaodi Huang[2], Allison Hein[2]

1. Department of Industrial Design, Ahmadu Bello University, Zaria, Nigeria.
2. Department of Minerals and Materials Engineering, Michigan Technological University, Houghton, Michigan, USA

ABSTRACT

In this study the compaction and sintering behavior of waste glass powder compacts of different particle sizes was studied. Compacts of waste soda lime silica glass and borosilicate glass were made at various proportions from 106 microns and minus 75 microns powders in combination with 5% bentonite as binder for each glass type, after several trials to determine compatibility of particle sizes. The compacts were sintered at a temperature range of $600^{\circ}C - 800^{\circ}C$ and tested using the American Standard Test Methods for shrinkage, warpage, and absorption. Results show that there was no significant difference as regards the ratio of coarse to fine grains contained in the various compacts and the compacts exhibited better sinterability at a temperature range of $700^{\circ}C - 750^{\circ}C$. The experimental findings suggest that these waste glass composites would be a potential alternative to clay for ceramic tile making.

INTRODUCTION

One of the most effective ways to produce thousands of relatively simple shapes of relatively small pieces is compaction and sintering[1]. Sintering involves the use of powders and the characteristics desired in a powder are largely dictated by the properties required of the product. The demand for impermeability therefore makes severe calls on the control of the sintering process and hence on the nature of the powder itself. All the detailed mechanisms of sintering predict that sintering rates will increase with decreasing particle size, other things being equal[2].

However the sintering temperature of any glass powder is dependent on a lot of factors so a fixed temperature value cannot be given. Such factors are particle size, composition, softening point of parent material, the heating rate, and method of compacting. Powder size and size distribution can have marked effect on the final structure produced as well as factors such as compacting density, efficiency and consolidating time. A reduction in particle size increases the driving force and reduces the mass transport distance for densification and achieves finer grain size in the sintered micro- structure[3], with smaller particles favoring grain boundary and surface diffusion and larger particles favoring bulk diffusion[4].

There is evidence that powders consisting of uniform spheres can yield high density ceramic materials at lower sintering temperatures[5]. Two sizes of glass 150 micron and 5 micron have been used to evaluate their size effect on clay firing at various clay and glass mixture ratios, which resulted in the reduction of the sintering temperature of the clay[6]. Another study showed the effect of initial compaction on the sintering of borosilicate glass matrix composites reinforced with 25 vol. % alumina (Al_2O_3) particles, this has been studied using powder compacts that were uniaxially pressed at 74, 200 and 370 MPa. The sintering behaviour of these samples heated in the temperature range 850–1150 °C showed increase in density, meaning increased in shrinkages[7]. So also composite glass parts produced with 10-20% alumina were

3

found to be sintered close to full density with porosity of less than 0.7% and very low amounts of water absorption (0.002% - 0.27%)[8].

This study sought to convert the glass waste of fluorescent tubes and window glass (soda-lime silica glass) and laboratory glass (borosilicate glass) to functional forms using the sintering method.

METHODOLOGY

The glass samples used in this study were fairly large recycled pieces of window glass and fluorescent tubes (Soda lime silica glass) and laboratory glassware (Borosilicate glass). They were crushed individually by passing them through a hammer mill. Using a hammer mill alone did not produce powders and so they were passed through a roll crusher and rod mill to reduce their sizes further. The crushed samples were sieved (Tyler mesh) with the aid of a mechanical shaker and a shaking time of 20 minutes. Sintering trials were made to determine which grain size would be suitable for subsequent experiments and it was quite evident from the preliminary sintering tests that less than 150 micrometer grain size would be required for subsequent experiments meaning that only meshes 150 to -200 would be needed.

Samples were made ranging from 70 – 30% of 106 microns mixed with 30 – 70%, -75 microns respectively making 5 samples of each type of glass. 5% bentonite was added and also 5% water as bonding agents. Compacts were made using the uniaxial press forming compacts approximately 29 mm in diameter by 9 mm in thickness with a pressing pressure of 10,000 psi.

The pressed pieces were left in the open over a period of 24hrs to dry and a set of samples was fired in an electric kiln at the following temperatures 600°C, 650°C, 700°C, 750°C and 800°C with holding time of two hours. 700 -750°C was determined as a suitable sintering temperature range and as such the subsequent samples were sintered within this temperature range at 700°C, 725°C and 750°C. The sintered pieces were tested for shrinkages, warpage and absorption. The microstructures of some specimens were observed using a JEOL JSM 820 Scanning Electron Microscope (SEM) to show the nature of fusion.

RESULTS AND DISCUSSION

Sintering

There was no compaction at 600°C and not enough fusion at 650°C either that could hold the compact completely together. But at 700°C there was good fusion and all compacts made were hard and had a matt surface. There was no distortion or warpage but shrinkage occurred.

Sintering at 725°C made the compact from fluorescent tube to begin to form a thin layer of glazing. There was also no warpage but further shrinkage. Sintering at 750°C left all compacts with thin layers of glazing, there was also no warpage but further shrinkage occurred. When the sintering temperature was raised to 800°C expansion and shape distortion occurred. Sintering temperature range of 700-750°C was therefore found suitable. The samples sintered at 750°C were very hard and they developed a thin layer of glazing, making it possible to produce a glazed tile all in one process, figures 1a -1c show some of the sintered compacts while figures 2a to 2c show shrinkage results, where S 1 is shrinkage of window glass, S 2 is shrinkage of fluorescent tubes and S 3 is shrinkage of laboratory glass.

Figure 1a: Window Glass Compact containing Bentonite as Binder fused at 700, 725 and 750 respectively

Figure 1b: Fluorescent Tubes Compacts containing Bentonite as Binder fused at 700, 725 and 750 respectively

Figure 1c Laboratory Glass Compacts Containing Bentonite as Binder fused at 700, 725 and 750 respectively

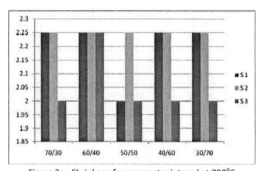

Figure 2a: Shrinkage for compacts sintered at 700°C

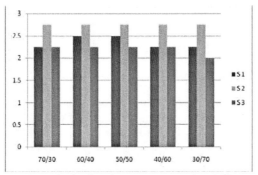

Figure 2b: Shrinkage for compacts sintered at 725°C

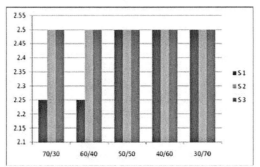

Figure 2c: Shrinkage for compacts sintered at 750°C

Water of Absorption

The effects of sintering temperature, glass type and glass particle size on water of absorption are illustrated in Figures 3a – 3f. In this discussion WA 1 represents water of absorption for window glass sample. WA 2 represents water of absorption for fluorescent tubes and WA 3 represents water of absorption for laboratory glass samples.

The water of absorption was higher at 24 hr absorption test for samples fused at 700°C (Figure 3b), laboratory glass L 30/70 had the highest water of absorption of 2.956% while fluorescent tube glass compact F 40/60 had the least water of absorption of 0%. The water of absorption was observed to decrease as the sintering temperature was raised to 725°C and 750°C (Figures 3c – 3f). The maximum water of absorption allowed for an individual tile is 9% at 24 hour cold water absorption as specified in ASTM 212, it can therefore be said that all tiles made fall within the ASTM standards.

The water of absorption for fluorescent tubes sintered at 725°C tended towards 0%, likewise the water of absorption for all samples sintered at 750°C. This means that the glazed layers that have been formed are pore free, however the SEM of the samples as shown in figures 4a – 4b exhibited some pores that were encased beneath the glazed layers.

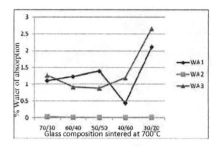

Figure 3a: 5 hr Water of Absorption

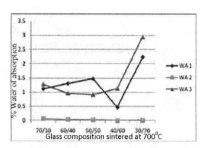

Figure 3b: 24 hr Water of Absorption

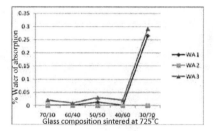

Figure 3c: 5 hr Water of Absorption

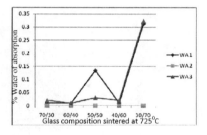

Figure 3d: 24 Hr Water of Absorption

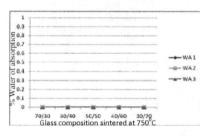

Figure 3e: 5 hr Water of Absorption

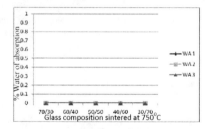

Figure 3f: 24 hr Water of Absorption

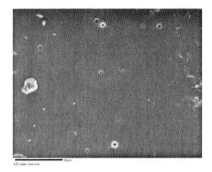
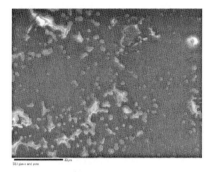

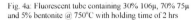

Fig. 4a: Fluorescent tube containing 30% 106µ, 70% 75µ
and 5% bentonite @ 750°C with holding time of 2 hrs

Fig. 4b: Fluorescent tube containing 30% 106µ, 70% -75µ
and 5% bentonite @ 750°C with holding time of 2 hrs

CONCLUSION

Sintered bodies above 90% glass powder from window glass, fluorescent tubes (soda lime silica glass and Laboratory glass (borosilicate glass) were achieved using PM method and the highest sintering temperature of 750°C attained is at least 300 degrees less than firing clay tiles, thereby saving energy and time if used as an alternative to clay in making ceramic tiles. And having achieved a glazing at 750°C, time can further be saved and glaze raw materials by eliminating a separate glazing stage.

REFERENCES

[1] Askeland D.R and Phule P.P (2003): The Science and Engineering of Materials. Brooks/Cole-Thomson Learning, Pacific Grove, United States p 635.

[2] Williams J (1965): Characterization of Fine Oxide Powders in Science of Ceramics Volume 2 Proceedings of the second conference of British Ceramic Society and Netherlandse Keramische Veremiging, edited by G.H Stewart Academic Press London pp 3 -4.

[3] Sacks et al (1992): Processing of Compopsite Powders, Fabrication of Ceramics and Composites by Viscous Sintering and Transient Viscous Sintering in Chemical Processing of Advanced Materials by Hench et al (eds) John Wiley and Sons Inc New York p 557.

[4] Barsoum M.W (2003): Fundamentals of Ceramics, Institute of Physics Publishing Ltd, Bristol and Philadelphia p 345.

[5] Matijevic E (1992): Control of Powder Morphology in Chemical Processing of Advanced Materials by Hench and West (eds), John Wiley and Sons Inc New York, p 513.

[6] Hwang J.Y, Huang X, Garkida A, Hein A (2006): Waste Colored Glasses as Sintering Aid in Ceramic Tile Production; Journal of Minerals & Materials Characterization & Engineering, Volume 5 no 2 2006, USA pp 119-129.

[7] Monteiro R.C.C and Lima M.M.R.A (2003): Effect of Compaction of on the Sintering of Borosilicate Glass/Alumina Composites, Journal of the European Ceramic Society, Volume 23, Issue 11 pp 1813-1818.

[8] Jiann-Yang Hwang et al (2007): Utilization of Waste Glass in Alumina-Glass Composite. Extraction and Processing Division (EPD) of The Minerals, Metals and Materials Society (TMS), United States pp 121-126.

INVESTIGATION OF EFFECTIVE PARAMETERS IN PRODUCTION OF A356/ TiB$_{2p}$ COMPOSITE USING TiB$_{2p}$/CMC/PPS MORTAR

M.Hizombor[1]; S.M.H. Mirbagheri[1]; A. Rezaie[2] and R. Abdideh[2]

1) Department of Mining and Metallurgical Engineering, Amirkabir University of Technology, Tehran, Iran, e-mail: hizombor@yahoo.com
2) Department of Metallurgical Engineering, Islamic Azad University(young researchers club), Ahwaz, iran.

ABSTRACT
Composite materials are among those new materials used for many industrial applications. Many Methods have been proposed to overcome the problem of poor wettability between ceramic reinforcement particles and molten aluminum for metal matrix composite (MMC) production in casting processes.

In this investigation, an innovation procedure has been proposed for casting of metal matrix composites by adding a mortar consist of expandable polystyrene beads, carboxy methyl cellulose paste, water and TiB$_{2p}$ particles as a mould pattern. This process was examined for A356/TiB$_{2p}$ composite. The use of pretreated TiB$_{2p}$ particles, 1wt% magnesium as a wetting agent and mechanical mold vibration while MMC$_s$ slurry is solidifying were found to promote wettability of TiB$_{2p}$ with molten matrix alloys. Produced composites were characterized using optical and scaning electron microscopy. Then mechanical properties of the composites, such as hardness, wear, tensile testing and porosity levels of produced Al/TiB$_{2p}$ composites were measured and results have been discussed. Results show the mechanical properties strongly dependent on the distribution of the TiB$_{2p}$ particles.

INTRODUCTION
The requirement for strong, light and stiff materials has extended an interest in metal matrix composites (MMC$_s$). During the past two decades, MMC$_s$ have received substantial attention because of their improved strength, high elastic modulus and increased wear resistance over conventional base alloys. The wide Scale introduction of MMC$_s$ has been increasing simultaneously with the technological development [1, 2].

Aluminum matrix composites posses many advantages such as low specific density, high strength and good wear resistance, with the development of some non- continuous reinforcement materials, whisker, fibres or particles. In particular, the particulate reinforced aluminum matrix composites not only have good mechanical and wear properties, but are also economically viable [3,4]. Therefore, TiB$_{2p}$-particulate reinforced aluminum composites have found many applications in aerospace and automotive industry [5]. There are many methods for fabrication of particulate reinforced metal matrix composites (MMC$_s$) Such as powder metallurgy, Squeeze casting, compocasting, and so on [5]. Among the variety of manufacturing processes available for discontinuous metal matrix composites, casting is generally accepted as a particularly promising rout, currently practiced commercially. Its advantages lie in its simplicity, flexibility and applicability to large quantity production. According to Skibo at al. [6], the cast of preparation composites materials using a casting method is about on- third to half of competitive methods, and for high volume production, it is projected that, the cast will fall to on- tenth[7].

In general, the solidification synthesis of metal matrix composites involves producing a melt of the selected matrix material followed by the introduction of a reinforcement ceramic into the melt, obtaining a suitable dispersion. The next step is the solidification of the melt containing suspended dispersions under selected condition to obtain the desired distribution of the dispersed phase in the cast matrix.

In preparing metal matrix composites by casting method, there are several factors that need considerable attention, including: The difficulty of achieving a uniform distribution, wettability between two main substance, porosity and chemical reaction between the reinforcement material and the matrix alloy.

In order to achieve the optimum properties of the metal matrix composites, the distribution of the reinforcement material in the matrix alloy must be uniform, and the wettability or bonding between these substances should be optimized. The porosity levels need to be minimized, and chemical reactions between the reinforcement materials and the matrix alloy must be avoided [7].

MATERIALS AND EXPERIMENTAL PROCEDURE

Materials
A commercial casting aluminum A356 alloy was utilized as the matrix materials, the chemical composition of A356 matrix alloy used in this investigation are given in table1. This alloy has been selected because of its good fluidity as well as the presence of silicon and magnesium. The Silicon content of A356 alloy is sufficiently high, therefore it can be casted in high temperature with out giving rise to the extensive formation of Al$_4$C$_3$ [8].

In the present study, composites containing 5, 7.5 and 10 %vol. TiB$_{2p}$ were manufactured, the average size of TiB$_2$ particles were 6μm. The TiB$_{2p}$ particles were oxidized in air at 1100°C for 4 hours to improve their wettability with molten A356 alloy. Fig.1 Show the SEM images of TiB$_2$ particles. EPS/ TiB2p performs were fabricated using pre- expandable polystyrene and carboxy methyle cellulose in infiltration methods has been mentioned in other refrences [9].

Fabrication of EPS/ TiB$_{2p}$ performs
In order to EPS/TiB$_{2p}$ fabrication, TiB$_2$ particles must be coated on the pre-expandable polystyrene. Solution of water and carboxy methyle cellulose was utilized as a binder. Suitable volume percentage of carboxy methyle cellulose soluble in water, was obtained experimentally (2-3) % wt., in the less amounts, there was no suitable adherence between TiB$_2$ particles and pre-expandable polystyrenes, and in above of 3% wt. carboxy methyle cellulose, because of high concentration solution, the coat of TiB$_2$ particles on the pre-expandable polystyrenes is not uniform.

Pre-expandable polystyrenes was poured into a container, mixture of binder and particles was added gradually and stirred continuously for about 5min, using an impeller at a speed of 400rpm. Coated pre-expandable polystyrenes were formed as a preforms with dimension of 12×3×1.6 cm. To annihilate humidity, the EPS/TiB$_{2p}$ performs was dried at 60°c for approximately 2 hours.

Production of Al/TiB$_{2p}$ composites
The composites were manufactured by molten metal of A356 alloy using an electric induction furnace, which is 30 KW power. The molten alloy was suitably degassed with hexachloroethane (C$_2$Cl$_6$) and fluxed with 45%wt. KCl-45%wt. NaCl-10%wt. NaF.

In this work, pouring temperature of A356 molten alloy was selected 780°c and pure magnesium was added to molten metal approximately 1, and 5 wt. % in order to wettability enhancement. EPS/TiB$_{2p}$ performs were put in the permanent mould (Fig.2), subsequently molten metal was poured into it and allowed to cool down to room temperature. In some experiments, mechanical mold vibration with 180 Cycle/sec and 2-3 mm amplitude was used to improve the wettability of molten aluminum and silicon carbide particles.

Chemical composition analysis

Content magnesium in A356 aluminum alloy is lost in high temperature by oxidation. Pure magnesium was added to molten A356 to restitute the magnesium and improvement the wettability of the alloy. The chemical composition analysis was carried out using D47533KELEVE coantimeter.

Microstructural characterization

The microstructures of the manufactured A356/TiB$_{2p}$ composites were investigated using Scanning electron microscope (SEM) to determine the distribution of the TiB$_2$ particles and presence of porosity. Sample for metallographic observation were sectioned and prepared by standard methods. Fabricated A356/TiB$_{2p}$ composites microstructures were invest perused by image analysis (IA) techniques to obtained the volume percentage and distribution of TiB$_{2p}$ in them.

Density measurements

The density of the A356/TiB$_{2p}$ samples was measured according to ASTM D797 standard and using Archimedes principle [10]. The samples were precision weighed in an electron balance to an accuracy of 0.1 mg. The theorical density of composites and alloy matrix specimens was then calculated according to the rule of mixtures [10]. The volume fraction porosity of the composites was also determined by use of equation 1.

$$\%vol.\,prosity = \frac{\rho_{cal} - \rho_{exp}}{\rho_{cal}} \tag{1}$$

Hardness and wear tests

The hardness of the Al/TiB$_{2p}$ composites and matrix alloy were preformed using Vicker method (30 sec. At a load of 31 kgf) according to ASTM E92 standard. In order to eliminate possible segregation effect, the mean of at least three tests was taken for each specimen.

Dry sliding wear tests were carried out according to the ASTM G99 standard using a pin-on-disc type apparatus [10]. The test material in the form of rings of diameter 5mm and length 15mm were slide against a steel disk equivalent to 3343 grade with hardness of 60HRC. Wear tests were preformed under normal pressure of 1Mp and with sliding distance of 100, 300, 600 and 1000 m.

In this study, the sliding distance of specimens was selected $0.5\,m/_s$, and in all tests, each experiment was repeated at least 3 times. After the tests, the specimens were cleaned with acetone, dried and weighted again to determine the weight loss due to wear. The weight loss was calculated from the difference in weight of the specimens measured before and after the test to the nearest 0.1 mg, using an electronic balance. Finally weight loss values were converted into volume loss and wear rate known density data.

RESULTS AND DISCUSSION

Investigation of effective parameters

Oxidation of TiB$_2$ particles, magnesium percentage in molten metal and mold vibration before and during solidification are important process parameters for fabrication of low volume fraction composites using EPS/TiB$_{2p}$ performs. The addition of magnesium to molten A356 alloy to promote the wetting of silicon carbide particles is particularly successful [11]. Indeed, the addition of magnesium to molten aluminum can modify the matrix metal alloy by generating a transient layer between the particles and molten metal. This transient layer has a low wetting angle, decrease the surface tension of liquid, and surround the particles with a structure that is similar to both the particle and matrix alloy [7]. The addition of magnesium to aluminum reduces its surface tension. The reduction is very sharp for initial 1%wt magnesium addition [10]. According to sukumaran et al. [12], increasing the magnesium content over 1%wt., increases the viscosity of matrix alloy and decrease the wettability, moreover, addition of magnesium over this value lead to the formation of a Mg$_5$Al$_8$ phase, which devastates the mechanical properties of composite because of its low melting point.

Heat treatment of TiB$_2$ particles before introduce into the melt annihilates absorbed gases from the particle surface, and alter the surface composition by forming an oxide layer on the surface [13].

In composites manufacturing, a mechanical force can usually be used to dominate surface tension and improve wettability between TiB$_{2p}$ and molten metal. In the experimental work of Zhou et al. [13] proposed that is necessary to break the gas layer surrounding the ceramic particles in order to achieve good wettability. Using mechanical and ultrasonic vibration of molten aluminum in order to improve the wettability of ceramic particles has been reported in other refrences [14].

Distribution of TiB$_{2p}$ in EPS/ TiB$_{2p}$ preforms

Fig.3 show image of coated pre-expandable polystyrenes with uniform TiB$_{2p}$ distribution. This uniformity aids uniform TiB$_{2p}$ distribution in final Al/TiB$_{2p}$ composites.

Microstructure of Al/TiB$_{2p}$ composites

The properties of the metal matrix composites depend not only on the matrix, particles and the volume fraction, but also on distribution of reinforcing particles and interface bonding between the particle and matrix. The SEM micrograph of the aluminum composite reinforced with approximately 5, 7.5 and 10 vol. % of TiB$_{2p}$ is shown in fig.4 (a), (b) and (c).

Density and porosity

The results of theorical and experimental densities and porosities of the composites according to the volume percentage of TiB$_{2p}$ particles are shown in table2. Table2 shows that, the theorical and experimental density of the composites increase linearly (as expected from the rule of mixture) [3]. The experimental value are lower than that of the theorical density, therefore, the density measurement showed that, these composites contained some porosity. The amount of porosity and density in the composites increased with increasing volume percentage of the TiB$_2$ particles. The increased porosity of the cast Al/TiB$_{2p}$ composites with increased TiB$_{2p}$ content has also been reported by other researchers

[15]. The porosity of the composites was found between 3.3 and 4.6 vol. %. These values indicate that, the porosities of the composites are at acceptable levels for low volume percentage of composites. Kok et al. [15] demonstrated that porosity level of metal matrix composites was found approximately 4%. SEM micrograph of porosity in TiB$_{2p}$ reinforced composite are shown in fig.5.

Hardness, Wear test
The hardness variation of the manufactured particle reinforced composite are shown in fig.6. Results indicate that, the hardness of the manufactured composites increase more and less linearly with volume percentage of TiB$_2$ particle phase.

The variation in weight loss with sliding distance for unreinforced matrix alloy and Al/TiB$_{2p}$ composites containing different volume percentage of TiB$_2$ particles are summarized in table3 and fig.7. It is clear that, the weight loss of the composites is lower than that of the unreinforced alloy. The weight loss also decreases with increasing the TiB$_{2p}$ volume percentage, which is in agreement with the previously reported results and can be attributed to the increased hardness of composites.

CONCLUSIONS
The following major conclusion can be drawn from the present investigation.

- Processing parameters such as magnesium content in the molten metal alloy, TiB$_2$ particle oxidation and mechanical mold vibration during and before solidification are among the important factures to be considered in the production of Al/TiB$_{2p}$ using EPS/TiB$_{2p}$ preforms.

- Microstructural investigation of EPS/TiB$_{2p}$ preforms showed that, there is a uniform distribution of TiB$_2$ particles on pre-expandable polystyrene, this uniformity aids uniform TiB$_{2p}$ distribution in fabricated Al/TiB$_{2p}$ composites.

- Al/TiB$_{2p}$ composites consisting of 5, 7.5 and 10 vol. % TiB$_2$ particles could be produced successfully by EPS/TiB$_{2p}$ preforms.

- SEM observation of the Al/TiB$_{2p}$ microstructures showed that, there is a uniform distribution of TiB$_2$ particles in manufactured composites.

- The density and the porosity of manufactured composites increased with increasing volume percentage of TiB$_2$ particles.

- The hardness of Al/TiB$_{2p}$ composites increased with increasing volume percentage of particulates.

- In the produced Al/TiB$_{2p}$ composites, the weight loss and wear rates are lower, compared those to
- the unreinforced matrix alloy

REFERENCES
1. M. Muratoglu, M. Aksoy,' The effects of temperature on wear behaviors of Al-Cu alloy and Al-Cu/SiC Composite', Mater.Sci.Eng., A282, 2000, pp. 91-99.

2. D. B. Miracle, 'Metal Matrix Composites – from Science to technological significance ' composite science and technology, 65, 2005, 2526-2540.

3. Y. Sahin,' Preparation and some properties of SiC particle reinforced aluminum alloy composites', Material&Design, 24, 2003, pp. 671-679.

4. Y. sahin, ' The effect of sliding speeed and microstructure on the dry wear properties of metal-matrix composites', Wear, 214, 1998, pp. 98-106.

5. J. R. Davis, Aluminum and Aluminum Alloys, ASM Specialty Handbook, 1998,160.

6. D. M. Skibo, D. M. Schuster, L. Jolla, 'Process for preparation of composite materials containing nonmetallic particles in a metallic matrix and composite materials', U. S. Patent No. 4786467, 1988.

7. J. Hashim, L. Looney, M. S. J. Hashmi,' Metal matrix composites: production by the stir casting method', Mat. Proc. Tech., 92, 1999, pp. 1-7.

8. S. Vaucher, O. Beffort, 'Bonding and interface formation in Metal Matrix Composites', EPMA report Nr 250, Thun, Switzerland, 2001.

9. A. Mortensen, J. Thomas,' Method for producing a microcellular foam', US. Patent, No. 5553658.

10. Y. Sahin, M. Acilar, ' Production and Properties Of SiC- reinforced Aluminum Alloy Composite ', Composite:Part A,34,2003,pp.709-715.

11. J. Hashim, L. Looney, M. S. J. Hashmi,' The enhancement of wettability of SiC particles in cast aluminium matrix composites', Mat. Proc. Tech., 119, 2001, pp. 329-335.

12. K. Sukumaran, S. G. K. Pillai, V. S. Kelukutty, B. C. Pai, K. G. Satyanarayana, K. K. Pavikumar, J. Mater. Sci. 30, 1995, pp. 1469-1472.

13. W. Zhou, Z. M. Xu,' Casting of SiC reinforced metal matrix composites', J. Mat. Proc. Tech., 63, 1997, pp.358-363..

14. Y. Tsunekawa, H. Nahaneshi, M. Okumia, N. Mohri, Key Eng. Mater., 104-107, 1995, pp.215-224.

15. M. Kok, ' Production and mwchanical properties of Al$_2$O$_3$ particle- reinforced 2024 aluminium alloy composites', Mat. Proc. Tech., 161, 2005, pp. 381-387.

Table1. Chemical Analyses of used A356 Alloy and standard element range in this alloy

	Si	Mg	Fe	Cu	Mn	Zn	Ti	Al
Used alloy	6.45	0.445	0.0293	0.0098	0.0027	0.0073	0.0049	Bal
Standard range	6.5-7.5	0.25-0.45	<0.5	0.25	0.35	0.35	0.25	Bal

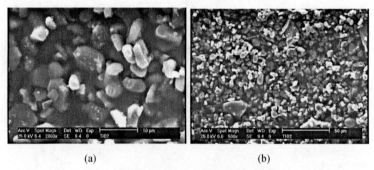

(a)   (b)

Fig. 1. SEM images of TiB$_{2p}$ a) Low magnification b) High Magnification

Mortar dried (EPS/TiB$_{2p}$) ——— Mould

Fig. 2. View of fabricated EPS/TiB$_{2p}$ after moulding and drying

3 mm

Fig.3. Images of coated pre-expandable polystyrenes Uniform TiB$_{2p}$ distribution

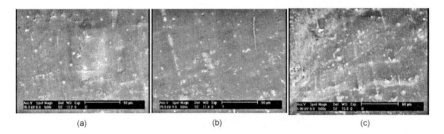

(a)   (b)   (c)

Fig.4. SEM micrograph of the aluminum composite reinforced with approximately a) 5 b) 7.5 and c) 10 vol. % of TiB$_{2p}$

Table2. Theorical and experimental densities and porosities of the composites and unreinforced A356 alloy

Samples	Theorical density (gr/cm^3)	Experimental density (gr/cm^3)	Porosity (%)
A356	2.70	2.61	3.33
A356-5% TiB$_{2p}$	2.74	2.62	3.94
A356-7.5%TiB$_{2p}$	2.79	2.68	4.38
A356-10%TiB$_{2p}$	2.84	2.71	4.57

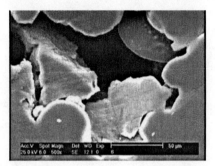

Fig.5. SEM micrograph of porosity in TiB$_{2p}$ reinforced composite

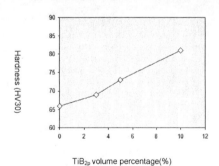

TiB$_{2p}$ volume percentage(%)

Fig.6. Hardness variation of the manufactured TiB$_{2p}$ reinforced composite

Table 3. Variation in weight loss with sliding distance for unreinforced matrix alloy and Al/TiB$_{2p}$ composites

Samples	Weight loss(mg)			
	100m	300m	600m	1000m
A356	0.7	3.4	4.4	5.2
A356-5% TiB$_{2p}$	0.6	2.1	3.5	4.1
A356-7.5%TiB$_{2p}$	0.5	1.9	3.3	3.9
A356-10%TiB$_{2p}$	0.48	1.7	2.4	3.1

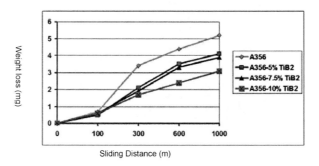

Fig.7. Weight loss variation according to the sliding distance and TiB$_{2p}$ volume percentage

CHEMICAL INTERACTION OF $Sr_4Al_6O_{12}SO_4$ WITH LIQUID ALUMINUM ALLOYS

José Amparo Rodríguez-García, Enrique Rocha-Rangel
Universidad Politécnica de Victoria, Avenida Nuevas Tecnologías 5902
Parque Científico y Tecnológico de Tamaulipas, Tamaulipas, México, 87138

José Manuel Almanza Robles, Jesús Torres Torres
CINVESTAV, Unidad Saltillo. Carretera Saltillo-Monterrey Km. 13.
Saltillo, Coahuila, México, 25900.

Ana Lilia Leal Cruz, Guillermo T. Munive
Universidad de Sonora, Boulevard Luis Encinas y Rosales S/N,
Zona Centro, Hermosillo, Sonora, México, 83000.

ABSTRACT
        The $Sr_4Al_6O_{12}SO_4$ compound was synthesized by a solid state reaction from a stoichiometric molar mixture of $SrCO_3$, Al_2O_3 and $SrSO_4$. The process of the formation was followed using DTA, TGA, XRD and SEM analyses. In order to study the interaction of this compound with liquid aluminum alloys, the samples were immersed in liquid aluminum. During the synthesis, it was found the formation of two transitory phases, $Sr_3Al_2O_6$ and $SrAl_2O_4$, between 900 and 1100°C. The strontium sulphoaluminate start to form around 900°C and it was completely formed at 1150°C. The $Sr_4Al_6O_{12}SO_4$ showed a slightly interaction with Al-Si alloy, with the formation of layer of alumina (less than 200μm), on the other hand, there was no interaction of this compound with Al-Mg alloy. The sulphoaluminate showed excellent corrosion resistance to aluminum alloys.

INTRODUCTION
        Strontium is one of the elements less known and studied. In nature it can be found as celestine ($SrSO_4$). Mexico is one of the main exporters of strontium worldwide contributing to ¼ of the world production. Recently it has been an increase in the use of this compound in several applications. Strontium carbonate ($SrCO_3$) is the most used of the strontium compounds, and it is used to the fabrication of kinescope for TV sets[1], in pyrotechnics as $Sr(NO_3)_2$[2], in the aluminum industry as phase modifier[3], etc. However the recent increase in the production of flat screens (plasma, LCD, led), the $SrCO_3$ demand decreased considerably. Due to this fact, there has been an interest in finding new applications for strontium compounds. Examples of such compounds have the strontium aluminates, which has several applications such as luminescence pigments[4], CO_2 sensors[5] and refractories[6]. However, there are some other compounds that has not been properly studied, such is the case of the strontium sulphoaluminate ($Sr_4Al_6O_{12}SO_4$), where there are scarce information related to this compound[7-8]. Aluminum industry is in growing demand and its production generally depends of refractory losses, these are due aggressive environments generated during the process. The degradation of refractories used in aluminum furnaces it is due to corrosion and erosion and wear, that leads to heat losses and decreased of the furnace thermal efficiency. Such issues can also lead to decreased product quality and production outages when repairs are needed. Additionally, large amounts of energy are lost when furnaces are cooled for refractory repair and are required for reheating the furnace[9]. To minimize the effect caused by the reaction between aluminum/refractory, resulting in formation of an alumina layer on the refractory, some authors have suggested the addition of non-wetting agents such as: $BaSO_4$[10], CaF_2, AlF_3[11-12] among others. Recently it was demonstrated[13] that the use of $SrSO_4$ increase the corrosion resistance of refractories in contact with liquid aluminum. In this work, the use of a strontium compound as a corrosion resistant to liquid aluminum was studied.

EXPERIMENTAL PROCEDURE

Stoichiometric mixtures, 3:3:1 molar, of $SrCO_3$, Al_2O_3 and $SrSO_4$, were homogenized for 4 hours with acetone in a plastic jar on spinning rolls. The mixture were dried at 100°C for 12 hours and then grounded to reduce agglomeration. Then disk pellets of 5 and 20mm in diameter were conformed at 100MPa. Powder samples of the mixture were characterized by TGA up to 1200°C at 10°Cmin⁻¹. Additionally the mixture (powder and pellets of 5mm in diameter) were analyzed by DTA up to 1200°C using 3, 5 and 10°Cmin⁻¹. Dick pellets of the mixture (20mm in diameter) were heat treated from 800 to 1200°C, with 50°C intervals, for 4 hours. Then the samples were analyzed by XRD and SEM.

The chemical interaction of the samples with liquid aluminum was studied using the immersion method. Disk pellets of the mixture (20mm in diameter) were double sinterized, first, the samples conformed at 100Mpa were heat treated at 1400°C for 4 hours. Then the disks were milled to -325 mesh and the conformed at 300Mpa (20mm in diameter) and sintered at 1400°C for 24 hours. This process was choosen in order to reduce at minimum porosity. The samples were fixed at the bottom of an alumina crucible and then 500g of solid aluminum was put on it. Then the crucible was put in a furnace were the temperature was rised up to 900°C for a period of 24 hours. The chemical composition of the alloys used were: Al-Si: 85.77%Al, 7.58%Si, 0.0403%Mg, 3.23%Cu, 0.951%Fe, 0.0403%Pb, 0.28%Mn, 1.85%Zn y 0.258% others; Al-Mg: 91.10%Al, 1.09%Si, 7.52%Mg, 0.052%Cu, 0.111%Fe, 0.0181%Pb, 0.0126%Mn, 0.0131%Zn y 0.083% others. The samples after the test were analyzed by SEM.

RESULTS AND DISCUSSION

Figure 1 shows a Thermal Gravimetric Analysis (TGA) curve of the $SrCO_3$-Al_2O_3-$SrSO_4$ mixture from room temperature to 1200°C. It was observed that there is a 15% weight loss between 750°C and 1050°C. This weight loss corresponded to the decomposition of $SrCO_3$ to SrO and CO_2.

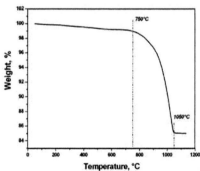

Figure 1. TGA curve obtained for a powder sample of the $SrCO_3$-Al_2O_3-$SrSO_4$ mixture heat treated to 1200°C with a heating rate of 10°Cmin⁻¹.

Figure 2 shows the results obtained of Differential Thermal Analysis (DTA) for the powder and pellet samples of the $SrCO_3$-Al_2O_3-$SrSO_4$ mixture from room temperature to 1200°C. First, an exothermic peak was observed around 850°C for the four curves (A: powder sample at 10°Cmin⁻¹, B: pellet sample at 10°Cmin⁻¹, C: pellet sample at 5°Cmin⁻¹ and D: pellet sample at 3°Cmin⁻¹), which corresponds to the formation of phases result of the partial decomposition of $SrCO_3$ to SrO and its combination with

Al$_2$O$_3$ and SrSO$_4$, resulting in the partial formation of Sr$_4$Al$_6$O$_{12}$SO$_4$, SrAl$_2$O$_4$ and Sr$_3$Al$_2$O$_6$. Between 1075 to 1105°C a change in slope was detected for all running conditions. This process corresponded to a phase transformation, specifically the reaction between Sr$_3$Al$_2$O$_6$ with strontium sulfate present in the system to form Sr$_4$Al$_6$O$_{12}$SO$_4$. At around 1110°C an endothermic peak was observed for the curves C and D, which corresponds to the disappearance of Sr$_3$Al$_2$O$_6$ in the ternary system due to its reaction with strontium sulfate to form larger amount of Sr$_4$Al$_6$O$_{12}$SO$_4$. Finally, in the same curves (C and D), an exothermic peak was observed at 1150°C that corresponded to the final formation of strontium sulphoaluminate. As the heating rate decreased and the same was compacted, it was observed in more detail all the events that occurred during heating. These events will be discussed more with X Rays Diffraction and Scanning Electronic Microscopy results. The X Rays Diffraction patterns (XRD) for the mixture at different temperatures are show in Figure 3.

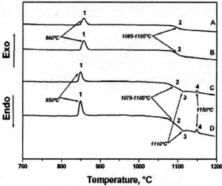

Figure 2. DTA curves for samples of the SrCO$_3$-Al$_2$O$_3$-SrSO$_4$ mixture. A: powder to 10°Cmin^{-1}, B: pellet at 10°Cmin^{-1}, C: pellet at 5°Cmin^{-1} and D: pellet at 3°Cmin^{-1}.

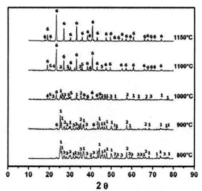

Figure 3. XRD patterns for samples heat treated at different temperatures with 4 hours of isotherm. 1: SrCO$_3$, 2: Al$_2$O$_3$, 3: SrSO$_4$, 4: Sr$_3$Al$_2$O$_6$, 5: SrAl$_2$O$_4$ and 6: Sr$_4$Al$_6$O$_{12}$SO$_4$.

The sample heat treated from 800°C shows the presence of the initial raw materials without the formation of the other phases. It was observed that the formation of Sr$_4$Al$_6$O$_{12}$SO$_4$, started at 900°C, as temperature increased (1000°C) the formation of two aluminates, Sr$_3$Al$_2$O$_6$ and SrAl$_2$O$_4$ was observed and the intensity of the peak corresponding to the Sr$_4$Al$_6$O$_{12}$SO$_4$ increased. As the temperature was increase (1100°C) the intensity of the peaks corresponding to the Sr$_3$Al$_2$O$_6$ and SrAl$_2$O$_4$ decreased with the increase in intensity of the peaks of Sr$_4$Al$_6$O$_{12}$SO$_4$. The complete formation of Sr$_4$Al$_6$O$_{12}$SO$_4$, was observed at 1150°C with no other phase present. The aluminates were formed from the reaction of SrO with alumina just after the decomposition of SrCO$_3$. These aluminates eventually reacted with more alumina and SrSO$_4$ to finally form the strontium sulphoaluminate.

The analysis of SEM (Figure 4), showed the evidence of the strontium aluminates formation at 1000°C.

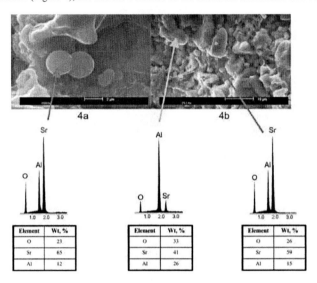

Figure 4. Micrographs of samples (pellets) of the mixture sintered at 1000°C for 4 hours. Note the formation of two strontium aluminates, the Sr$_3$Al$_2$O$_6$ and SrAl$_2$O$_4$.

The Sr$_3$Al$_2$O$_6$ corresponded to the spherical particles according to the EDS analysis. The SrAl$_2$O$_4$ has placket morphology. These two aluminates as stated before were transitory phases, which acted as nuclei sites to the final formation of Sr$_4$Al$_6$O$_{12}$SO$_4$. XRD and SEM results confirm that there are the formation of aluminates at relatively low temperature and then the transformation to another phase as it was observed in DTA results, however more studied is necessary.

Figure 5 shows the morphology of the strontium sulphoaluminate heat treated at 1150°C for 4 hours. In Figure 5a semispherical particles can be observed formed by agglomerates that small spheres of around 10μm according to the EDS analysis. Figure 5b corresponds to the increase in the image to see in detail the agglomerates of spherical particles; it shows the process of sintering between particles. No other phase was found according to the EDS analysis.

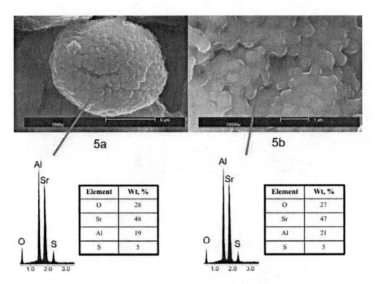

Figure 5. Micrographs of a sample (pellets) of the mixture sintered at 1150°C for 4 hours. Note the formation of a strontium compound, the Sr₄Al₆O₁₂SO₄.

The sample of $Sr_4Al_6O_{12}SO_4$, after the immersion in the Al-Si alloy (Figure 6), showed a small reaction layer of less than 200μm. According to the EDS analysis this layer corresponded to alumina formation, it was not clear if the aluminum reduced the strontium phase since no Sr was detected in the corrosion layer. Since the layer is so small it difficult to establish the corrosion mechanism using this technique. It was reported[13] that a mullite sample, in an similar experiment, showed a corrosion layer of about 2mm, therefore the $Sr_4Al_6O_{12}SO_4$ can be considered as an excellent corrosion resistant material to liquid aluminum.

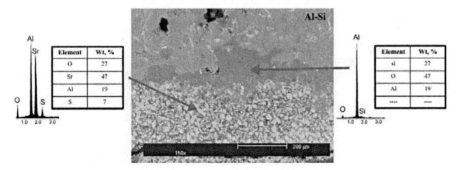

Figure 6. Micrograph of a sample (pellet) of the strontium compound $Sr_4Al_6O_{12}SO_4$ exposed to contact with an Al-Si alloy at 900°C for 24 hours.

As can be observed in Figure 7, there was no evidence of any interaction of the strontium sulphoaluminate with liquid Al-Mg alloy. There are several reports of corrosion test with some aluminum alloys, for example it was found that the addition of BaSO$_4$ to mullite substrates increased the corrosion resistance to an Al-Mg alloy[14-15]. It was reported that the corrosion of mullite substrates can reach a corrosion speed up to 6mm for hour[16]. Some other strontium compounds have been tested such as SrAl$_2$O$_4$ and Sr$_2$Al$_2$SiO$_7$, and it was found that only a small reaction layer (100-200μm) was found after a test with and Al-Si alloy[17]. The results in this work corroborated that strontium compound are indeed resistant to liquid aluminum corrosion. An important point is that strontium sulphoaluminate does not contain SiO$_2$, which corrodes rapidly with Al, and contains SrSO$_4$, which has been demonstrated that is corrosion resistant additive.

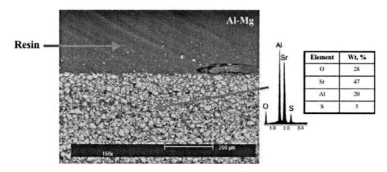

Figure 7. Micrograph of a sample (pellet) of the strontium compound Sr$_4$Al$_6$O$_{12}$SO$_4$ exposed to contact with an Al-Mg alloy at 900°C for 24 hours.

CONCLUSIONS

o   It was found that Sr$_4$Al$_6$O$_{12}$SO$_4$ can be synthesized at low temperature (1150°C) as a pure phase. After SrCO$_3$ decomposition two transitory phases, Sr$_3$Al$_2$O$_6$ and SrAl$_2$O$_4$, were formed, these phases acted as a nuclei for the formation of the strontium sulphoaluminate. The morphology of the Sr$_3$Al$_2$O$_6$ corresponds to spherical particles (1 - 2μm); the SrAl$_2$O$_4$ has placket morphology (5 - 10μm) and the Sr$_4$Al$_6$O$_{12}$SO$_4$ corresponds to the spherical particles formed of agglomerates of small particles (5- 10μm).

o   There was found that there was no chemical interaction of Sr$_4$Al$_6$O$_{12}$SO$_4$ with the Al-Mg alloy. On the other hand, a layer of less than 200μm was found after the immersion in Al-Si alloy. This layer corresponded to the alumina formation. The corrosion mechanism was not clear due to the small layer and the technique used to identify the phases. It can be concluded that the Sr$_4$Al$_6$O$_{12}$SO$_4$ has an excellent corrosion resistance to liquid aluminum.

ACKNOWLEDGMENT

Authors would like to thank to National Council of Science and Technology of Mexico for financial support for the development of this work as well as the Politechnique University of Victoria, CINVESTAV – Campus Saltillo and University of Sonora for the technical support.

REFERENCES

[1]Kira, R. E. and Othmer, D. F., Compuestos de estroncio: Enciclopedia de la Tecnología Química, Tomo VII, Ed. UTHEA, 472-476 (1962).

[2]Collins, R. K. and Andrews, P. R. A., Celestite in Canada: CIM Bulletin, 84, 130-250 (1961).

[3]Massonne, J., Technology and uses of barium and strontium compounds, Industrial Mineral International Congress, 117, 25-29 (1982).

[4]Katsumata, T.; Sasajima, K.; Nabae, T.; Komuro, S. and Morikawa, T., Characteristic of the strontium aluminates crystal used for long duration phosphor, J. A. Ceram Soc., 81, 413-416 (1998).

[5]Goto, T.; He, G.; Narushima, T. and Iguchi, T., Application of $Sr\beta$-alumina solid electrolyte to CO_2 gas sensor, Solid State Ionic, 156, 329-336 (2003).

[6]Capron, M. and Douy, A., Strontium dialuminate $SrAl_4O_7$: synthesis and stability, J. A. Ceram. Soc., 85, 3036-3040 (2002).

[7]Gilioli, C.; Massazza, F. and Pezzuoli, M., A new compound, strontium sulphoaluminate and its relationship with calcium sulphoaluminate, Cement and Concrete Research, 1, 621-629 (1971).

[8]Teoreanu, I.; Georgescu, M.; Puri, A. and Badanolu, A., Mechanims at formation of some unitary strontium and barium sulphoaluminates, International Congress Chemistry Cement, 3, 250-255 (1992).

[9]Hemrick J. G. and Peters K.-M., Advanced ceramic composites for molten aluminum contact applications, 11th Biennial Worldwide Conference on Refractories, ORNL Editor. Salvador, Brazil (2001).

[10]Aguilar-Santillan, J., Wetting of Al_2O_3 by molten aluminum: the Influence of $BaSO_4$ additions. Journal of Nanomaterials, 12 (2008).

[11]Allahevrdi, M; Afshar, S. and Allaire, C., Additives and the corrosion resistance of aluminosilicate refractories in Molten Al-5Mg. Journal of the Minerals, Metals and Materials Society, 50(2), 30-34 (1998).

[12]Koshy, P., Effect of chemical additives on the interfacial phenomena of high alumina refractories with Al-Alloys, School of Materials Science and Engineering, University of New South Wales: Syndey, Austalia. 284 (2009).

[13]Ibarra Castro, M. N. Almanza Robles, J. M., Cortés Hernandez, D. A., Escobedo Bocardo, J. C. and Torres Torres, J., The effect of $SrSO_4$ and $BaSO_4$ on the corrosion and wetting by molten aluminum alloys of mullite ceramics. Ceramics International 36, 1205–1210 (2010).

[14]Afshar, S. and Allaire, C., Furnaces improving low cement castables by non-wetting additives, JOM, 24-27 (2001).

[15]Oliveira, M. I. L. L.; Agathopoulos, S. and Ferreira, J. M. F., The influence of BaO on the reaction of oxides ceramics by molten aluminum alloys, Key Engineering Materials, 13, 1711-1714 (2002).

[16]Taiken, J. and Krietz, L., Refractories for the next millennium, additive effects on physical proprieties of castables systems for aluminum contact applications, The 9th symposium on refractories for the aluminum industry, The Minerals Materials Society, 5-207 (2000).

[17]Hernández Valdez, A. Efecto de la adición de aluminato y solicoaluminato de Ba y Sr sobre la mojabilidad de refractarios base sílice por aluminio líquido. Tesis de maestría, CINVESTAV, Unidad Saltillo. (2009).

EFFECT OF TEMPERATURE ON THE HYDRATION OF ACTIVATED GRANULATED BLAST FURNACE SLAG

Enrique Rocha-Rangel
Universidad Politécnica de Victoria, Avenida Nuevas Tecnologías 5902
Parque Científico y Tecnológico de Tamaulipas, Tamaulipas, México, 87138

M. Juana Martínez Alvarado, Manuela Díaz-Cruz
Departamento de Metalurgia y Materiales, ESIQIE-I. P. N.
UPALM-Zacatenco, México D. F., 07738

ABSTRACT
     In this work it was carried out a study about the hydration products formed by the activation with NaOH of granulated blast furnace slag at different temperatures (25, 50, 100 and 150°C), as a function of time (1, 3, 7, 14 and 28 days). The progress of hydration reactions was assessed by X-ray diffraction and infrared spectroscopy. From these analyses, they were found that the main product of hydration was the calcium silicate hydrate gel (CSH-gel), whose formation is favored with the increments of the temperature of hydration. Also, the hydrotalcite compound was formed in significant amounts for the different studied conditions. At high temperatures (100 and 150°C) they were formed in small amounts the compounds tobermorite and ettringite. Monosulphoaluminate and brucite phases were formed from 50°C. Results show the presence of calcite due to the absorption of atmospheric CO_2 by the slag. High amount of MgO in the slag provokes the formation of dolomite at high temperatures.

INTRODUCTION
     During the recent years, researches related to monitoring of hydration of granulated blast furnace slag (GBFS)-alkaline activated have been increased[1-3]. However, even today the knowledge about the chemical reaction and formation of hydration products is not fully understood, because the mechanism of GBFS hydration continues remaining confused. Since then the lack of GBFS hydration, is that it has been possible to identify possible applications of the same. GBFS is considered to be a total or partial replacement of Portland cement and can be used in the construction of various civil works such as; dams, bridges and buildings[3]. With the advantage over Portland cement that has good resistance to high temperatures, high values of compressive strength, good chemical resistance and sulfate resistance, low heat of hydration, does not emit pollutants into the atmosphere, energy saving, use of industrial sub product confined for years, between others. Due to these good properties if they knew more about the mechanism of GBFS hydration, its application could be considered better, as example in cementing oil wells, where it requires a cementitious material that has to work under extreme pressure and temperature. The aim of this work is to establish the effect on GBFS hydration, which would have the temperature of hydration (25, 50, 100 y 150°C) as function of time (1, 3, 7, 14 and 28 days) for the system GBFS 100% using NaOH as an alkaline activator, this in order to expand the knowledge that we have about the progress of GBFS hydration.

EXPERIMENTAL PROCEDURE
     GBFS from Altos Hornos de Mexico Company was crushed and ground mechanically until less than 75 microns sizes. The powders obtained from milling were granulometrically analyzed to determine size distribution. Furthermore, its specific surface was evaluated by the method Blaine using an H-3810, Humboldt equipment. The chemical composition of the slag was determined by wet chemical analysis. Structural characterization of GBFS before and after hydration was carried out by

X-ray diffraction and infrared spectroscopy by Fourier transform in the following equipments; Bruker AXS D8 and Fourier 2000FTRI Perkin Elmer System respectively. For alkaline activate the slag it was used NaOH in the proportions suggested in the literature[2, 4-5]. The water/slag ratio was fixed at 0.5. Monitoring of the hydration reactions was performed at 1, 3, 7, 14 and 28 days. Hydration temperatures studied were 25, 50, 100 and 150°C.

RESULTS AND DISCUSSION
Starting materials
        Prior to GBFS hydration, slag has received a conditioning treatment as follows: size reduction in a hammer crusher and grinding in a ball mill with the purpose to carry sizes suitable for study. After crushing and milling, GBFS has particle size < 43 microns, presenting a specific surface area of $3,562 cm^2/g$, the density of the slag was 2.882g/ml. Also, the slag was carried out in quantitative wet chemical analysis for determine its composition, the result of this analysis are presented in Table 1.

Table 1. GBFS chemical composition.

Compound	Composition (wt. %)
CaO	33.32
SiO_2	38.00
Al_2O_3	10.73
Fe_2O_3	1.65
MgO	13.40
MnO	0.47
TiO_2	2.43

        Figure 1 shows the infrared spectra of the non-hydrated GBFS. The main bands present in this figure are the next: the broadband, which is between 800 and 1100 cm^{-1} that is characteristic of the gehlenite. The broad and little intense band from 3000 to 3700 cm^{-1} is due to the presence of moisture in the sample with (-OH) vibration. At 1647 cm^{-1} there is a small band due to H-O-H vibrations from moisture. Next observed band is at 1432 cm^{-1} this is narrow and present low intense and is due to CO_3. The most intense band is at 973 cm^{-1} and indicates the presence of silicates in the sample with symmetric stretching vibrations of the Si-O-Si and Al-O-Si types. At 713 cm^{-1} there is a little intense peak which reflects the asymmetric stretching bond Al-O-Al from the oxides of aluminum present in the slag. Finally, there is at 506 cm^{-1} a narrow band of medium intensity which is also due to the presence of silicates bond Si-O and a very small shoulder at 480 cm^{-1} assigned to SiO_4 group.

        In agreement with Taylor[6] and taking in consideration these results, it is concluded that slag presents the characteristics of a cementitious material and therefore it can be used for the studied proposed.

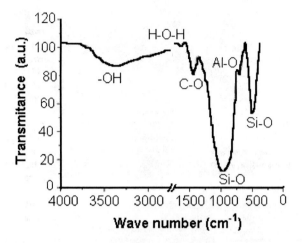

Figure 1. Infrared spectra of the non-hidrated slag.

Characterization of hydrated products

Hydrated design

The idea of this study on the GBFS hydration is with the purpose of using it in the manufacture of cements for oil used according to the Mexican standard NRF-069-PEMEX-2006[7-8]. The design of hydration is widely described in the thesis of J. Martinez[9]. The characteristics of the GBFS slurry used for the hydration study proposed are listed in Table 2.

Table 2. Characteristics for the GBFS hydration.

GBFS	860 g GBFS
Water	43.63 wt. % 375.2 g H_2O
NaOH	5 wt. % Na_2O, 55.5 g NaOH
Specific surface	3562 cm^2/g
Consistence	11 Bc
Water/slag ratio	375.2/ 860 = 0.4363

X-ray diffraction

Figures 2 and 3 show X-ray diffraction patterns GBFS hydrated during 1 and 28 days respectively, as a function of temperature. It is noteworthy that the study was conducted to see the sequence of hydration at different days of the slag between 1 and 28 days, however, due to situations of space, patterns corresponding to 3, 7 and 14 days are not presented here, these can be found in the work of J. Martinez[9]. As shown in these two figures, all diffraction patterns remain the same tendency of amorphous material that is a characteristic of GBFS non-hydrated, although there are observed some hydration products of low crystallinity. The main products of GBFS hydrated are: CSH gel, hydrotalcite, hydrogranate, AFm phases (C_4AH_{13} y C_2ASH_8) and ettringite. There is also the presence of carbonates in the form of calcite and dolomite[10]. The most abundant product is the CSH gel, it has been documented that its presence is very important because it provides good mechanical properties[1,2,6].

In Figure 2, for a temperature of 25°C are present as hydration products the CSH gel, hydrotalcite, strätlingie (C_2ASH_8) and in addition calcite, gehlenite and merwinite, the two latter are the crystalline phase of GBFS and remain in the diffractograms for all time and temperature of hydration. Among the products of hydration, it has hydrotalcite as is observed in the XRD patterns the corresponding peaks to the same at 22°, 35°, 43° and 53°. The hydrotalcite is a hydration product of all systems with GBFS.

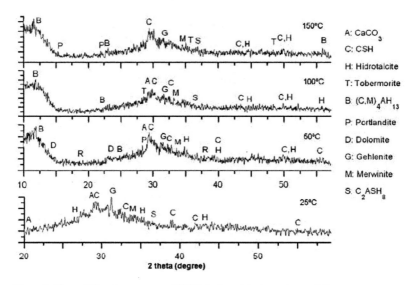

Figure 2. X-ray diffraction patterns of GBFS hydrated during a day at different temperatures.

Diffraction patterns for the hydration time of 28 days as a function of studied temperatures are shown in Figure 3. In this figure it is observed that the strätlingite (C_2ASH_8) peak located at 36.1° is not very intense at all temperatures, the presence of ($C,M)_4AH_{13}$ is well observed at 11.5° and there is another peak at 22°. Finally, in this same figure at 150°C are present as hydration products; the CSH gel, hydrotalcite, tobermorite, the AFm phases (($C,M)_4AH_{13}$ and C_2ASH_8), portlandite, and in addition dolomite, calcite, merwinite and gehlenite.

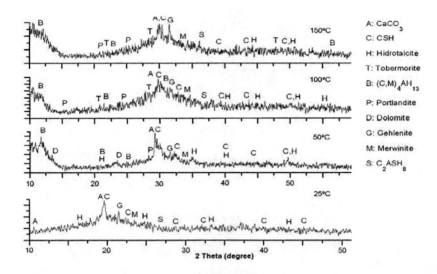

Figure 3. X-ray diffraction patterns of GBFS hydrated during 28 day at different temperatures.

Fourier transform infrared (FTIR)

Figure 4 shows the majority of bonds that were observed in all infrared spectra analyzed. The purpose of this figure is to visualize the main bonds that are found more frequently during GBFS hydration at different temperatures. The spectrum of this figure corresponds to the experimental conditions of 25°C and 14 days of hydration. Bands observed in all spectra are: the band of 3465 cm^{-1} corresponds to the vibrations of the O-H bond of water. Similarly, the band of 1650 cm^{-1} corresponds to water. The bands at 1480 cm^{-1} and 1416 cm^{-1} are due to the C-O bond in calcite, and so is the shoulder at 870 cm^{-1}. The band at 950 cm^{-1} is due to Si-O-Si of the CSH gel. The two small bands at 710 and 665 cm^{-1} are due to link Al-O-Al of aluminates. The small band begins to form at 594 cm^{-1} is due to the S-O bond. Finally, the band at 460 cm^{-1} and 483 cm^{-1} are due to the link TO$_4$ (where T = Si, Al, Mg, etc.). According to the FTIR spectra increments in the temperature favors the polymerization of the CSH gel and therefore its rapid formation, which has already mentioned is a very important hydration product due to good mechanical properties that can give to the cement manufactured with this slag.

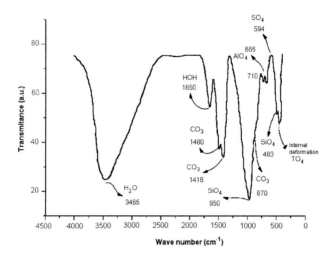

Figure 4. Main links in the GBFS hydrated.
Temperature and time of hydration 25°C and 14 days respectively.

CONCLUSIONS
- Due to the results of chemical composition and infrared analysis, granulated blast furnace slag (GBFS) presents the characteristics of a cementitious material and therefore it can be used in the construction of cementing oil wells.

- Hydration products observed were: the CSH gel, hydrotalcite, tobermorite, the AFm phases $((C,M)_4AH_{13}$ and $C_2ASH_8)$, portlandite, and in addition dolomite, calcite, merwinite and gehlenite, the two latter are the crystalline phase of GBFS

- The most abundant hidration product is the CSH gel, it presence is very important because it provides good mechanical properties.

- As the temperature increases the hydration products also increase, on the other hand, as the hydration time progresses, some hydration products intensify his band of infrared and X-ray diffraction peaks.

ACKNOWLEDGMENT
Authors would thank Departamento de Metalurgia y Materiales, ESIQIE-I. P. N. and Universidad Politécnica de Victoria for technical support.

REFERENCES

[1]J.I. Escalante García, Materiales Alternativos al Cemento Portland, *Revista Avance y Perspectiva*, **21**, march-april (2002). In Spanish.

[2]J.I. Escalante-García, A.F. Fuentes, A. Gorokhovsky, P.E. Faire-Luna and G. Mendoza-Suárez, Hydration Products and Reactivity of Blast-Furnace Slag Activated by Various Alkalis, *Journal of American Ceramic Society*, **86**, 2148-2153 (2003).

[3]M.G.P. Silva, C.R. Miranda, A.R. Almeida, G. Campos and M.T.A. Becerra, Slag Cementing Versus Conventional Cementing Comparative Bond Results, *PETROBRAS Latin American and Caribbean Petroleum Engineering Conference*, 30 August-3 September, Rio de Janeiro, Brazil, 1-14 (1997).

[4]R. Mejia de Gutiérrez and J. Maldonado, Propiedades de Morteros de Escoria Activada Alcalinamente, *Universidad del Valle, Escuela de Ingeniería de Materiales*, Cali, Colombia. (2003). In Spanish.

[5]M. Georgescua, C.D. Vointchib and L. Radub, Hardening Properties of Some Alkali Activated Blast Furnace Slag Binders, *Polytechnic University of Bucarest, Faculty of Industrial Chemistry*, Romani, (2000).

[6]H.F.W. Taylor, Cement Chemistry, 2nd Edition, Thomas Thelford, *American Society of Civil Engineers*, USA, (2004).

[7]ASTM C-114, Standart test method for chemical analysis of hydraulic cement (2006).

[8]NRF-069-PEMEX, Norma Mexicana para los Cementos clase H y H-Pemex Empleados en Pozos Petroleros, (2006).

[9]M.J. Martínez Alvarado, Estudio de la Hidratación de la Escoria Granulada de Alto Horno a Diferentes Temperaturas, *Master Thesis ESIQIE-IPN* (2009).

[10]W. Chen and H.J.H. Brouwers, The Hidratation of Slag, Part 1: Reaction Models for Alkali-Activated Slag, *Journal Materials Science*, **42**, 428-443 (2007).

IN SITU FORMATION OF WC PLATELETS DURING THE SYNTHESIS OF WC-Co NANO-POWDER

Yang Zhong[1], Angel L. Ortiz[2] and Leon L. Shaw[1]

[1] Department of Chemical, Materials and Biomolecular Engineering, University of Connecticut, Storrs, CT 06269, USA
[2] Departamento de Ingeniería Mecánica, Energética y de los Materiales, Universidad de Extremadura, Badajoz 06006, Spain

ABSTRACT
    WC platelets are found for the WC-Co nano-powder synthesized from the mixture of WO_3, Co_3O_4 and graphite. XRD and EDXS analysis show that W atoms are partially substituted by Co atoms in the WC crystal causing subtraction of the lattice. It is found that the formation of ternary carbides during reduction stage is important for the formation of WC platelets. The formation of WC platelets is a defect-assisted transformation process from the W-containing source to WC.

INTRODUCTION
    Cemented tungsten carbide with cobalt binder (WC/Co) has wide applications as machining, drilling and cutting tools in both military and civilian fields[1-4]. In all of the applications, good wear resistance derived from a combination of high hardness and high toughness is required. The properties of cemented carbides depend primarily on the binder content and grain size of WC[5].

    It is generally known that the hardness of WC/Co can increase by decreasing the grain size of WC to submicron or nanometer scales[6-9]. Grain growth inhibitors (e.g., VC, Cr_3C_2, NbC, TaC and LaB_6)[10-13] additions have already been proven to be useful to limit the growth of WC crystal in cemented carbides. These ultrafine cemented carbides possess high hardness values. However, the fracture toughness is reported to be decreased as the hardness increases[5]. Thus, the challenge is to obtain high values of both hardness and fracture toughness in order to improve the performance of the cemented carbide based tools.

    Novel design of engineered microstructures is one of the approaches to break this unfavorable general trend. In the 1960s, du Pont de Nemours and Company filed a series of patent applications on WC plate formation[14]. In the 1990s, Toshiba Tungalloy Co., Ltd filed another patent on the processing of plate-crystalline WC containing hard alloys[15]. Since then, several groups have studied the formation of WC plates during the sintering of WC/Co[11, 16-21]. There are mainly two modes of WC plate formation during sintering, i.e. (i) dissolution/reprecipitation process[14, 16] and (ii) reaction sintering process[15, 18]. In the first process, plates are formed through dissolution/reprecipitation of nano-sized WC on heating, while for the reaction sintering process, the WC plates are formed during sintering via the transformation of a tungsten containing source (e.g., Co_3W_3C, $W_9Co_3C_4$) to WC. It has been observed that platelet-nanocrystalline WC containing hard alloys exhibit a superior hardness/fracture toughness combination[19, 22]. It is proposed that the high hardness of WC/Co is due to a limited mean free path for slip, while the toughness depends on the aspect ratio of WC platelets[22]. The higher the aspect ratio, the more crack deflection and hence the higher toughness.

    Thus it is technologically important to study the platelet-nanocrystalline WC containing hard alloys. Up to now, this structure has only been reported as a product of sintering as described earlier. In the present study, for the first time, we report that the platelet-nanocrystalline WC containing WC-Co powders can be achieved *in situ* during synthesis without sintering. These powders can be used as the starting powders for coating and sintering process to achieve superior mechanical properties due to the ultrafine size of the powder and the platelet structures of WC crystals.

EXPERIMENAL

WO$_3$ (99.8%) and Co$_3$O$_4$ (99.7%) purchased from Alfa Aesar and graphite (99.99%) purchased from Aldrich were used as the starting materials. To reduce its large particle size, graphite powder was ball milled 6 h before mixing with other powder. These WO$_3$, C, and Co$_3$O$_4$ powders were mixed in a molar ratio of 1:2.4:0.123 to form a final product of WC-10 wt% Co. The mixed powders were subsequently ball milled using a modified Szegvari attritor[23, 24]. The ball milled powders were then treated in a mixture of H$_2$ (P$_{H2}$ = 0.5 atm) and Ar (P$_{Ar}$ = 0.5 atm) at 670°C for 2 h for the reduction process. The as reduced powders were subjected to the carburization at 1000°C for 1 h in flowing Ar atmosphere. For comparison purposes, the as milled powders were also heated at 1000°C for 1 h in flowing CO atompshere. Detailed experimental parameters can be found in the previous study[23].

Direct observation of morphologies and crystal structures of the powders was made using scanning electron microscope (FESEM, JEOL, JSM 6335F) and transmission electron microscope (TEM, JEOL 2010 and TEM, Fei Technai). Elemental analysis was taken by the EDXS equipped with the TEM. Phase identification was carried out by employing X-ray diffraction (XRD) with Cu Ka radiation (Bruker Axs D5005D X-ray Diffractometer). Rietveld analysis[25] of the XRD patterns was employed to analyze the structure of the WC platelets.

RESULTS AND DISCUSSIONS

Figure 1 shows the XRD patterns for the WO$_3$ + Co$_3$O$_4$ +C powder mixture in the as received form and after different treatments. The comparison between Figure 1(a) and (b) clearly shows the peak broadening and the decrease of the intensities of the WO$_3$ peaks, indicating the nano-crystallite formation and/or introduction of the internal strain. It is calculated that the crystallite size for WO$_3$ after milling is ~22 nm. The fact that Co$_3$O$_4$ peak is hardly seen after milling is due to the small amount of powder in the mixture. Figure 2(a) shows that particles are around 30~300nm after 6 h milling, creating large surface specific area and a uniform mixture. Thus the powder after milling is nano-crystalline with large surface specific areas, which will enhance the reaction kinetics and improve the diffusivity. After heating in H$_2$ for 2 h at 670°C, W and Co$_3$W$_3$C are formed based on the XRD pattern shown in Figure 1(c). The WO$_3$ and Co$_3$O$_4$ are reduced by H$_2$ and/or graphite, forming W and Co. W and Co react to form Co$_3$W, which will further react with graphite to form Co$_3$W$_3$C. This kind of reaction pathway is identified in the previous study via residual gas analysis[23]. Shown in Figure 2(b) and (c) are the bright-field and dark-field TEM images for the powder after heating at 670°C, it can clearly be seen that the crystallites are in the range of 10~40 nm, indicating the formation of nano-crystalline W/ Co$_3$W$_3$C.

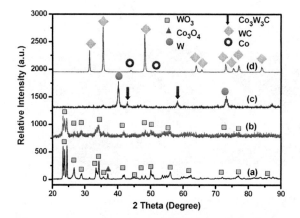

Figure 1. XRD patterns for $WO_3 + Co_3O_4 + C$ powder mixture (a) in as-received form, (b) after milling for 6h, (c) after milling for 6h and heating in H_2 at 670°C for 2h and (d) after milling for 6h, heating in H_2 at 670°C for 2h and heating in Ar at 1000°C for 2h.

From Figure 1(d), it is observed that after heating in Ar for 2 h at 1000°C, the powder mixture comprises of WC and Co, indicating that the ternary phases as well as W have been transformed to WC phase, while Co have been expelled from the ternary phases. The comparison between the calculated crystalline size of WC (~40 nm) and the crystallite size of W/ Co_3W_3C after reduction suggest that the formed WC crystals have little or no growth. Figure 3(a) shows the morphologies of the as-synthesized WC-Co. It is important to note that many platelets structures are found in the images, indicating the formation of WC plates during the synthesis process. Our prior first principles study[26] shows that the equilibrium structures of WC are "bulky" rather than platelet. Hence, the structure formed is a non-equilibrium structure. It is suggested by Shatov, et al. that the shape of WC crystals in cemented carbide is determined by a balance of shape relaxation (towards equilibrium structure) and carbide crystal growth processes (towards growth-determined structure)[16]. We also found that the growth of WC in WC-Co will follow different fashions at different heating conditions[27]. In our case, as the temperature is low, the growth and relaxation process still have little effect on the final structure of the WC crystal. Hence, the platelet structure of WC is formed *in situ* during the synthesis process.

To better understand the structure of the WC plates formed in this study, Rietveld refinement analysis[25] was carried out for the XRD pattern. It is found that the lattice parameters (0.290129(22) nm for *a* and 0.283328(23) nm for *c*) are smaller than the theoretical numbers in the PDF card 00-051-0939 (0.290631 nm for *a* and 0.283754 for *c*), indicating a subtracting distortion exists in the WC crystal. Kobayashi, et al. observed that the WC platelets formed during sintering process possess a higher *c/a* ratio than the theoretical value [15] and they also noticed the elongated distortions of the lattice in both *a* and *c* directions. In our case, the c/a ratio (0.9765) is slightly larger than the theoretical numbers (0.9763), while the lattices are subtracted rather than elongated. It is also found by the Rietveld analysis that some W atoms are substituted by Co atoms, and W to Co occupation ratio is 4.48, meaning ~18 at.% W atoms are substituted by Co atoms. Hence, it is reasonable to argue that the subtraction of lattice parameters is due to the smaller atomic size of Co compared with W[28].

Figure 2. (a) SEM image of $WO_3+Co_3O_4+C$ powder mixture after milling for 6h, (b) Bright field TEM image and (c) Dark field TEM image of $WO_3+CO_3O_4+C$ powder mixture after milling for 6h and heating at 670°C in H_2 for 2h.

Figure 3. (a) SEM image and (b) TEM image of 6 h ball milled $WO_3+Co_3O_4+C$ powder mixture after heating at 670°C in H_2 for 2h and at 1000°C in Ar for 2h. The labels in figure (b) show the region of the EDXS analysis.

Elemental analysis is carried out using energy dispersive X-ray spectroscopy (EDXS) equipped in the HRTEM to study the chemical composition of a single WC particle. Figure 3(b) shows a high resolution TEM image of WC crystal, as we can notice that there are some defects in the crystal. The EDXS for the whole crystal shows the W to Co atomic ratio is 82:18, which agrees well with the Rietveld analysis. To clearly prove the Co atoms are embedded in the WC crystal rather than forming a thin film on the WC crystal, EDXS patterns were taken for four spots (i to iv) shown in Figure 3(b). If one argues the Co atoms are from the thin film rather than in the WC crystal, a higher Co ratio should be expected that from the spots at the edges of the crystal (ii to iv) than the center (spot i). The results for all the four spots show constantly that the W to Co atomic ratio is around 82:18. Thus, it proves that some W atoms are substituted by Co atoms in the WC crystal, leading to the subtraction of the lattice. In our study, as described earlier, the oxide particles are reduced to form W and also Co_3W_3C in the reduction step. The further carburization step of the W and ternary carbides will form WC. During the carburization process, the formed Co_3W_3C react with graphite to form WC and release Co atoms from the ternary carbides. In this study, some Co atoms are retained in the WC solid solution. These retained Co atoms caused the subtraction of the WC lattice.

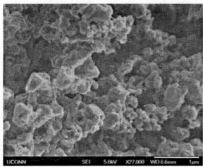

Figure 4. SEM image of 6 h ball milled $WO_3+Co_3O_4+C$ powder mixture after heating at $1000^{\circ}C$ in Co for 2h.

In all the studies of WC platelets[11, 16-21], ball milling process is employed to create defects. It is also believed that the defects, especially stacking faults are related to the formation of WC platelets. In our study, stacking faults are clearly seen in the TEM image. Also, the Co dissolution in the WC crystals leads to the subtraction of WC crystals. Thus, the formation of WC platelet is a defect-assisted process.

As is mentioned earlier, WC platelets can be formed during the sintering of WC-Co[17-19]. During the sintering, WC is decarburized forming some ternary carbides and the ternary carbides react with graphite to transform to WC. It is believed that the transformation of ternary carbides to WC during sintering leads to the formation of WC platelets[17]. The reaction sintering process of W containing sources and C can also promote the formation of WC platelets by the reaction of W containing sources and C to form WC[15-18]. Comparing these reports with the present study, it can clearly be seen that the transformation of ternary carbides to WC is a critical process for the formation of WC platelets. To further prove this, the $WO_3 + Co_3O_4 +C$ powder mixture after milling for 6 h is directly heated at $1000^{\circ}C$ for 2 h in CO atmosphere. In this process, the WO_3 is directly carburized to WC without the formation of ternary carbides. The morphologies of the powders after heating are shown in Figure 4. It is clearly seen that the powders exhibit "bulky" rather than platelet structure. Thus, the reduction step of oxide to form ternary carbides is a critical step for the formation of WC platelets. The powders after ball milling cannot proceed directly to platelet morphologies without the reduction stage to form ternary carbides.

CONCLUSION

WC platelets are formed *in situ* during the synthesis of nano-crystalline WC-Co powder. The comparison of WC morphologies between two step synthesis and one step synthesis clearly indicates that the reduction step is critical for the formation of WC platelets. XRD and Elemental analysis proves the Co dissolution into WC crystals and the subtraction of the lattice. TEM also show the defects in the WC crystal. The in situ formation of WC platelets is a defect assisted transformation process of the W containing source (ternary carbides) to WC.

ACKNOWLEDGEMENT
This research was sponsored by the U.S. National Science Foundation (NSF) under the contract number CMMI-0856122. The support and vision of Dr. Mary Toney is greatly appreciated.

REFERENCES

[1]P Schwarzkopf, R. Kieffer, Refractory Hard Metals. New York: The Macmillan Company; p. 138-161 (1953).

[2]D.H. Jack, In: Schwartz MM, editors. Engineering Applications of Ceramic Materials: Source Book. Materials Park, OH: American Society for Metals; p. 147-153 (1985).

[3]Metals Handbook, Vol. 3: Properties and Selection: Stainless Steels, Tool Materials and Special-Purpose Metals, 9th edition. Materials Park, OH: American Society for Metals (1980).

[4]S.W.H. Yih and C.T. Wang, Tungsten Sources, Metallurgy, Properties and Applications. New York: Plenum Press (1979).

[5]Z.Z. Fang ZZ, X. Wang, T. Ryu, K. Hwang and H. Sohn, Synthesis, sintering, and mechanical properties of nanocrystalline cemented tungsten carbide – A review. Int. J. Refract. Met. Hard Mater., 27, 288-99 (2009).

[6]Z.Z. Fang and J.W. Eason, Study of nanostructured WC-Co composites, Int. J. Refract. Metal Hard Mater., 13, 297-303 (1995).

[7]I. Azcona, A. Ordonez, J.M. Sanchez and F. Castro, Hot isostatic pressing of ultrafine tungsten carbide-cobalt hardmetals, J. Mater. Sci., 37, 4189-95 (2002).

[8]J.M. Densley and J.P. Hirth, Fracture toughness of a nanoscale WC-Co tool steel, Scripta Mater., 8, 239-44 (1997).

[9]K. Jia, T.E. Fischer and B. Gallois, Microstructure, hardness and toughness of nanostructured and conventional WC-Co composites, Nanostruct Mater., 10, 875-91 (1998).

[10]C.W. Wills, D.J. Morton and K. Stjernberg, The temperature ranges for maximum effectiveness of grain growth inhibitors in WC–Co alloys, Int. J. Refract. Met. Hard Mater., 23, 287-93 (2005).

[11]G.H. Lee and S. Kang, Sintering of nano-sized WC–Co powders produced by a gas reduction–carburization process, J. Alloys Compd., 419, 281–9 (2006).

[12]S.G. Huang, L. Li, O. Van der Biest and J. Vleugels, VC- and Cr_3C_2-doped WC–NbC–Co hardmetals, J. Alloys Compd., 464, 205–11 (2008).

[13]S.S. Shen, D.H. Xiao, X.Q. Ou, M. Song, Y.H. He, N. Lin and D.F. Zhang, Effects of LaB6 addition on the microstructure and mechanical properties of ultrafine grained WC–10Co alloys, J. Alloys Compd., 509, 136-43 (2011).

[14]G. Meadow, Anisodimensional tungsten carbide platelets bonded with cobalt, US Patent Application 418808, United States Patent Office (1964).

[15]M. Kobayashi, K. Kitanura and S. Kinoshita, Plate-crystalline tungsten carbide-containing hard alloy, compostion of forming plate-crystalline tungsten carbide and process for preparing said hard alloy, European Patent 0759480A 1, European Patent Office (1995).

[16]A.V. Shatov, S.A. Firstov and I.V. Shatova, The shape of WC crystals in cemented carbides, Mater. Sci. Eng. A 242, 7-14 (1998).

[17]M. Sommer, W.D. Schubert, E. Zobetz and P. Warbichler, On the formation of very large WC crystals during sintering of ultrafine WC-Co Alloys, Int. J. Refract. Met. Hard Mater., 20, 41-50 (2002).

[18]B. Reichel, K. Wagner, D.S. Janisch and W. Lengauer, Alloyed W-(Co, Ni, Fe)-C phases for reaction sintering of hardmetals, Int. J. Refract. Met. Hard Mater., 28, 638-45 (2010).

[19]V. Bonache, M.D. Salvador, D. Busquets, P. Burguete, E. Martinez, F. Sapina and E. Sanchez, Synthesis and processing of nanocrystalline tungsten carbide: Towards cemented carbides with optimal mechanical properties, Int. J. Refract. Met. Hard Mater. 29, 78-84 (2011).

[20]F.A. da Costa, F.F.P. de Medeiros, A.G.P. da Silva, U.U. Gomes, M. Filgueira and C.P. de Souza, Structure and hardness of a hard metal alloy prepared with a WC powder synthesized at low temperature, Mater. Sci. Eng. A 485, 638-42 (2008).

[21]S. Lay S, M. Loubradou, W.D. Schubert, Structural analysis on planar defects formed in WC platelets in Ti-doped WC-Co, J. Am. Ceram. Soc., 89, 3229-34 (2006).

[22]L. Shaw L, L. Hong and Y. Zhong, submitted to Mater. Sci. Eng. A.

[23]Y. Zhong and L. Shaw, A study on the synthesis of nanostructured WC- 10wt% Co particles from WO_3, Co_3O4 and graphite, J. Mater. Sci., **46**, 6323-31 (2011).

[24]Y. Zhong, L. Shaw, M. Manjarres, and M.F. Zawrah, Synthesis of Silicon Carbide Nanopowder using Silica Fume, J. Am. Ceram. Soc., **93**, 3159-3167 (2010).

[25]H.M. Rietveld, A profile refinement method for nuclear and magnetic structures, J. App. Crystallography, **2**, 65–71 (1969).

[26]Y. Zhong, H. Zhu, L. Shaw and R. Ramprasad, The Equilibrium Morphology of WC Particles - A Combined Ab Initio and Experimental Study, Acta. Mat., **59**, 3748-57 (2011).

[27]Y. Zhong and L. Shaw, Growth mechanisms of WC in WC–5.75 wt% Co, Ceramics International. **37**, 3591-3597 (2011).

[28]N.N. Greenwood, A. Earnshaw, Chemistry of the Elements. 2nd ed. Oxford: Butterworth-Heinemann (1997).

SYNTHESIS, SHAPED AND MECHANICAL PROPERTIES OF HYDROXYAPATITE–ANATASE BIOMATERIALS

Roberto Nava-Miranda, Lucia Téllez-Jurado
ESIQIE-I. P. N., UPALM-Zacatenco, 07738, D.F., México

Enrique Rocha-Rangel
Universidad Politécnica de Victoria, Av. Nuevas Tecnologías 5902, Parque Científico y Tecnológico de Tamaulipas, 87138, Cd. Victoria, Tamaulipas, México.

ABSTRACT
Hydroxyapatite powder was synthesized by the precipitation technique, starting from aqueous solutions of phosphoric acid and calcium hydroxide. To the hydroxyapatite powders were added 5, 10 or 20 wt % titanium oxide (anatase) as a reinforcing agent, powders were mixed by high energy mechanical milling during different times (0.5, 0.75 and 1h) in order to obtain homogeneous mixtures. With the powder mixture it was manufactured a paste, same that was extruded to obtain a solid cylinder using a stainless steel die. Finally, extruded cylinders were sintered in an electrical furnace at 1100°C during 1 h. Synthesized materials were characterized by IR, XRD, SEM and mechanically. XRD results show that in the sintered products are present mainly the crystalline phases; hydroxyapatite and titanium oxide in anatase form, these phases also could be identified in the infrared analysis. After milling, powders present sizes smaller than 5 microns. Microhardness and fracture toughness was similar to the values of human bone reported in literature of these same properties.

INTRODUCTION
The calcium phosphate most commonly used in the manufacture of implants is hydroxyapatite $Ca_{10}(PO_4)_6(OH)_2$, because it is the main inorganic constituent of bone tissue[1-3]. The main characteristics of hydroxyapatite are its biocompatibility, non-toxicity, chemical stability, osteoconduction and bioactivity[4]. On the other hand, titanium oxide (TiO_2) is a ceramic material whose major holdings for use as a biomaterial are its biocompatibility and resistance to compression. The studies and experiments with hydroxyapatite have shown that possess properties which make it be the base material used in the scope of biomaterials, however its most significant limitations are its low mechanical properties, because of this, it has been used other accompanying materials to improve its mechanical properties. The most interesting results that have been found in the production of hydroxyapatite-anatase bioceramics are the development of a biomaterial that has better properties than hydroxyapatite pure, since it increases the hardness, compressive strength, modulus of rupture, and not alter or modify the biocompatibility[5]. There are several techniques to form ceramic materials (bioceramics) with different geometries and shapes. Extrusion is one of these forming techniques, in which a mass is used as a mixture of different compounds among which are ceramic powders, plasticizers, solvents and dispersants. This mass is plastically deformed by passing it through a given cross section of which give the dough the desired final geometry. After extrusion, the product was dried and sintered in a controlled manner to obtain the desired physical properties[6]. In this work, hydroxyapatite is manufactured using a chemical synthesis method starting from aqueous solutions. The hydroxyapatite powders obtained were mixed with anatase powders, through high energy milling varying the weight percentages of TiO_2, for later conform samples to cylindrical shape and well established dimensions. Finally, they were characterized and evaluated some mechanical properties (fracture toughness and hardness) to compare the results with the same mechanical properties of human bone reported in the literature.

EXPERIMENTAL PROCEDURE

The synthesis of materials study was carried out at different stages: a first stage consisted in to obtain hydroxyapatite through the precipitation method. In a second stage, hydroxyapatite and anatase powders were characterized. In the third stage hydroxyapatite-anatase mixtures were obtained using high energy milling. The fourth stage consists in the characterization of powders produced by high energy milling. In the fifth stage, the powders will be conformed using the extrusion method; finally some mechanical properties were assessed and compared with mechanical properties of human bone.

Hydroxyapatite synthesized by the precipitation technique

The synthesis of hydroxyapatite was carried out using the method of precipitation. Table 1 describes the specifications of raw materials used for the synthesis of hydroxyapatite. As precursors of hydroxyapatite were used phosphoric acid and calcium hydroxide.

Table 1. Raw materials characteristics for hydroxyapatite synthesis by the precipitation method.

Raw materials	Chemical formula	Purity %	Indications	Commercial brand
Phosphoric acid	H_3PO_4	99.999 (85% in water)	Corrosive	J.T. Baker
Calcium hydroxide	$Ca(OH)_2$	95	Irritant	Sigma – Aldrich

Titanium oxide (anatase)

The characteristics of TiO_2 used in the production of study materials are given in Table 2. From this information it should be noted that powder is present as anatase phase, its purity and size is over 99% and very fine respectively.

Table 2. Characteristics and properties of titanium oxide.

Commercial brand	SIGMA – ALDRICH
Name of the chemical compound	Titanium oxide (anatase)
Purity (%)	≥ 99
Size particle (mesh #)	-325
Density (gcm^{-3})	3.9
Melting temperature (°C)	1825

Obtaining of hydroxyapatite–anatase mixtures (high energy milling)

The hydroxyapatite–anatase mixtures were obtained in a high energy ball mill (Simoloyer, CM01-2L Zoz GmbH, Germany). The grinding chamber is lined with sheets of Si_3N_4, designed for grinding ceramic to avoid powders contamination by wear of the grinding chamber. The grinding media used were commercial stainless steel balls with diameter of 5mm. Milling of the powders (hydroxyapatite + anatase) was performed during 0.5, 0.75 and 1h in air at room temperature, with a ball-to-powder weight ratio of 30:1

Forming of powder through the extrusion technique

The technique used for forming and fabrication of the ceramic pieces (solid cylinders) was direct extrusion combined with a sintering process. The extrusion device consists of an extrusion container, plunger, die and the die support. Every part was made of stainless steel. The system was assembled to a universal testing machine. The ceramic bodies were prepared according to Table 3.

Table 3. Compositions of materials used in the manufacture of extruded ceramics.

Hydroxyapatite (wt. %)	TiO$_2$ (anatase) (wt. %)	Ceramic paste		
		Ceramic powders HA-TiO$_2$ (wt. %)	plasticizer (methocel) (wt. %)	Distilled water (wt. %)
95	5	80	5	15
90	10	75	7	18
80	20	70	10	20

The constituents of the ceramic paste were kneaded manually until get a homogeneous and consistent paste. Then paste was placed in the extrusion container and it was applied a preload of 100 kgf on the past to remove air trapped in it. The paste was extruded by applying pressures of 200 kgf and speeds of 1cm/min. They were obtained green solid cylindrical samples of 1.5cm in diameter. The extruded samples were air-dried during 72h. The next step was the sintering of samples at 1100°C during 1h in an electrical furnace (Carbolite RHF17/3E). Heating and cooling rates were 5 °C/min.

Characterization and mechanical properties evaluation

Through infrared spectroscopy (IR) it were determined the types of bonds and functional groups of hydroxyapatite and anatase powders. X-ray diffraction (XRD, SIEMENS D5000 with Cu Kα) was used to identify phases in the materials. Scanning electron microscopy (SEM, Philips 6300) allowed to analyze the microstructure. Hardness measurement was performed using a Vickers hardness tester (Nikon, DS-L2, Japan), with a pyramidal diamond indenter, and using loads of 100g during 15s on the outer surface. They were performed a minimum of ten indentations for each measure of hardness[7]. To measure the fracture toughness (K$_{IC}$) it was used the indentation fracture method proposed by Evans[8].

RESULTS AND DISCUSION

X-ray diffraction analysis of synthesized hydroxyapatite [Ca$_{10}$(PO$_4$)$_6$(OH)$_2$]

The pattern of X-ray diffraction of the hydroxyapatite dry powder obtained by the precipitation method from aqueous solutions (phosphoric acid and calcium hydroxide) is presented in Figure 1. Figure 1 shows the characteristic X-ray diffraction peaks of hydroxyapatite. During its synthesis they were occurred chemical reactions between aqueous solutions of phosphoric acid (H$_3$PO$_4$) and calcium hydroxide [Ca(OH)$_2$]. A phase that accompanied the hydroxyapatite formation was the β-tricalcium phosphate [β-Ca$_3$(PO$_4$)$_2$]. Similarly, figure 1 shows that most of the peaks are broad; this feature predicts that the particle size of the powder samples obtained by the precipitation method is about (10-20 microns). In the same figure 1, it is show at an angle of 2Θ = 31.774° the most intense peak that is characteristic of the hydroxyapatite.

X-ray diffraction analysis of titanium oxide (anatase) commercial powders

The pattern of X-ray diffraction in Figure 2 corresponds to titanium oxide dry powder. This pattern shows well-defined peaks, characteristic of the TiO$_2$ anatase phase, as present in the diffractogram, peaks are well defined and correspond to those presented in the (00-021-1272) diffraction of anatase phase chart. It is also seen clearly that all peaks shown in the diffractogram are fully defined and narrow; this is because the powders of the sample have a fully crystalline structure. We also analyzed the X-ray spectrum using the 00-021-1272 diffraction chart, looking for the presence of another possible phase of titanium oxide as rutile, but with the analysis performed was discarded the presence of a second phase in the titanium oxide powder. Narrow peaks indicate that the size of particles in the powder is of the order ≤ 1micron. The most intense peak observed in the diffraction pattern in figure 2, which is in the position angle 2Θ = 25.281° is the characteristic peak of TiO$_2$ anatase phase.

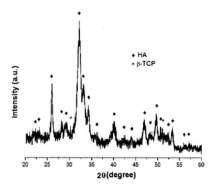

Figure 1. X-ray diffraction pattern of sample in powder of hydroxyapatite
prepared by the precipitation method at room temperature.

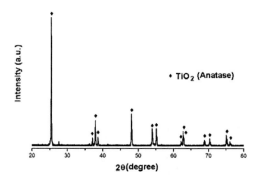

Figure 2. X-ray diffraction pattern of titanium oxide powder.

Infrared spectroscopy analysis of obtained hydroxyapatite

Figure 3 shows the infrared spectrum of hydroxyapatite obtained by the precipitation method, in which are observed frequencies of the bands of the bonds and functional groups of the hydroxyapatite. In the spectrum of figure 3 the vibration band frequencies at 633 and 3578 cm^{-1} are due to O-H bonds belong to the functional group hydroxyl (OH^{1-}), this functional group is characteristic of hydroxyapatite. The vibration band frequencies at 474, 574, 600, 962 and 1046 cm^{-1}, occur due to O-P bonds belong to the phosphate functional group (PO$_4^{3-}$), also this characteristic functional group is always present in the hydroxyapatite. Finally, the vibration band frequencies at 875, 1419 and 1465 cm^{-1} belong to the O-C bonds of the functional group carbonate (CO$_3^{2-}$). The presence of this functional group is due to the partial replacement of the phosphate group (PO$_4^{3-}$) in the crystalline structure of hydroxyapatite.

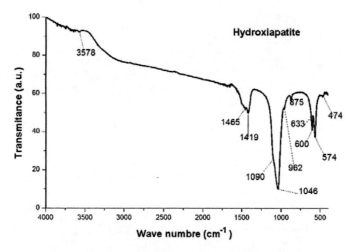

Figure 3. Infrared spectra of hydroxyapatite [$Ca_{10}(PO_4)_6(OH)_2$] obtained by the precipitation method.

Infrared spectroscopy analysis of titanium oxide commercial powder in anatase form

Figure 4 shows the infrared spectrum obtained from the commercial anatase phase of titanium oxide powder. It can be seen in the spectrum, the appearance of 4 peaks, the peaks that is located at the vibration frequencies of 500 and 700 cm^{-1} corresponds to bond Ti-O-Ti, the peak at 1630 cm^{-1} corresponds to the presence of water (H_2O), finally the frequency of vibration that appears at 3400 cm^{-1} corresponds to link hydroxyl (OH^{-1}).

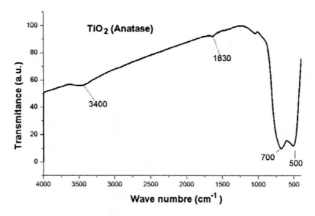

Figure 4. Infrared spectra of titanium oxide commercial powder in anatase form.

X-ray diffraction analysis of milled (0.5, 0.75 y 1 h) powder mixtures

Figure 5 shows the X-ray diffraction spectra of hydroxyapatite, anatase and mixtures of these two ground components at different times (0.5, 0.75 and 1h). In the diffractograms of the three mills, are observed the peaks belonging to hydroxyapatite and anatase, because there is no movement or appearance of peaks at different 2θ angular positions. None of the mixtures made by high-energy grinding are presenting chemical reactions, and formation of new phases, or the appearance of different compounds with hydroxyapatite or anatase. It is noted that high energy milling, allowed the homogenization of the powders, thus decreasing the particle size.

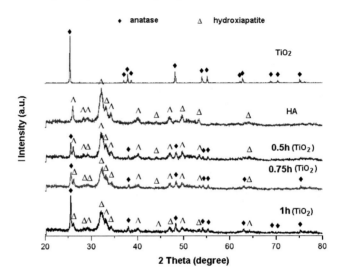

Figure 5. X-ray diffraction patterns of anatase, hydroxyapatite, milled for 0.5, 0.75 and 1h.

Scanning electron microscopy analysis milling 0.5 h

The micrograph in Figure 6, shows small clusters of hydroxyapatite particles and anatase, these mixes are about of 20 microns, it is clear that high-energy milling that was performed the hydroxyapatite + anatase powders, reduced the size particle, grinding time for this mixture was 0.5h and milling powders of hydroxyapatite was carried out with 95 wt. % hydroxylapatite + 5 wt. % anatase, is not as significant the presence of anatase particles in the matrix (hydroxyapatite) of the mixture due to the small amount of that ceramic powder. However, it is observed small amounts reach the anatase particles on the surface of some agglomerates of hydroxyapatite.

With this milling (0.5h) it was possible to reduce the particle size of anatase from 0.5 microns to 0.25-0.3 microns, also were able to reduce the number and size of the agglomerated particles. We did not observe chemical reactions or diffusion between the particles of hydroxyapatite + anatase, which itself is remarkable is that the particles of anatase were found on the surface of the agglomerates and some particles that are larger than the particles of anatase. Anatase particles shows spherical morphology and large agglomerates, but the distribution of these anatase particles is not homogeneous.

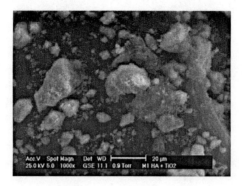

Figure 6. Resulting powder from the milling during 0.5h.

In the micrograph of Figure 7 it is show the grinding powder mixture during 0.75h, the milling was performed in a mixture of powder with 90 wt. % hydroxyapatite + 10 wt. % anatase, there is a homogeneous mixture of both powders, anatase particles are distributed over the surface of larger particles (hydroxyapatite). The morphology of anatase particles is in the form of flakes, with high-energy milling, the particle size of anatase decreased from 0.5 microns to 0.26 microns. The smaller particles seen in the micrograph correspond to anatase, it is clear that in this zone are seen the anatase particles, because in the grinding it was increased the amount (wt. %) of anatase.

Figure 7. Resulting powder from the milling during 0.75h.

The micrograph of Figure 8 shows the mixture powders milling during 1h in a high energy mill. This micrograph shows a good distribution of hydroxyapatite and anatase particles, the larger particles correspond to hydroxyapatite, and smaller particles to anatase. Hydroxyapatite particles are observed with spherical morphology and sizes on the order of 3-5 microns. Anatase particles have morphology of flakes and small spheroids of sizes on the order of 0.15 microns.

Figure 8. Resulting powder from the milling during 1h.

X-ray diffraction analysis of sintered samples at 1100°C during 1h

The diffraction patterns presented in Figure 9 correspond to samples sintered at 1100°C during 1h, with different contents of titanium oxide. Unlike the XRD analysis of the compounds non-heat treated these patterns show a gradual change in both phases (hydroxyapatite and titanium oxide). In the anatase pattern it is shows the beginning of the transformation from anatase to rutile phase. In the pattern of the hydroxyapatite there is a different phase belonging to tricalcium phosphate. It is noteworthy that tricalcium phosphate is presented in the three compounds (5, 10 and 20 wt. % TiO_2) and its intensity peak increases with increasing TiO_2 content and milling time, these results were also observed for some researchers[9], they mentioned that tricalcium phosphate is obtained for both following reasons: by mixing hydroxyapatite and titanium oxide in its anatase phase and when the compound is exposed to a very high sintering temperature.

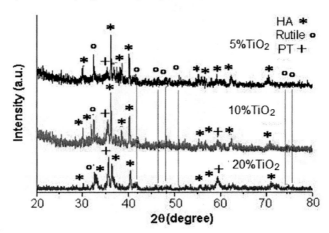

Figure 9. X-ray diffraction analysis of different compounds with ratios HA/TiO2: 95/5, 90/10 and 80/20 sintered at 1100 °C during 1h.

Microstructure

Figure 10 shows the micrograph of the sample prepared with 20 wt. % anatase and sintered at 1100°C during 1h. There is a fine and homogeneous microstructure with equiaxed shape and grain sizes smaller than 10 microns. Also, there is some isolated porosity in the sample. The presence of anatase phase is not observed here, however, is thought to be found in intergranular positions and for that reason its size should be much less than 1 micron. In general, it has a dense and well-sintered body.

Figure 10. Microstructure of sample prepared with
20 wt. % anatase and sintered at 1100°C during 1h.

Mechanical properties

Table 4 presents the values of hardness and fracture toughness for hydroxyapatite-anatase composites sintered at 1100°C during 1h. It also presents data of the same properties measured in the bone of a human femur. From the data presented in this table, is that there is a slight decrease in the hardness of the composites as well as the content of anatase in the same increases, this in turn results in a slight increase in fracture toughness of the composites with higher content of anatase. On the other hand, we have that hardness and toughness results of materials manufactured here, are very similar to data reported for these same properties evaluated in a human femur bone[10].

Table 4. Hardness and fracture toughness of femur bond and of the hydroxyapatite-anatase composites sintered at 1100°C during 1 h.

Hydroxyapatite (wt. %)	Titanium oxide (anatase) (wt. %)	HV (GPa)	K_{IC} (MPa·m$^{-1/2}$)
95	5	5.277 +/- 0.013	0.97 +/- 0.08
90	10	5.273 +/- 0.012	1.0 +/- 0.09
80	20	5.270 +/- 0.015	1.02 +/- 0.09
Femur bond		5.272	1.0

CONCLUSIONS

Hydroxyapatite-titania was obtained by the precipitation method at room temperature through the reaction between mixtures of phosphoric acid and calcium hydroxide presented in aqueous solution. The results of the characterization by XRD and IR showed the formation of the desired compound in addition to small amounts of tricalcium phosphate. The analysis of powders by SEM allowed the observation of homogeneous mixtures of the two main phases present in the hydroxyapatite-anatase compounds. During sintering at 1100°C there is a slight transformation of titanium oxide because the anatase phase changes to rutile phase. Also it is observed the formation of tricalcium phosphate at this stage in those samples with higher content of titanium oxide. The microstructure of the composites is homogeneous and has fine and equiaxed grains with sizes smaller than 10 microns. Finally, it has that hardness and toughness of materials manufactured here are very similar to data reported for these same properties evaluated in a human femur bone.

ACKNOWLEDGMENT

Authors would thank Departamento de Metalurgia y Materiales, ESIQIE-I.P.N. and Universidad Politécnica de Victoria for technical support. As well as LTJ and ERR the support from SNI and RNM the support from CONACyT.

REFERENCES

[1]A. Osaka, Y. Miura, K. Takeuchi, M. Asada, K. Takahashi, Calcium apatite prepared from calcium hydroxide and orthophosphoric acid. *J of Materials Science: Materials in Medicine* **2**, 51-55 (1991).

[2]K. Niespodziana, K. Jurczyk, J. Jakubowicz, M. Jurczyk, Fabrication and properties of titanium–hydroxyapatite nanocomposites, *Materials Chemistry and Physics* **123**, 160–165 (2010).

[3]W. Que, K.A. Khor, J.L. Xu, L.G. Yu, Hydroxiapatite/titania nanocomposites derived by combining high energy ball milling with spark plasma sintering processes. *J. of Eur. Ceram. Soc.* **28**, 3083-3090 (2008).

[4]M. E. Londoñ, A. Echavarría, F. de la Calle, Características cristaloquímicas de la hidroxiapatita sintética tratada a diferentes temperaturas, *Revista EIA* **5**, 109–118 (2006).

[5]E. Fidancevska, G. Ruseska, J. Bossert, A. Boccaccini, Fabrication and characterization of porous bioceramic composites based on hydroxyapatite and titania, *Materials Chemistry and Physics* **103**, 95–100 (2007).

[6]E. Rocha, R. H. López, E. R. García, Producción de cerámicos triangulares base alúmina por la técnica de extrusión, *Ingenierías* **VI**, 19-23 (2003).

[7]J.J. Saura, K. A. Habib, C. Ferrer M. S. Damra, I. Cervera, E. Giménez, L. Cabedo Propiedades mecánicas y tribológicas de recubrimientos alúmina/titania proyectados por oxifuel (spray llama), *Bol. Soc. Esp. Ceram.* 47, 1 7-12 (2008).

[8]A. G. Evans and E. A. Charles, Fracture Toughness Determination by Indentation, *J. Am. Ceram. Soc.*, **59**, 371-372 (1976).

[9]M. E. Santos, M. de Oliveira, L. P. de F. Souza, H. S. Mansur, W. L. Vasconcelos, Synthesis control and characterization of hydroxyapatite prepared by wet precipitation process, *Materials Research*, **7**, 4 625–630 (2004).

[10]Y. Chun, G. Ying-kui, Z. Mi-lin, Thermal decomposition and mechanical properties of hydroxyapatite ceramic, *Trans. Nonferrous Met. Soc. China*, **20**, 2 95-99 (2010).

Combustion Synthesis and SHS Processing

COMBUSTION FORMATION OF Ti₂AlC MAX PHASE BY ELECTRO-PLASMA PROCESSING

Kaiyang Wang, Jiangdong Liang, P.G. Zhang and S.M. Guo
Mechanical Engineering Department, Louisiana State University, Baton Rouge, LA 70803

ABSTRACT

The formation of high-purity MAX Ti₂AlC powders from titanium, aluminum and graphite powder compact was directly obtained by electro-plasma process (EPP) and the following combustion reaction. The theoretical density of the powder compact was found to play an important role for the initiation of the combustion reaction. For the compacts with a higher than 70% of theoretical density, the combustion reaction occurred and the Ti₂AlC powders were obtained in-situ. Because of the occurrence of combustion reaction in the electrolyte liquid, the resultant particles experienced rapid quenching and resulted in the formation of high-purity Ti₂AlC powders. The resultant particles which were not experienced the rapidly liquid quenching were found to have less Ti₂AlC purity. X-ray diffraction (XRD), and scanning electron microscopy (SEM) were used to characterize the products after the processing of electro-plasma processing. The mechanism of formation of Ti₂AlC powders was also discussed.

INTRODUCTION

MAX Ti₂AlC ternary carbide belongs to a special class of materials known as nano-layered ternary compounds or "machinable ceramics". It has unique physical and mechanical properties, such as high melting point, good thermal and electrical conductivity, high strength and modulus, and machinability by both electrical discharge method and conventional cutting tools [1-3]. Right now there is a growing interest in layered ternary compounds since they have a variety of potential applications, which include gas turbine engine components, heat exchangers, heavy duty electric contacts, etc.

Self-propagating high-temperature synthesis (SHS) or combustion synthesis (CS) was first developed by the Institute of Structural Macrokinetics and Materials Science (ISMAN) in Russia in the late 1960's [4]. This technique employs exothermic reaction processing, which circumvents the difficulties of long processing time and high energy consumptions associated with the conventional sintering method.

In general, there are two modes by which combustion synthesis may occur: self-propagating high-temperature synthesis (SHS) and volume combustion synthesis (VCS) [5]. In both cases, reactants may be pressed into a pellet, typically cylindrical or parallelepiped – shaped. The samples are then heated by an external source (e.g. electric arc [6], tungsten coil [7], etc.), either locally (SHS) or uniformly (VCS), to initiate an exothermic reaction. Łopacinski et al. [8] also reported the formation Ti-Al-C compounds (Ti₂AlC and Ti₃AlC₂) by SHS method using powder mixtures of Al, Ti, C and compound TiAl.

Electrochemical discharges also known as contact glow discharge electrolysis, is a well known phenomenon and has been systematically investigated [9, 10,11] in the past. Oishi et al. [12] showed the fabrication of metal oxide nano-particles by anode discharges between a metal anode and molten salt electrolyte in argon atmosphere at 700 K by applying 500 V DC. Toriyabe et al. [13] reported the controlled formation of metallic nanoballs during plasma electrolysis. Liang et al. [14] reported to use EPP for surface cleaning. A gas/vapor sheath is formed at the solid electrode/electrolyte interface when the applied voltage is high enough to induce discharge plasma.

More recently, we already synthesized Al_2O_3 and TiC nanocomposites using this method [15]. In this paper, in order to synthesize Ti_2AlC particles, a same process, which involves electro-plasma, combustion reaction, and liquid quenching, was developed. The metals-carbon composites powder compact was used as a cathode. Through controlling electrolyte concentration, applied DC voltage and the compacting pressure, electro-plasma discharge occurred around the titanium-aluminum-graphite composites surface and induced the combustion reaction to produce Ti_2AlC particles.

EXPERIMENTAL PROCEDURES

The reactants were powders of titanium (99.8% pure, average particle size of 45 μm), aluminum (99% pure, atomized, average particle size of 30 μm), and graphite fibers (99% pure, diameter size of 5 μm). Appropriate amounts of these powders were mixed in accordance with the stoichiometric molar ratios of Ti:Al:C = 2:1:1, respectively. Mixing was done in a stainless steel ball-mill using stainless steel balls for 20 minutes in an argon atmosphere and the blends were then pressed into compacted electrode samples. The dimension of compact electrode was 20 x 5 mm² and 2.5 mm in thickness, respectively.

The electro-plasma apparatus was set-up as shown in Figure 1. The water based electrolyte solution was prepared with 75g/L sodium carbonate, which gave an equivalent electrical conductivity of 43±0.5 mS/cm. During the experiment, there is no preheating to the electrolyte bath. A DC power source (Magna-power PQA) was connected to the powder compact cathode and an AISI 304 stainless steel plate anode to form a complete circuit. The input voltage was set to be 150 V, which took 2±0.5 seconds to reach from zero. The cathode was fed by an electric motor at a constant speed (1.0 mm/min) toward the electrolyte bath. As the cathode just made contact with the electrolyte, electro-plasma formed between the electrolyte and the contacting cathode surface. Due to electrical heating, electrolyte turns into liquid-vapor two phases and instability occurs. High potential between the electrodes leads to concentration of positive ions that are present in the electrolyte, in the close proximity of the cathode, mostly on the surface of the gas bubbles. This results in high localized electric field strength between the cathode and the positive charges. When the electric field strength reaches $\sim 10^5$ V/m or higher, gas space inside the bubbles is ionized and a plasma discharge takes place. In this process, the cathode was continually fed toward the electrolyte bath; cathode kept solid shape at the contact by electro plasma; and reaction happened in the solid cathode in the electrolyte bath. After the reaction, the solid and porous product was then diluted and purged with distilled water to remove sodium and carbonate ions. For the sample without quenching, the sample feeding will be stopped when the combustion occurred. Resultant was collected and dried in air for microstructure characterization.

The phases in the product were characterized using a powder X-ray diffraction (XRD) (CuKα, λ=0.154nm, Rigagu). Field emission scanning electron microscopy (FESEM) (Model Quanta 3D FEG, FEI Company, USA) was used to characterize the powder microstructures. The as-combusted and fractured sample was also examined by XRD and FESEM methods.

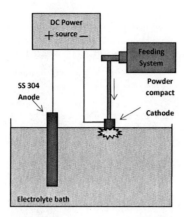

Figure 1 Electro-plasma processing (EPP) set-up

RESULTS AND DISCUSSION

For the composite powder compacts, made under different compacting pressures thus with different apparent densities, electro-plasma processing was applied. No reaction was found for the compact samples with a density less than the 70% of the theoretic value. Instead, for the samples with lower densities, the powder compacts were disintegrated under the electro-plasma processing. In contrast, for samples with an apparent density over 70% of the theoretic density, electro-plasma was successfully maintained around the tip of the compact cathode. Under a proper feeding rate, the plasma envelope makes high temperature on the cathode/electrolyte interface and induces the reaction of the powder compact, which is subsequently quenched immediately by the surrounding electrolyte liquid. By controlling the electrical current, the temperature of the cathode core may be limited to a safe low value, thus preventing the induced combustion reaction in the cathode.

Figure 2 (a) shows the X-ray diffraction (XRD) pattern of the initial as-milled powder mixtures. From Figure 2(a), diffraction peaks of titanium, aluminum are clearly seen. Figures (b) and (c) show the products after electro-plasma processing, with induced reactions, no quenching and quenching. After electro-plasma processing without quenching, Ti$_2$AlC and titanium carbonitride and trace of Al$_3$Ti are formed in a complete reaction, Figure 2(b). Compared with the diffraction peaks of Ti$_2$AlC and titanium carbonitride, one can see that the amount of titanium carbonitride is more than that of Ti$_2$AlC. In contrast, the sample after electro-plasma processing with following quenching, Ti$_2$AlC and titanium carbonitride and trace of Al$_3$Ti are also formed in a complete reaction, Figure 2(c). High purity of Ti$_2$AlC was obtained from this pattern. Through the adjusting of stoichiometric composition, higher purity of Ti$_2$AlC can be obtained by this method [16].

FESEM images of the EPP products without quenching are shown in Figure 3 (a), which shows the typical nano-layer structures which is belong to the Ti$_2$AlC phases. Micro-elemental (EMPA) analyses on these microstructures show that Ti(C,N) is partially dissolved within the

MAX phase after the reaction. Due to the limitation of EDAX resolution, the interfaces between Ti(C,N) and Ti₂AlC cannot be identified. Figure 3 (b) shows the FESEM image of the other part, which indicated as Al₃Ti phase. From the above results, we can suggest that the Al₃Ti was in the molten state and Ti(C,N) was in the solid state during this reaction stage.

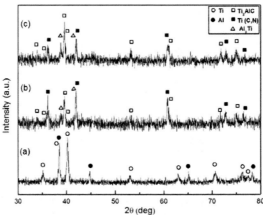

Figure 2 XRD Patterns of (a)As-received powder mixtures 2Ti+Al+Carbon fibers (b) EPP treatment without quenching (c)EPP treatment following quenching .

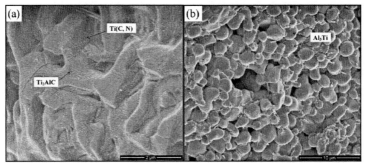

Figure 3 FE-SEM images of the sample after EPP treatment without quenching

Figure 4 (a)-(d) show the FESEM images of the sample after EPP treatment with following quenching. Figures 4(a), (b) and (c) are typical nano-layer structures, Ti₂AlC phase. Figure 4 (d) shows the surface images of the sample after EPP and following quenching. One can see the typical wave propagation after combustion reaction. Detailed analysis will be carried out in the future.

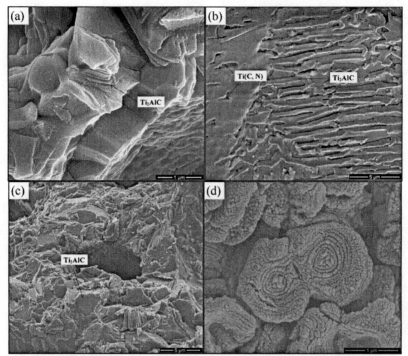

Figure 4 FE-SEM images of the sample after EPP treatment following quenching

The present experiment is different from those previously reported SHS processes induced by either electric arc or a tungsten wire [6, 7]. Figure 5 shows the schematic reaction mechanism of forming MAX Ti$_2$AlC particles during the EPP treatment. In this process, the bonding strength of the powder compact plays an important role for the completion of the combustion reaction in the ablated particles. For samples with high apparent density (70%), the combustion reaction can occur and generate Ti$_2$AlC powders. Once a highly densified powder compact (cathode) surface makes close contact with electrolyte, large current is introduced through the contact surface due to good electrical conductivity. Consequently, local joule heat evaporates the electrolyte solution. As such, an unstable gas layer, mainly water vapor, is formed around the cathode/electrolyte interface. High voltage application through such gas layer induces localized plasma, which could be as high as 3000-6000 K [12, 13, 17]. Gidalevich et al. [17] reported a theoretical analysis of a radial plasma column induced hydrodynamic effects under pulsed low current arc discharges. It was found that during the initial 10^{-8} s the velocity of the plasma column expansion exceeds many times of the speed of sound, creating a shock wave in the water with a pressure discontinuity of ~10 GPa, and a high temperature in the order of 10^4 to 10^5 K. Timoshkin et al. [18] reported that a high voltage spark discharge could induce a pressure up to several GPa in the plasma bubble. High temperature in plasma bubbles leads to the melting of localized surfaces on the cathode and induce the occurrence of combustion. In this process, the cathode was continually fed toward the electrolyte bath; the compact cathode exfoliated due to the discharge impact and expansion of residual gas in the pellet during rapid heating, while the synthesis and rapid quenching occurred as the temperature threshold reaches and the liquid-solid contact forms. The sample is quenched by the surrounding electrolyte, leading to the formation of high purity Ti₂AlC [19]. For the samples with low theoretical densities (i.e. 55% and 65%), the compact pellets were easily crashed due to the strong localized pressure waves in the electro-plasma process, thus no following combustion reaction occurred.

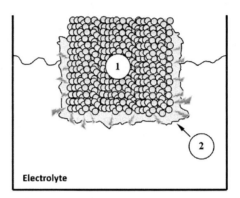

Figure 5 Reaction mechanism of formation of Ti₂AlC particles by EPP
(1) powder compact (Cathode); (2)plasma

Unlike the traditional electro-plasma process where metallic solid electrodes are used, a combustion synthesis of MAX Ti₂AlC particles is obtained using titanium - aluminum - graphite powder compacting electrodes in an electro-plasma process. In this process, the compacting electrode is locally partial melted and ejected by the localized micro electro-plasma arcs, induced by electrothermal instability. Combustion reactions happen quickly in the solid and porous composite electrode. With cold surrounding solution, the solid and porous composite electrode, due to localized shock wave from plasma discharge, experience rapid quenching and result in the formation of high purity MAX Ti₂AlC. Comparing to the traditional process of combustion reaction, this EPP-assisted combustion reaction provides fast cooling from ambient solution, by which combustion reaction could be controlled.

CONCLUSION

Ti₂AlC particles can be directly synthesized through electro-plasma processing and following combustion reaction and quenching using titanium, aluminum and graphite powder compacts. The apparent density of the powder compacts played an important role for the onset of electro-plasma induced combustion/quenching reactions. For the compacts with theoretical density larger than 70%, the combustion/quenching reactions can occur. Because of the occurrence of combustion reaction in the EPP ejected small particles in a liquid environment, the resultant particles experienced rapid quenching and resulted in high purity of Ti$_2$AlC particles. Comparing to other methods, the electro-plasma discharge method presented in this paper can be used to produce Ti$_2$AlC powders easily, quickly, and inexpensively. However, more studies are needed to optimize complicated interaction between nonisothermal temperature fields and condensed phase reactions.

ACKNOWLEDGMENTS

This research is sponsored by NASA under Cooperative Agreement Number NNX09AP72A, and Louisiana Board of Regents NASA/LEQSF (2009-12)-Phase3-03.

REFERENCE

[1]   M.W. Barsoum, T. El-Raghy, Synthesis and Characterization of a Remarkable Ceramic: Ti$_3$SiC$_2$, J. Am. Ceram. Soc., 79, 1953-1956 (1996).

[2]   Y.C. Zhou, Z.M. Sun, Electronic Structure and Bonding Properties of Layered Machinable Ti$_2$AlC and Ti$_2$AlN Ceramics, Phys. Rev., B, 61, 12570-12573 (2000).

[3]   C.J. Rawn, M.W. Barsoum, T. El-Raghy, A. Procipio, C.M. Hoffmann,C.R. Hubbard, Structure of Ti$_4$AlN$_3$-a Layered M$_{n+1}$AX$_n$ Nitride, Mater. Res. Bull., 35, 1785-1796 (2000).

[4]   V. I. Nikitin, A. I. Chmelevskich. A. P. Amosov, A. G Merzhanov, Book of Abstracts of the 1st International Symposium on SHS, Alma-Ata. Russian Federation, (1991).

[5]   A. Varma, A.S. Mukasyan, Combustion Synthesis of Advanced Materials: Fundamentals and Applications, Korean J. Chem. Eng., 21, 527-536 (2004).

[6]   T.D. Xia, Z. A. Munir, Y.L. Tang, W.J. Zhao, T.M. Wang, Structure Formation in the Combustion Synthesis of Al$_2$O$_3$-TiC Composites, J. Am. Ceram. Soc., 83, 507-512 (2000).

[7]   J.H.Lee, S.K.Ko, C.W.Won, Sintering Behavior of Al$_2$O$_3$-TiC Composite Powder Prepared by SHS Process, Mater. Res. Bull., 36, 989-996(2001).

[8]   Michał Łopacinski, Jan Puszynski, Jerzy Lis, Synthesis of Ternary Titanium Aluminum Carbides Using Self-Propagating High-Temperature Synthesis Technique, J. Am. Ceram. Soc., 84, 3051-3053 (2001).

[9]   C. Guilpin, J. Garbaz-Olivier, Analyse de la lumière émis aux électrodes pendant les Effects d'électrode, dans des solutions aqueuses d'électrolyte, Spectrochim. Acta B, 32, 155-164 (1977).

[10]   H. Vogt, Contribution to the Interpretation of the Anode Effect, Electrochim. Acta, 42, 2695-2705 (1997).

[11]   R. Wüthrich, C. Comninellis, H. Bleuler, Bubble Evolution on Vertical Electrodes under Extreme Current Densities, Electrochim. Acta, 50, 5242-5246 (2005).

[12]   T. Oishi, T. Goto, Y. Ito, Formation of Metal Oxide Particles by Anode-Discharge Electrolysis of a Molten LiCl-KCl-CaO System, J. Electrochem. Soc., 149, D155-D159 (2002).

[13]   Y. Toriyabe, S. Watanabe, S. Yatsu, T. Shibayama, T. Mizuno, Controlled Formation of Metallic Nanoballs during Plasma Electrolysis, *Appl. Phys. Lett.*, 91, 041501-3 (2007).

[14]   J.D. Liang, M.A. Wahab, S.M. Guo, Surface Cleaning and Surface Modifications through the Development of a Novel Technology of Electrolytic Plasma Process (EPP), *the World Journal of Engineering*, 7.3, 54-61(2010).

[15]   Kaiyang Wang, Peigen Zhang, Jiandong Liang, S.M.Guo, Synthesis of Al₂O₃-TiC Nano-Composite Particles by a Novel Electro-Plasma Process, Processing and Properties of Advanced Ceramics and Composites III: Ceramic Transactions, Edited by N.P.Bansal et al., 225, 51-58(2011).

[16]   Kaiyang Wang, S.M.Guo, to be published

[17]   E. Gidalevich, R. L. Boxman, S. Goldsmith, Hydrodynamic Effects in Liquids Subjected to Pulsed Low Current Arc Discharges , J. Phys. D: Appl. Phys., 37, 1509-1514 (2004).

[18]   I.V. Timoshkin, R.A. Fouracre, M.J. Given, S.J. MacGregor, Hydrodynamic Modelling of Transient Cavities in Fluids Generated by High Voltage Spark Discharges, J. Phys. D: Appl. Phys., 39, 4808-4817 (2006).

[19]   A. Hendaoui, M. Andasmas, A. Amara, A. Benaldjia, P. Langlois, D. Vrel, SHS of High-Purity MAX Compounds in the Ti–Al–C System, International Journal of Self-Propagating High-Temperature Synthesis, 17, 129-135 (2008)

PROPERTIES OF HOT-PRESSED Ti$_3$AlC$_2$ OBTAINED BY SHS PROCESS

L. Chlubny and J. Lis

AGH - University of Science and Technology
Faculty of Material Science and Ceramics, Department of Technology of Ceramics and Refractories
Al. Mickiewicza 30, 30-059, Cracow, Poland

ABSTRACT

Some of ternary materials in the Ti-Al-C system are called MAX-phases and are characterised by heterodesmic layer structure. Their specific structure consisting of covalent and metallic chemical bonds influence their semi-ductile features locating them on the boundary between metals and ceramics. These features may lead to many potential applications, for example as a part of ceramic armour. Ti$_3$AlC$_2$ is one of these nanolaminate materials. Self-propagating High-temperature Synthesis (SHS) was applied to obtain sinterable powders of Ti$_3$AlC$_2$. Intermetallic compounds at Ti-Al system were used as precursors in the synthesis. For densification of obtained powders hot-pressing technique was used. Phase composition of dense samples was examined by XRD method. Properties such as hardness, bending strength, relative density and elastic properties were determined.

INTRODUCTION

Among many covalent materials such as carbides or nitrides there could be found an interesting group of ternary compounds called in a literature as H-phases, Hägg-phases, Novotny-phases or thermodynamically stable nanolaminates. These compounds have a M$_{n+1}$AX$_n$ stoichiometry, where M is an early transition metal, A is an element of A groups (mostly IIIA or IVA) and X is carbon and/or nitrogen. Heterodesmic structures of these phases are hexagonal, P63/mmc, and specifically layered. They consist of alternate near close-packed layers of M$_6$X octahedrons with strong covalent bonds and layers of A atoms located at the centre of trigonal prisms. The M$_6$X octahedral, similar to those forming respective binary carbides, are connected one to another by shared edges. Variability of chemical composition of the nanolaminates is usually labeled by the symbol describing their stoichiometry, e.g. Ti$_2$AlC represents 211 type phase and Ti$_3$AlC$_2$ – 312 typ. Structurally, differences between the respective phases consist in the number of M layers separating the A-layers: in the 211's there are two whereas in the 321's three M-layers [1-3]. Their specific, layered structure leads to an outstanding set of properties, combining those characteristic for ceramics, like high stiffness, moderately low coefficient of thermal expansion and excellent thermal and chemical resistance with low hardness, good compressive strength, high fracture toughness, ductile behavior, good electrical and thermal conductivity characteristic for metals. Among their potential application can be found ceramic based armors for tanks, armored personal carriers etc., on functionally graded materials (FGM) or as a matrix in ceramic-based composites reinforced by covalent phases.

The SHS is a method that allows obtaining lots of covalent materials such as carbides, borides, nitrides, oxides and intermetallic compounds. The base of this method is utilization of exothermal effect of chemical synthesis, which can proceed in powder bed of solid substrates or as filtration combustion. An external source of heat is used to initiate the process and then the self-sustaining reaction front is propagating through the bed of substrates. This process could be initiated by local ignition or by thermal explosion. The form of synthesized material depends on kind of precursor used for synthesis and technique that was applied. Typical feature of this reaction are low energy consumption, high temperatures obtained during the process, high efficiency and simple apparatus. The lack of control of the process is the disadvantage of this method [4].

The objective of this work was obtaining dense samples of Ti$_3$AlC$_2$ materials manufactured from powders synthesized by SHS method and sintered by hot-pressing method and examining of some of their mechanical properties.

PREPARATION

Following the experience gained while synthesizing ternary materials in Ti-Al-C-N system [5, 6, 7], intermetallic material, namely Ti$_3$Al, was used as a precursor for synthesis of Ti$_3$AlC$_2$ powders. Ti$_3$Al was selected basing on previous research for the highest content of MAX phase in a product, after SHS process [10].

Due to relatively low availability of commercial powders of intermetallic materials from Ti-Al system, it was decided to synthesize it by SHS method initiated by a thermal explosion [5]. Titanium hydride powder - TiH$_2$ (Sigma-Aldrich powder no.209279, grain size < 45 μm, +98% pure, density 3.91g/cm^3), and metallic aluminium powder (ZM Skawina recovered from electrofilter system, grain size ca. 20 μm, +99% pure) were used as a substrates mixed with a molar ratio 3:1 (equations 1).

$$3TiH_2 + Al \rightarrow Ti_3Al + 3H_2 \qquad (1)$$

Obtained product was ground in the homemade rotary-vibratory mill for 8 hours in isopropanol, to the grain size ca. 10 μm [11].

In the next step synthesis of MAX phase, namely Ti$_3$AlC$_2$ was conducted by SHS method with a local ignition system and with use of intermetallic precursor obtained in previous process. Synthesis was conducted in the self-designed high-pressure reactor. The precursors used for a synthesis were commercially available aluminium powder (ZM Skawina recovered from electrofilter system, grain size ca. 20 μm, +99% pure), graphite powder as a source of carbon (Merck no. 1.04206.9050, 99,8% pure, grain size 99.5% < 50μm) and SHS derived Ti$_3$Al powder (characterized in section Result and Discussion). The mixtures of precursors for a SHS synthesis were set in appropriate stoichiometric ratio and are presented in equation 2[11].

$$Ti_3Al + 2C \rightarrow Ti_3AlC_2 \qquad (2)$$

Following the synthesis, X-ray diffraction analysis method was applied to determine phase composition of the synthesised materials. The basis of phase analysis was data from ICCD [8]. Quantitative phase analysis was performed by the Rietveld refinement method [9]. The apparatus used for XRD analysis was Philips X'Pert Pro equipped with X'Celerator counter. The density of obtained powders was determined by use of Micromeritics Accu-Pyc 1330 helium pycnometer. The specific surface of powder was established by BET method and also grain size was measured with use of Micromeritics ASAP 2010.

In the next step powder was densified by hot-pressing method. Samples were sintered in the Thermal Technologies 2000 hot press, at temperature 1350°C, annealing time in maximum temperature was 1 hour, sintering was conducted in constant flow of argon and maximum pressure applied was 25MPa.

Sintered samples were examined to estimate phase quantities of particular sintered bodies. Scanning electron microscopy (FEI Europe Company Nova Nano SEM 200) was applied to examine morphology of the samples. The relative density was measured by the hydrostatic weighing. Elastic properties were established by ultrasonic measurement wit use of UZP-1 INCO-VERITAS defectoscope. Hardness was measured by the Vickers indentation method (FV-700 / Future Tech) and bending strength by the three point bending test (Zwick/Roell BTC-FR2.5TS.D14).

RESULTS AND DISCUSSION

Phase analysis proved that Ti_3Al synthesised by SHS method was pure, single phase powder (Fig. 1) [10]. In case of Ti_3AlC_2 synthesised by SHS method mixture of three phases was obtained Ti_3AlC_2 (36.9 wt.%), Ti_2AlC (35.1 wt.%) and titanium carbide (28.1 wt.%). The XRD pattern of obtained powder is presented on Figure 2. The parameters of MAX phase powder are presented in Table I.

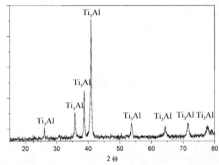

Figure 1. XRD pattern of the Ti_3Al powders obtained by SHS

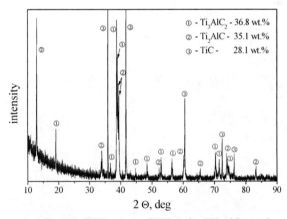

Figure 2. XRD pattern of SHS derived Ti_3AlC_2 powder.

Table I. Characteristic of SHS derived Ti_3AlC_2 powders

Parameter	Value
Density	4.317 ± 0.004 g/cm³.
BET specific surface	2.027 ± 0.018 m²/g
Average grain size	$4.25\mu m$

In case of hot pressed sample, sintered at 1350°C, dense polycrystalline MAX phase material containing 95.8 wt.% of Ti₃AlC₂ and 4.2 wt.% of TiC was obtained. The XRD results are presented on Figure 3. Above this temperature decomposition of MAX phases takes place and hexagonal structure changes into regular structure which is characteristic for lot of MAX phases, for example Ti₃SiC₂.

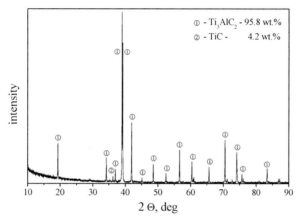

Figure 3. XRD pattern of Ti₃AlC₂ hot pressed at 1350°C.

The SEM pictures of sintered materials presents characteristic for MAX phases plate-like grains are presented on Figure 4.

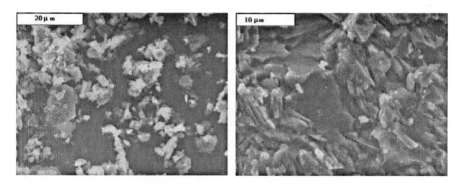

Figure 4. SEM picture of SHS derived powder (left) and dense Ti₃AlC₂ hot pressed at 1350°C (right).

Results of examination of relative density, elastic properties, Vickers hardness and bending strength of hot pressed Ti₃AlC₂ sample compared to the literature data of typical MAX phase material (Ti₃SiC₂) presented in Table II[12, 13].

Table II. Characteristic of hot-pressed Ti$_3$AlC$_2$ materials.

Parameter	Ti$_3$AlC$_2$ (obtained by HP process)	Ti$_3$SiC$_2$
Relative density	4.36 ± 0.02 g/cm^3	4.53 g/cm^3
Young's modulus	357.8 ± 6.2 GPa	333 GPa
Shear modulus	156.2 ± 3 GPa	139 GPa
Poisson ratio	0.145 ± 0.015	0.2
Vickers hardness	6.47 ± 0,60 GPa	4 GPa
Bending strength	597 ± 34 MPa	-

CONCLUSIONS

It was shown that Self-propagating High-temperature Synthesis (SHS) is effective and efficient method for obtaining of sinterable Ti$_3$AlC$_2$ precursors. Intermetallic compound, namely Ti$_3$Al had to be used as a precursor for SHS of Ti$_3$AlC$_2$. Hot pressing of these powders led to obtaining almost single phase (96 wt.%) Ti$_3$AlC$_2$ dense material, proving that further chemical reaction takes place during sintering process. These phenomena were also observed in case of other MAX phases such as Ti$_3$SiC$_2$, Ti$_2$AlN and Ti$_2$AlC.

Morphology of obtained samples is characteristic for this specific group of materials and consists of plate-like grains which influence cracking mechanism of these materials. Elastic and mechanical properties are also comparable to other materials from this group. Further researches on fracture toughness and mechanism of strains within the material as well as some of thermal and electrical properties will be conducted.

ACKNOWLEDGMENTS

This work was supported by the National Science Centre under the grant no. 2472/B/T02/2011.

REFERENCES

[1] W. Jeitschko, H. Nowotny, F.Benesovsky, Kohlenstoffhaltige ternare Verbindungen (H-Phase). *Monatsh. Chem.* **94**, p 672-678, (1963)

[2] H. Nowotny, Structurchemie Einiger Verbindungen der Ubergangsmetalle mit den Elementen C, Si, Ge, Sn. *Prog. Solid State Chem.* **2** , p 27, (1970)

[3] M.W. Barsoum: The MN+1AXN Phases a New Class of Solids; Thermodynamically Stable Nanolaminates- *Prog Solid St. Chem.* **28**, p 201-281, (2000)

[4] J.Lis: Spiekalne proszki związków kowalencyjnych otrzymywane metodą Samorozwijającej się Syntezy Wysokotemperaturowej (SHS) - *Ceramics 44* : (1994) (*in Polish*)

[5] L. Chlubny, M.M. Bucko, J. Lis "Intermetalics as a precursors in SHS synthesis of the materials in Ti-Al-C-N system" *Advances in Science and Technology*, **45**, p 1047-1051, (2006)

[6] L. Chlubny, M.M. Bucko, J. Lis "Phase Evolution and Properties of Ti$_2$AlN Based Materials, Obtained by SHS Method" Mechanical Properties and Processing of Ceramic Binary, Ternary and Composite Systems, *Ceramic Engineering and Science Proceedings*, Volume **29**, Issue 2, 2008, Jonathan Salem, Greg Hilmas, and William Fahrenholtz, editors; Tatsuki Ohji and Andrew Wereszczak, volume editors, p 13-20, (2008)

[7] L. Chlubny, J. Lis, M.M. Bucko: Pressureless Sintering and Hot-Pressing of Ti$_3$AlC$_2$ Powders Obtained by SHS Process, MS&T'10 Houston: Material Science and Technology 2010, p 2339-2345, (2010)

[8] "Joint Commitee for Powder Diffraction Standards: International Center for Diffraction Data"

[9] H.M. Rietveld: A profile refinement method for nuclear and magnetic structures - *J. Appl. Cryst.*, **2**, p. 65-71, (1969).

[10] L. Chlubny: New materials in Ti-Al-C-N system. - PhD Thesis. AGH-University of Science and Technology, Kraków 2006. (*in Polish*)

[11] L. Chlubny, J. Lis, M.M. Bucko: Preparation of Ti$_3$AlC$_2$ and Ti$_2$AlC powders by SHS method MS&T Pittsburgh 09: Material Science and Technology 2009, p 2205-2213, (2009)

[12] N. V. Tzenov, M. W. Barsoum: Synthesis and Characterization of Ti$_3$AlC$_2$ - *J. Amer. Cer. Soc.,* **83**, (2000)

[13] P. Finkel, M. W. Barsoum, T. El-Raghy: Low Temperatures Dependencies of the Elastic Properties of Ti$_3$Al$_{1.1}$C$_{1.8}$, Ti$_4$AlN$_3$ and Ti$_3$SiC$_2$ – *J.Appl.Phys.*, 87, p 1701-1703 (2000)

SHS DIE-CASTING (SHS-DC) OF MAGNESIUM METAL MATRIX COMPOSITES (MMCs)

I. Jo[1], J. Nuechterlein[1], W. Garrett[1], A. Munitz[1], M.J. Kaufman[1], K. Young[2], A. Monroe[3] and J.J. Moore[1]

1) Advanced Combustion Synthesis and Engineering Laboratory (ACSEL), Department of Metallurgical and Materials Engineering, Colorado School of Mines, Golden, Colorado, USA

2) VForge Inc, Denver, Colorado, USA

3) North American Die Casting Association, Wheeling, Illinois, USA

ABSTRACT

SHS die-casting (SHS-DC) of metal matrix composites (MMCs) is a novel process that couples rapid ceramic particle synthesis in-situ within the molten metal during casting process. Processing of Mg metal matrix composites using self-propagating high temperature synthesis (SHS) was investigated and implemented to produce Mg MMC billets with both different processing parameters and different volume fractions of the selected reinforcement (TiC in this study). Powder chemistries were optimized to reduce the ignition temperature of the SHS reaction so that it could be coupled effectively with typical molten Mg processing methods.

Billets of Mg-TiC MMCs were successfully semi-solid die cast into lightweight automotive parts as well as wedge test specimens, which were used for mechanical test specimens. X-ray diffraction (XRD) and scanning electron microscopy (SEM) clearly show that stable TiC reinforcement particles in the range of $5 \sim 10 \ \mu m$ in particle size were uniformly distributed throughout the Mg matrix. In addition, the mechanical testing results indicate that the TiC-reinforced Mg-MMC samples display superior mechanical properties, such as high hardness (~2.2 GPa vs. 0.6 GPa for Mg), elastic modulus (~100 GPa vs. 45 GPa for Mg), ultimate tensile strength (~300 MPa), ultimate compressive strength (~510 MPa), and wear resistance, compared with unreinforced Mg.

INTRODUCTION

Significant progress in the development of lightweight metal matrix composites (MMCs) has been achieved in recent decades resulting in specific applications in the automotive and aerospace industries. Magnesium and its alloys are receiving renewed attention in the area of "lightweighting" for structural applications since they have good damping properties, high strength and stiffness, excellent castability, 30 % lower density than aluminum and are potentially cost competitive. However, compared to other structural metals, magnesium and magnesium alloys have a relatively low absolute strength and wear resistance, especially at elevated temperatures. A possible solution to these drawbacks is to incorporate reinforcements such as titanium carbide (TiC) particulates into the Mg alloy matrix. TiC reinforcements are of interest because of their thermodynamic and thermal stability, high hardness and elastic modulus, and superior wear resistance. [1-5]

Three major disadvantages of metal matrix composites include (1) high cost of the reinforcement materials, (2) high fabrication costs and (3) non-uniformity in structure and properties. Cost-effective processing of MMC's is, therefore, an essential element for expanding their application. Die-casting is one of the most cost effective high-speed production processes for making complex mechanical parts out of light metals such as aluminum, magnesium and zinc alloys. Recently, self-propagating high temperature synthesis (SHS) has been extensively studied because in-situ production of the ceramic reinforcement offers significant advantages: (1) the clean interfaces between the reinforcement and matrix introduces strong bonding; (2) the high reaction temperatures (1500 < T < 5000 °C) remove volatile impurities; (3) the high exothermic reaction and reaction rate (0.1 < v < 25 cm/s), lower operating and processing costs; and (4) fine and well dispersed ceramic reinforcements can be achieved in the metal matrices by controlling the reaction parameters. The coupling of these two technologies can lead to rapid processing techniques for high quality net shape metal matrix composite components. [6-11]

The SHS processing method is based on the fact that a high exothermic reaction occurring during the formation of a compound can propagate spontaneously without additional heat supply to the reactants. It is also commonly referred to as combustion synthesis (CS) or reactive synthesis (RS). The adiabatic temperature is calculated assuming no heat loss during the entire reaction and complete conversion to the products using data of heat capacities and enthalpies of formation. In practice, the adiabatic temperature needs to be above about 1500°C for the reaction to be self-sustaining. Using this approach, more than 500 compounds have been synthesized over the past 30 years. Among these compounds, titanium carbide ceramic reinforcements show favorable thermodynamic stability within magnesium matrices. [8-10, 12, 13]

In the present study, the SHS process was combined with high pressure die-casting techniques to produce in situ Mg-TiC MMC components with 10-30 volume percent of fine and well dispersed titanium carbide reinforcing phase. Particular attention has been paid to the characterization of the microstructural development and mechanical properties of the SHS Die-Casting (SHS-DC) products.

EXPERIMENTAL

10-30 volume percent TiC reinforced MMC billets of either 1.75 inch diameter and 3.5 inch height (Fig. 1a) or 2.5" diameter by 3.5" height were produced by injecting Ti + C pellets into the magnesium matrix with 9 wt % aluminum as a diluent. The smaller billets were then reheated and semi-solid die cast at VForgeTM into wedge samples (Fig. 1b) and the larger billets into automotive belt tensioning brackets. (Fig. 1c)

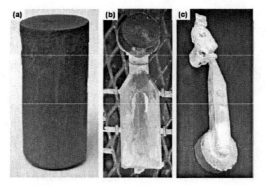

Figure 1. (a) MMC billet for semi-solid die casting, (b) Wedge sample made from billet and (c) Automotive belt tensioning bracket made from billet.

Titanium and graphite powders were mixed at a ratio corresponding to that of stoichiometric TiC. The green powders were mixed by ball milling for 24h with spherical alumina balls, and then were uniaxially pressed into 0.5 inch diameter cylindrical pellets. The average green density of the pellets was approximately 60% of the theoretical density. The aluminum powder as a diluent material was obtained from commercial powders which has ~44μm size and 99.5% purity. The raw materials for TiC pellets were made from commercial powders of titanium (99.7% purity, ~44μm) and graphite (99.5% purity, 1~5 μm).

To make Mg-9 wt % Al matrix, a pure Mg ingot was cut into pieces and 9 wt % Al powder were placed into a clean graphite crucible that was then located inside a 300 lb induction furnace containing an alumina crucible. In an effort to minimize oxidation, a protective argon "blanket" was used to cover the melt throughout the entire process to prevent oxidation of the MMC. In addition, the top of the furnace was covered with Fiberfrax inside a steel lid to prevent the melt from contact with the atmosphere.

When the crucible and protective assembly were in place, the induction furnace was used to increase the temperature to ~750 °C. The molten magnesium was held at 750 oC and the desired volume percent of the reinforcement was introduced by submerging the cold pressed pellets into the melt. After their insertion, the "melt" was mechanically stirred for ~5 min using steel rod to facilitate the incorporation and uniform distribution of the TiC reinforcement into the matrix. During the stirring, the temperature was maintained between 750 °C and 800 °C. After stirring, the stirrer was removed and the graphite crucible was extracted and then quenched into water.

The resulting billets containing 10, 20, and 30 vol % TiC were reheated above the liquidus of the Mg matrix and then placed into the shot sleeve of the HPDC machine at V-forge where they were die cast into either wedge samples (Fig. 1b) or automotive belt tensioning brackets (Fig. 1c). To avoid oxidation and burning during the reheating process, a commercial 2-part (epoxy / TiB$_2$ powder) coating was applied to the billets.

The phases in the samples were determined using X-Ray diffraction (Philips X'pert diffractometer) with Cu-Kα radiation. Scanning electron microscopy (ESEM, FEI Quanta 600) was used to investigate the microstructures of the samples.

The mechanical properties of the samples were evaluated by Rockwell hardness (Instron Wilson Rockwell Series 2000), Vickers hardness (Leco micro hardness tester-MHT220), tensile testing (conducted at Worcester Polytechnic Institute) and compression testing (MTS instru-Met A30-33). For each of the castings, Rockwell scale hardness values were determined by taking 20 measurements per sample coupled with a statistical evaluation. Also, Vickers microhardness tests were carried out under 10N load for 10 seconds and the indent diagonals were measured using an optical microscope. Twenty indents were made on each sample and a distance of at least 50 times greater than the indent depth was maintained between the center of each indent and the edge of the specimen. Abrasion wear tests were also conducted using an ASTM G-65 standard with a dry sand / rubber wheel apparatus.

RESULTS AND DISCUSSION

Fig. 2 shows an X-ray diffraction (XRD) pattern of the MMC sample containing 20 vol % TiC. From the XRD data, it was confirmed that the SHS reaction had successfully converted the Ti and C into TiC and the resulting TiC-MMC sample consisted of two main phases corresponding to the magnesium and titanium carbide in the sample. While Mg also forms carbide in the Mg-C system, it appears that the TiC is thermodynamically more stable in the Mg liquid as supported by the lack of any additional peaks in the XRD patterns. It should be noted that the conventional melting of magnesium using relatively high temperatures results in the oxidation of the Mg. Fortunately, the XRD analysis in this work indicates that there are no peaks corresponding to magnesium oxide phases. This indicates that the processing precautions taken in this study were effective in minimizing the oxidation of the molten magnesium. It also appears that the SHS process with its rapid and high-temperature reaction produces relatively high purity TiC-MMC samples with minimum secondary phases or impurities.

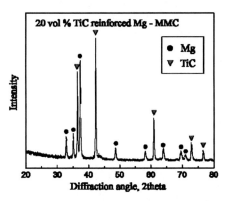

Figure 2. XRD pattern of the wedge sample with 20 vol % TiC reinforced MMC. (Black solid circle indicates pure metal Mg phase and red solid triangle indicates TiC phase)

Scanning electron microscopy (SEM) images of the wedge samples with 20 and 30 volume percent TiC are shown in Fig. 3. Using EDS analysis, it was determined that the light color phase is the TiC reinforcement and the dark grey color phase is the magnesium matrix. As shown in Fig. 3, 5~10 μm sized TiC particles were dispersed throughout the magnesium matrix and appear to have a "clean" interface (Fig. 3d). However, as the volume percent of the TiC reinforcement was increased to 30 vol %, some of the TiC reinforcements remained as clusters in the magnesium matrix (Fig. 3b) in spite of the stirring and the shearing action associated with the SS die casting. The average size of the agglomerates was in range from 50 to 100 μm, i.e., about 10X the individual TiC particle size. Furthermore, a few particles, appeared bright due to charging (Fig. 3c) in the 30vol % TiC reinforced MMC sample; EDS indicated that these particles are magnesium oxide (MgO). The MgO phase was not observed in the XRD data because of its apparent small volume fraction in the samples.

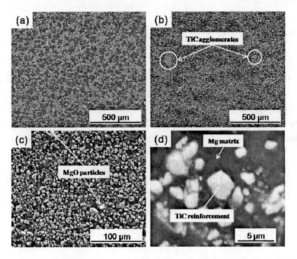

Figure 3. SEM micrographs of wedge samples with (a) 20 vol % TiC and (b-d) 30 vol % TiC.

Rockwell and Vickers hardness values of the wedge samples with various volume fractions of the TiC reinforcement are presented in Fig. 4. The results show a similar trend of hardness change in both tests, i.e, with increasing volume percent of TiC in the MMC samples, the hardness gradually increased from 30 to 90 HRB and 60 to 220 HV consistent with the rule of mixtures on the composite materials. The average Vickers hardness of the 30 vol % TiC-MMC sample is about 220 HV (~ 2.2 GPa) which is 3.6 times higher than that of the Mg-9 wt % Al matrix matrix. (62.7 HV)

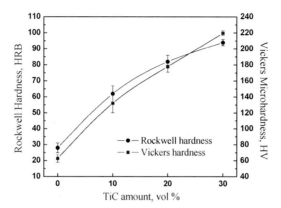

Figure 4. Rockwell and Vickers hardness values as a function of the volume percent of the TiC reinforcement.

Tensile properties of the MMCs were measured at room temperature with ASTM standard sub-sized tensile specimens that were machined from the wedge samples (Fig. 5). As expected, the reinforced MMC samples with higher volume fractions of TiC have superior tensile strengths and elastic moduli compared to the unreinforced samples. Compared with the unreinforced matrix, the 30 vol % TiC reinforced wedge sample exhibited a 140% higher elastic modulus and 47% higher ultimate tensile strength. It should be noted, however, that the UTS and YS increases are accompanied by a corresponding reduction in tensile ductility.

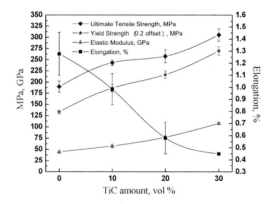

Figure 5. Ultimate tensile strength (UTS), yield strength (YS), elastic modulus and elongation, obtained from SHS-DC Mg-TiC MMCs as a function of TiC vol %.

In order to measure the compressive properties of the SHS-DC MMCs, the samples containing different volume fractions of TiC reinforcement were machined into rectangular specimens (5mm x 5mm x 8mm) and tested at room temperature. Stress-strain curves of the MMCs samples were plotted under compressive loading at room temperature, as shown in Fig. 6. A minimum of three specimens were tested for each system to obtain typical results. All compression curves can be divided into three regions: (1) linear elastic region before yielding, (2) plastic deformation region and (3) collapse region. As expected, the MMCs with higher volume percent of TiC display higher elastic moduli, higher yield strengths and higher strengths prior to "buckling".

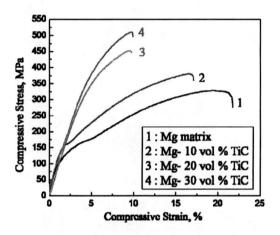

Figure 6. Compressive stress-strain curves of the SHS-DC MMCs with different TiC volume fractions.

Table 1 shows the values obtained from the compression tests. The ultimate compressive strength (UCS) of the sample with 30 vol % TiC was about 510 MPa or ~35% higher than that of the unreinforced sample whereas the compressive yield strength (CYS) was ~3.6 times higher.

Table 1.Summary of the compression tests for the MMCs samples.

Mechanical Properties	Compression test samples			
	Mg matrix	Mg – 10 vol % TiC	Mg – 20 vol % TiC	Mg – 30 vol % TiC
0.2 % offset CYS (MPa)	105	152	320	380
UCS (MPa)	328	380	452	510
Fracture Strain (%)	16.8	12.3	5.7	4.5

Abrasive wear resistance tests were performed by standard methods using a dry sand/rubber wheel apparatus based on ASTM G-65. 14 All tests were carried out at room temperature without cooling. Samples were weighed before and after tests to calculate wear volume loss. Volume losses of the MMC samples (Fig. 7) indicate that, as the TiC content is increased from 0 to 30%, the wear volume loss of the samples gradually decreases from 670 to 250 mm3. From the hardness and wear test results (Fig. 7), it is clear that the volume losses of the MMC samples are inversely proportional to hardness.

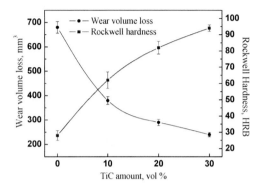

Figure 7. Wear volume loss during abrasion wear test on the wedge samples as a function of TiC volume percent.

CONCLUSION

10 – 30 volume percent TiC reinforced Mg - MMC billets were successfully produced by in-situ self-propagating high-temperature synthesis (SHS), and then semi-solid die cast into wedge test samples and automotive belt tensioning brackets. Microstructural evaluations clearly indicated that stable TiC reinforcements in range of 5 ~ 10 μm in particle size could be distributed somewhat uniformly in the Mg matrix. The TiC reinforced Mg - MMCs samples have higher hardnesses (~2.2 GPa), elastic moduli (~100 GPa), ultimate tensile strengths (~300 MPa), and ultimate compressive strengths (~510 MPa), and wear resistance, compared to the unreinforced Mg. It is concluded that the combination of SHS and die-casting has potential to be used to produce light weight Mg MMC components with enhanced mechanical properties for various applications in a rapid and economical way.

*Acknowledgments

This research was supported by the US Department of Energy under Award Number DE-EE0001100, the North American Die Casting Association and the China Scholarship Council of the Chinese Ministry of Education. The authors would like to thank Vforge Inc. and WPI for performing the semi-solid die casting trials and mechanical tests respectively.

REFERENCES

[1]Tresa M. Pollock, *Science*, **328**, 986 (2010).

[2]B.L. Mordike, T. Ebert, *Mater. Sci. Eng.* **A302**, 37 (2001).

[3]A.B. Pandey, R.S. Mishra, Y.R. Mahajan, *Mater. Sci. Eng.* **A206**, 270 (1996).

[4]S. K. Bhaumik, C. Divakar, A. K. Singh, G. S. Upadhyaya, *Mater. Sci. Eng.* **A279**, 275 (2000).

[5]L. Lu, M.O. Lai, Y. Su, H.L. Teo, C.F. Feng, *Scripta Mater.* **45**, 1017 (2001).

[6]H.Z. Ye, X.Y. Liu, *J. Mater. Sci*, **39**, 6153 (2004).

[7]X.C. Tong, *J. Mater. Sci.* **33**, 5365 (1998).

[8]J.J. Moore, *The Minerals, Metals & Materials Society*, 817 (1994).

[9]J.J. Moore, H.J. Feng, *Progress in Materials Science*, **39**, 243 (1995).

[10]Z.A. Munir, Anselmi-Tamburini, *Mater. Sci. Reports*, **3**, 277 (1989).

[11]Y.F. Yang, H.Y. Wang, J.G. Wang, Q.C. Jiang, *J. Am. Ceram. Soc.*, **91**, 3813 (2008).

[12]S.K. Roy, A. Biswas, S. Banerjee, *Bull. Mater. Sci.*, **5**, 347 (1993).

[13]A. Contreras, C.A. Leon, R.A.L. Drew, E. Bedolla, *Scripta Metall.*, **48**, 1625 (2003).

[14]ASTM G-65 Standard Test Method for Measuring Abrasion Using the Dry Sand/Rubber Wheel Apparatus

Microwave and Milli-Meter Processing and Its Field Effects

EVALUATION OF MICROWAVE-SINTERED TITANIUM AND TITANIUM ALLOY
POWDER COMPACTS

Arne W. Fliflet[1], Spencer L. Miller[2], and M. Ashraf Imam[1]

[1] Materials Science and Component Technology Directorate, Naval Research Laboratory,
  Washington, DC, USA
[2] University of Notre Dame, South Bend, IN, USA

ABSTRACT
Titanium (Ti) has many attractive attributes for naval applications including high strength, no
magnetic signature, and excellent corrosion resistance; however, its use has been limited by high
costs. Microwave sintering of titanium powder compacts can potentially lower the cost of
titanium parts. This approach benefits from recent low cost metal-from-ore production methods
that produce powder instead of sponge. Current experiments utilize an over-moded, variable
atmosphere, S-Band processing chamber and are evaluating the optimum preform density and
microwave processing conditions for consolidating titanium powder into complex, near-net-
shape parts. These experiments indicate that microwave sintering leads to shorter processing
times and lower processing temperatures compared to the processing in a conventional furnace.
As-received commercially pure titanium powder was also milled to reduce particle size in order
to improve densification but it did not improve the densification of the sintered product. The as-
received pure titanium alloy (Ti-6Al-4V) achieved higher densification compared to the as-
received pure titanium powder with similar particle size.

INTRODUCTION

Titanium (Ti) has many attractive attributes for naval applications including high specific
strength, no magnetic signature, and excellent corrosion resistance. Titanium has the highest
strength to weight ratio of any metal and is 45% lighter than steel and over twice as strong as
aluminum.[1] It can be easily alloyed with other metals such as aluminum and vanadium to
produce stronger light weight alloys. Ti-6AL-4V (6 wt% aluminum and 4 wt% vanadium) is the
most commonly used alloy in aerospace, medical, marine and chemical processing. However,
titanium's use has been limited by high costs. In the past few years the availability of low cost
titanium powders, produced by a number of novel processes, have resulted in broad interest in
utilizing titanium for military, industrial, and aerospace applications. Powder metallurgy is an
effective, cost reducing way to produce high-quality, near-net-shape products by sintering at a
fraction of the cost of melting and casting. Titanium is very sensitive to oxygen and is used as
oxygen-getter at high temperatures. Oxidization of titanium by 0.1 wt% will raise the hardness of
commercially pure (CP) titanium by a factor of three.[1] Because of its affinity to oxygen, titanium
must be processed in vacuum or inert atmosphere to maintain a low oxidization state.
Traditionally, sintering is performed in a conventional vacuum furnace and overall processing
times can be many hours to days leading to a costly process. Microwave sintering of titanium
metal/alloy powders is of interest as being a potentially quicker, lower cost process. Generally,
the direct heating of metals by microwaves is not effective because the high conductivity of
metals limits penetration of electric fields. This is not the case with powder metal compacts
having densities less than the theoretical value. These should be more properly treated as

artificial dielectrics – a composite of the metal powder and gas/vacuum. Recently, experiments have shown that metal powder compacts can be effectively heated and sintered by microwave fields and that, unlike ceramics, coupling to the magnetic component of the microwave field is important.[2,3] Theoretical and experimental studies have confirmed the effectiveness of microwave heating of conductive powder compacts.[4-6] A comparison of microwave and conventional sintering of titanium powder compacts was recently reported by Luo *et al.*[7]

In this work the emphasis was on the microwave sintering properties of compacts of pure and alloyed titanium produced in a high-pressure Cold Isostatic Press (CIP). The effect on sintering of milling to reduce the particle size of the as-received powders was also investigated. Previous experiments on compacts produced using a Uniaxial Press (UP) had shown that the final density depends greatly on the initial (green) density of the compact but not that much on the sintering temperature or the hold time. We were interested to see if improved results could be obtained by sintering compacts produced by the higher compaction pressures obtainable using the CIP. We were also interested to investigate the effects of temperature and hold time. An advantage of the CIP process is that it can be applied to more complexly-shaped compacts than can be produced in a UP. To test this feature, we sintered CIP'ed valve-like compacts as well as simple disk-shaped compacts. Very high densities (>90% theoretical density) were obtained for all of the sintered compacts produced by CIPing. Interestingly, milling the powders did not improve the post-sintered density. It was also noted that the titanium alloy (Ti-6AL-4V) achieved slightly higher densities than the pure titanium powder.

EXPERIMENTAL SETUP AND PROCEDURE

Titanium powders produced using the Armstrong Process were obtained from Cristal Global also known as International Titanium Powder (ITP). Three different powders types were used in the experiments. The first type used was a commercially pure (CP) as-received powder. The second type was a CP powder that had been milled to three different tap densities: 28%, 24% and 20% of Theoretical Density (TD). A higher percentage tap density generally refers to smaller particle size. The third type was the most commonly used titanium alloy (Ti-6Al-4V) powder in as-received condition.

Disk-shaped Titanium compacts were pressed in a 40-mm die at 10 tons axial load corresponding to a pressure of 10-ksi. This uniaxial load densified the compacts to 50–60%TD. They were then CIPed at 100 ksi to achieve higher green density (~ 87%). The compacts were sintered for various times at different temperatures. Compacts with a more complex shape were formed using a polymer mold made from a commercial available valve. The initial density of the pre-CIPed complex shape was limited to the 20–30% of theoretical density. Initial green densities of the disk compacts were measured using a technique based on Archimedes' Principle. The samples were wrapped in plastic tape to avoid contact with the water. The temperature of the water was measured to find its density and the tape's density was measured and subtracted to obtain a more accurate measurement. After the green density was measured and initial data on the sample was collected, the samples were put in an insulating heating casket, then placed in a chamber for sintering, and pumped down overnight to about 10-mT using an oil-based mechanical pump. The pump-down time was impacted by the out-gassing of the fiberboard materials in the insulation casket, particularly when the casket had been exposed to air for several days.

Sintering was carried out in the Naval Research Laboratory's (NRL) S-Band 2.45 GHz Microwave Materials Processing system shown in Fig. 1. Microwaves are generated in a 2.45 GHz Cober Electronics Model S6F Industrial Microwave Generator and transported to the heating chamber via S-Band waveguide. A three-stub tuner is used for impedance matching into the chamber. Forward and reflected powers are monitored using calibrated 60-dB couplers located upstream from the three-stub tuner. A liquid-nitrogen sorption pump capable of producing submillitorr pressures is connected to the chamber during the sintering process to improve the vacuum (to minimize the risk of oxygen contamination) and to reduce carbon contamination from the oil-based mechanical pump. The samples were brought up to temperature at a ramp rate of about 30°C/min. The chamber pressure increased as the casket and compact were heated, usually reaching several hundred millitorr at a temperature of ~500°C. As heating continued to higher temperatures, the pressure slowly decreased to <100-mT at sintering temperatures. The increase in pressure during heating is attributed mainly to outgassing of the fiberboard materials used the casket as well as some outgassing of the compact. The ramp rate was occasionally affected by the formation of plasma either inside or outside the casket. Plasma formation, especially outside the casket, was generally unfavorable to workpiece heating; however, plasma formation was not an issue at sintering temperatures when microwave coupling to the titanium compact is most efficient. The surface temperature of the compact was measured with an Omega OS3750 two-color optical pyrometer. Compacts were held at the maximum hold temperatures for either 10 min. or 60 min. Typically, three disk compacts were stacked and sintered together in order to obtain data for different powders sintered under similar conditions.

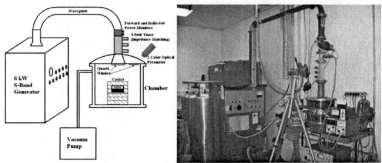

Figure 1. Schematic and photograph of the NRL 6 kW, 2.45 GHz microwave system for titanium sintering experiments.

To improve microwave heating efficiency during initial heating a lossy "susceptor" material is added to the insulting casket, a technique called hybrid heating. In previous experiments, the titanium compacts were placed in a densified zirconia crucible and the relatively lossy crucible improved initial heating. However the solid crucible was prone to cracking due to thermal gradients so for this work it was replaced by a combination of a silicon carbide plate and zirconia fiberboard. This system provided better heating at subsintering temperatures and was could be used repeatedly. The casket/hybrid heating system is shown in Fig. 2.

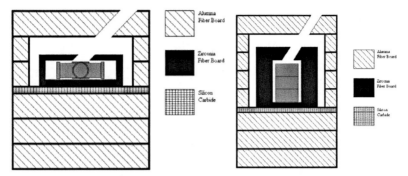

Figure 2. Schematics of the insulation casket and hybrid heating system for different compacts.

The goal of these experiments was to see if microwaves could sinter titanium powders effectively to near theoretical density and to find what temperatures and hold times are needed to obtain the highest final density. Experiments involving the disk compacts were done at temperatures ranging from 900°C to 1300°C for hold times of 10 minutes or 60 minutes. Compacts CIPed into valve shapes were sintered at 1200°C with holding times of 30 minutes and 60 minutes. Disk compacts pressed from different types of powder were sintered three at a time by stacking them one on top of the other. Loose titanium powder was placed between the samples (made from milled 28%, as-received, and as-received Ti-6Al-4V powders) to keep them from bonding together. The stacks of samples were sintered at temperatures of 900°C, 1100°C, and 1300°C at 10 minutes and 60 minutes. Samples were ramped at a steady increase of power and held at temperature (measured by a two-color optical pyrometer) by adjusting the power as needed. The only limiting factor for increasing the power was the formation of plasma. Plasma was no longer a problem after the sample had reached a temperature of ~600°C.

Figure 3. Left: three different disk compacts cut in half. Right: valve-shaped compact.

RESULTS

It was found that all the disk-shaped powder compacts made from as-received and milled ITP CP Ti powder, and as-received ITP Ti-6AL-4V powder could be sintered to near theoretical density at temperatures ranging from 900°C to 1300°C. The best results came from the unmilled Ti-6AL-4V which achieved a sintering density of ~99% TD. The as-received ITP CP Ti powder achieved an average final density of ~98%, and the ITP milled 28% Ti powder achieved an average final density of~96%. Increasing the hold time from ten to sixty minutes had a negligible effect on the final density. However, the temperature ramp-up time of several hours was similar in all cases and probably reduced the sensitively to temperature hold time. These results are shown in Fig. 4. The final density was found to be insensitive to the sintering temperatures in the range of 900°C to 1300°C. The surface of the compacts usually turned black during the sintering process probably from the oil vapor produced by the mechanical vacuum pump. It was found that the compact on the top of the stack picked up the most carbon. Loose Ti powder was therefore placed on top of the stacked compacts to minimize the amount of carbon deposited on the samples.

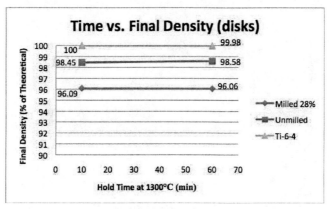

Figure 4: Comparison of final densities for two hold times at 1300°C.

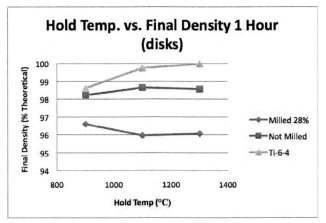

Figure 5. Sintered density vs. hold temperature for compacts made from as-received and milled CP Ti powders, and as-received Ti-6Al-4V powder.

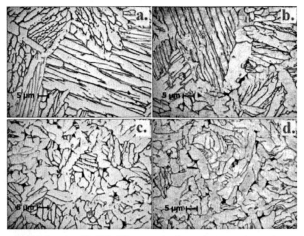

Figure 6. Micrographs showing Ti-6Al-4V compacts sintered at a maximum temperature of: a) 1300°C for 60 minutes, b) 1300°C for 10 min. c) 1100°C for 60 minutes, and d) 900°C for 60 min.

To investigate the microstructure of the sintered compacts, the surface of a densified compact of as-received Ti-6AL-4V powder was etched using a solution of 100 mL H_2O, 2 mL hydrofluoric acid, and 5 mL H_2O_2 (50% dilute) to show the grain boundaries. Micrographs were taken at different locations on the sample using a metallograph. It can be determined by visual inspection that higher temperatures and longer hold times resulted in larger grains. The results, shown in Fig. 6 for Ti-6AL-4V compacts sintered at maximum temperatures of 900°C to 1300°C with final densities of ~99%, show the presence of both the α and β phases.

The etchant caused many large and small etching pits that made it difficult to tell how many pores are in the samples. In order to see the pore structure, the samples had to have the etched surface removed and then be re-polished and photographed. The left micrograph in Fig. 7 shows the same area as shown in Fig. 6(a) after polishing off the etched surface to reveal the etching pits. The right micrograph in Fig. 7 shows the same surface after re-polishing. Some of the etching pits are still visible after polishing. The compacts were found to have minimal porosity.

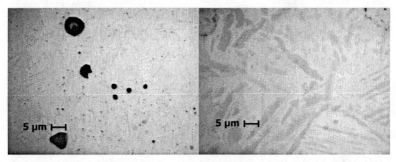

Figure 7: Micrographs of the compact shown in Fig. 6(b). Left micrograph shows with etched surface layer removed. The dark areas are etching pits. Right micrograph shows surface after re-polishing.

To look at the microstructure of the CP Ti compacts, a solution of 100 mL H_2O, 2 mL hydrofluoric acid, and 4 mL H_2O_2 (50% dilute) was applied for 30-60 seconds to show grain boundaries. Micrographs were taken at different locations on the sample using a metallograph. The results yielded the α-phase (close-packed hexagonal). Figs. 8 and 9 show the micrographs of the grain structures for milled 28% CP Ti and as-received CP Ti powders at different magnifications. The larger "pores" or "pits" are etching pits and can be removed by polishing.

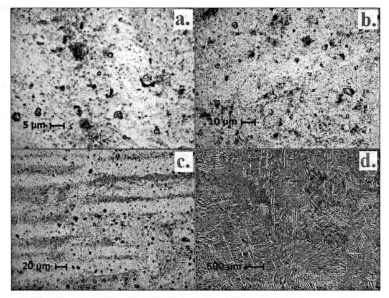

Figure 8. Micrographs of 28% milled CP Ti sintered at 1300°C for 1 hr. shown at different magnifications.

Hardness measurements were taken using a Vickers hardness tester. Hardness results for a sintered ITP CP 28% milled Ti powder compact are shown in Fig. 10. It was found that the hardness increases near the surface of the titanium compacts. This indicates that the surface absorbed more oxygen than the interior. To minimize the pickup of oxygen during sintering, there must likely be a better vacuum than in the present experiments during which the pressure increased to >100-mT. There were variations in the hardness of the samples. Average hardesses were comparable to industry standards. The hardness is an important property and has to be appropriate to the intended application (structural, mechanical, medical, plumbing, aerospace, etc.). Through surface treatment, the hardness can be adjusted for the intended application.

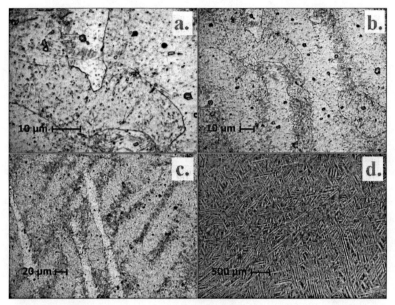

Figure 9. Micrographs of as-received CP Ti sintered at 1300°C for 1 hr. shown at different magnifications.

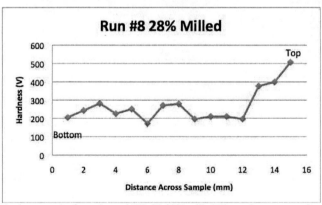

Figure 10. Hardness results for a sintered ITP CP 28% milled Ti powder compact showing higher hardness near the surface.

CONCLUSIONS

It was found that compacts of all three powder type sintered to near theoretical density at temperatures ranging from 900°C to 1300°C. The best results came from the as-received alloy of Ti-6Al-4V powder with an average sintering density of 99% TD. The as-received CP Ti powder achieved an average final density of 98.5% and the CP milled 28% CP Ti powder achieved an average final density of 96%. The results indicate that varying the hold time from 10 minutes to 60 minutes did not have a significant effect on the final density of the sintered compact. The powder type (milled CP Ti, as-received CP Ti, and as-received Ti-6Al-4V) and the initial compaction pressure determine the final density. Compaction pressure can be adjusted to achieve the desired final sintered density. The samples were found to be harder near the surface than near the center. This is because of oxidization of the compact near the surface. To avoid the oxidation, the vacuum must be improved. Future research will be targeted to optimize different parameters for controlling the final density and controlling the amount of shrinkage during the sintering for different applications. Research will also be focused on creating complex shapes by sintering.

ACKNOWLEDGEMENT

This work supported by the Naval Research Laboratory Base Program.

REFERENCES

[1]M. A. Imam, F. H. Froes and K. L. Housley. *Titanium and Titanium Alloys*. Kirk-Othmer Encyclopedia of Chemical Technology (2010) 1-41.
[2]R. Roy, D. Agrawal, J. Cheng, and S. Gedevanishvili, *Nature (London)* **399** (1999) 668.
[3]J. Cheng, R. Roy, and D. Agrawal, *J. Mater. Sci. Lett.* **20** (2001) 1561.
[4]J. Luo, C. Hunyar, L. Feher, G. Link, M. Thumm, and P. Pozzo, *Appl. Phys. Lett.* **84** (2004) 5076-5078.
[5]K.I. Rybakov, V.E. Semenov, S.V. Ergorov, A.G. Eremeev, I.V. Plotnikov, and Yu, V. Bykov, *J. Appl. Phys.* **99** (2006) 023506.
[6]T. Galek, K. Porath, E. Burkel, and U. van Rienen, *Modelling Simul. Mater. Sci. Eng.* **18** (2010) 025015.
[7]S.D. Luo, M. Yan, G.B. Schaffer, and M. Qian, "Sintering of Titanium in Vacuum by Microwave Radiation," *Metallurgical and Materials Transactions* **42A** (2011) 2466-2474.

MICROWAVE-ASSISTED SYNTHESIS OF TiC BY CARBOTHERMAL REDUCTION

Rodolfo F. K. Gunnewiek, Pollyane M. Souto and Ruth H. G. A. Kiminami

Federal University of São Carlos, Dept. of Materials Engineering
13565-905 São Carlos, SP, Brazil
rodolfo_foster@yahoo.com.br, ruth@ufscar.br

ABSTRACT

The synthesis of TiC by carbothermal reduction process has been carried out to evaluate the effect of microwave energy as a alternative source. The factors influencing the powder synthesizing process, such as power and vacuum, are discussed. TiC powders with particle sizes less than 600 nm were produced at 2.1kW for 20 min in an argon atmosphere. The combination carbothermal reduction reaction and microwave processing makes this route a promising route relatively simple, and very fast to synthesize titanium carbide powders.

INTRODUCTION

Titanium carbide is transition metal carbide widely used in several environmental applications that requires high temperature and chemical stability allied to extremely resilient mechanical properties. TiC is interstitial carbide with a NaCl-type fcc structure, whose carbon atoms occupy all the fcc octahedral interstitial sites of the lattice when in the stoichiometric condition. However, it is not always a truly stoichiometric compound since it can exist in the form of TiC_{1-x}[1]. TiC is a very hard material with Vickers hardness of 28 to 35 GPa. It is stoichiometry-sensitive, reaching maximum values when it is stoichiometric, x=0.2, with good thermal conductivity, which makes it suitable for cutting tools and hard coatings. The elevated bond energy of 14.66eV of Ti-C implies a high melting point of 3067°C as well as chemical stability when attacked by HNO_3, halogens and HF, oxidizing gradually above 800°C. These characteristics make TiC suitable for high temperature applications in extreme environmental conditions, since it is a light material with a density of $4.91g/cm^3$, which is slightly higher than that of metallic titanium. TiC can dissolve many carbides and nitrides and is suitable for composites with the optimal characteristics of TiC and other carbides/nitrides such as titanium carbonnitrides (which are completely intermiscible and form titanium carbonitrides, $Ti(C_xN_{1-x})$, with x ranging from 0 to 1). TiC/Cr_2C_3 and other materials are of great interest today as reinforcement materials for cutting tools and wear-resistant applications. When combined with other carbides and metals (cermets), the properties of TiC can be enhanced to improve its fracture toughness and activate its densification processes during sintering, enabling its deposition as a coating[3-11].

The literature describes numerous methods for synthesizing TiC. These methods include *in situ* coating[12], solid state reactions[7], carbon coating to produce submicrometric particles[13], combustion reaction and self-propagating high-temperature synthesis (SHS) using metallic titanium, carbon or graphite[1,9], and high energy mechanical milling of TiO_2 and C to produce nanosized TiC[14]. Additional methods are laser pyrolysis[15] and carbothermal reduction by conventional heating (using carbon[16] or using polymer as carbon source[17])and by microwave heating. Microwave plasma-assisted synthesis has been used to synthesize TiN and other nitrides, demonstrating the feasibility of microwave-assisted non-oxide synthesis[18]. The process is influenced by several parameters, such as microwave power, reaction time, temperature, and the purity and proportions of the raw materials, and presence of controlled atmosphere. The kinetics of the reaction and thermodynamically favored products are determined by process temperature[17]. Microwave-assisted synthesis is attractive for processing and synthesizing ceramics because this route is faster than the conventional one, requiring less energy since it provides volumetric and uniform thermal absorption and heating, as well as greater absorption in the case of some materials such as carbon. Carbon interacts very well with 2.45GHz microwaves, even at low temperatures, and absorbs almost all the energy, which enhances heating of powders mix until it reaches the critical temperature and the reaction begins. Microwave-assisted carbothermal reduction is a promising route to synthesize carbides[19,20]. TiC and TiN can be obtained from Ti[21] or TiO_2[22,23,24]. Therefore, microwave-assisted carbothermal reduction of transition metal oxides is very fast and efficient. Few papers describing microwave carbothermal reduction have been published so far. However, with respect to the synthesis of TiC, many years ago Hassine et al[23] reported the production of TiC and TaC under conditions that yielded particle sizes exceeding 18μm in up to 120 min of reaction dwell time.

In this context, the aim of this work was to use microwaves as the source of the energy to carry out carbothermal reduction to synthesize submicrometric TiC starting from TiO_2 and carbon black. The use of microwave energy makes this process economically promising, relatively simple, and very fast.

EXPERIMENTAL PROCEDURES
Firstly, pure titanium oxide (MERCK, >99%, with an average grain size of 160 ± 40 nm) and carbon black (with an average grain size of 115 ± 30 nm) in a proportion of 1:2 mol were stirred into ethanol at 50°C for one hour and then partially dried. The mass was extruded, yielding pellets. About 1.5g of pellets was weighed in a mullite/alumina reaction boat, which was placed in the reaction chamber of a multimodal microwave applicator (2.45GHz Cober, USA). The chamber was first cleaned by an outlet vacuum and inlet argon flow for 10 min and then preheated at 600°C for 10 min to 0.6kW to avoid thermal shock of the reaction chamber components. The reactions, which were performed at 1.8kW and 2.1kW for 20 min, are identified here as RT-1 and RT-2, respectively. To determine the efficiency of the reaction without the vacuum outlet, the vacuum pump was disconnected and a reaction (RT-3) was performed at 1.8kW for 20 minutes. The synthesized powders were disagglomerated in a mortar.

The crystallinity and phases of the powder were determined from X-ray diffraction patterns recorded by a Siemens D5005 diffractometer operating with Kα-Cu radiation from 30° to 90° (2θ) and a scan step of 2°/min. The powder's microstructure was examined by field emission gun scanning electron microscopy (FEG-SEM) (Philips XL30 FEG). Samples were prepared by dispersing the powders in acetone and droplets of the suspension were deposited on an aluminum sample holder. The droplets in the sample holder were then coated with an ultrathin gold layer. The BET method was employed to measure the surface area in a Micromeritics ASAP 2020 analyzer, using N_2 adsorption/desorption at liquid nitrogen temperature.

RESULTS AND DISCUSSION
Microwave carbothermal reduction is an effective method for producing well crystallized titanium carbide, as indicated by the X-ray diffraction patterns of RT1 and RT2 powders depicted in figure 1. Applying 1.8kW and 2.1kW of power and a reaction time of 20 min, the reaction between carbon and TiO_2 was completed and no evidence was found of any oxygenated phase such as Ti_2O_3. The 1:2 ratio of TiO_2 and carbon raw materials suggests the evolution of carbon dioxide during the reaction, yielding TiC as the only solid phase (JCPDS 32-1383), as expressed by equation 1.

$$TiO_2 + 2C \rightarrow TiC + CO_2 \qquad (1)$$

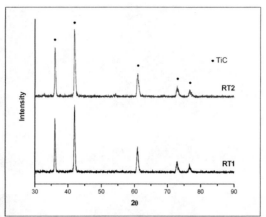

Figure 1. X-ray diffraction patterns of RT1 and RT2 powders synthesized under 2.45GHz microwave radiation, showing well defined and crystallized TiC phase (JCPDS 32-1383).

The broadening of peak widths suggests small particles. In fact, an analysis of the powders' morphologies (see figure 2) reveals a bimodal particle size distribution. The RT1 powder comprises a group with an average particle size of 500 nm and another with sizes smaller than 100 nm, while the RT2 powder consists of a group with an average particle size of 650nm and another with particle sizes smaller than 100nm. As this figure indicates, the particles of the RT1 powder are smaller but show a higher standard deviation (± 240nm) than those of the RT2 powder (± 180nm). This indicates that increasing the microwave power augments the homogeneity of the material but increases its particle size due to the growth of particles in close contact with each other. The particles of the RT2 powder appear to be more connected to each other than those of the RT1 powder and may be more difficult to disagglomerate.

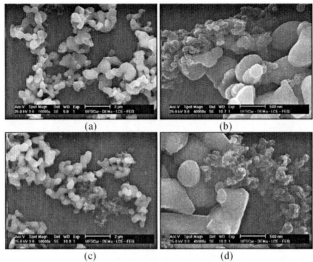

Figure 2. Morphologies of RT1 ((a) and (b)) and RT2 ((c) and (d)) powders, showing bimodal particle size distribution

In 1995, Hassine et al[23] synthesized TiC, starting from stoichiometric mixes of TiO_2 and carbon black using microwaves as the heating source but, for a long time thereafter, no relevant research was published about microwave-assisted carbothermal reduction to produce pure TiC. Our work resumes where that of the aforementioned authors stopped, but uses a different approach involving very small particles, in the carbothermal synthesis of TiC employing 2.45GHz of microwave power. In an earlier study, Hassine et al[23] synthesized TiC with an average particle size of 18μm in 120min of reaction dwell time. These authors applied varying temperatures and reaction times and produced particles with sizes no smaller than 10 μm and with a TiC content of about 20% after a long dwell time of synthesis controlled by temperature and power, from 500W to 5kW and up to 1600°C. Their dwell time and particle sizes far exceeded those obtained at this work, which were 650 nm particle size in less time at 2.1kW and 500 nm particle size in less time and lower power, i.e., 1.8kW.

TiC can with very small particles[14] can be produced by other techniques such as mechanochemical synthesis, but they require very long dwell times exceeding 30 hours and temperatures above 1200°C, in contrast to the 20 min required by our technique.The BET analysis indicated that both powders had a specific surface area of about 2.2 m^2/g and and an average particle size of 500 nm, according to the BET equivalent spherical particle diameter.

Considering oxidation of TiC at temperatures above 800°C, even if slow, contamination prevention of the reaction chamber by oxygen is necessary when is expected to produce only TiC. Disconnecting the vacuum pump from the outlet caused air borne contamination, leading to TiC oxidation and the formation of oxygenated phase Ti_2O_3, and the pressure of argon was insufficient to prevent air from entering the reaction chamber, as indicated in the X-ray diffraction patterns in figure 3. TiC is highly sensitive to oxidation at high temperatures. Although well crystallized TiC was formed, secondary oxygenated phase Ti_2O_3 was also formed, probably due to the oxidation of TiC synthesized at the high temperatures reached during the reaction, even at 1.8kW.

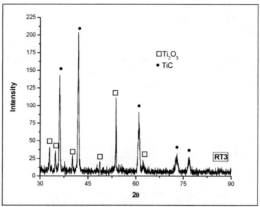

Figure 3. X-ray diffractogram of RT3 powder: note the presence of well crystallized TiC (JCPDS 32-1383) and Ti$_2$O$_3$ (JCPDS 10-63) phases.

CONCLUSIONS

Microwave-assisted carbothermalreduction synthesis is a rapid and economical method for producing well crystallized bimodal TiC in just 20 min of reaction, using simple raw materials, TiO$_2$ and carbon black. The SEM micrographs show smaller particle sizes when using lower power and a high particle size distribution of 500 nm at 1.8kW and 650 nm at 2.1kW. Although the particle size is higher at 2.1kW, the specific surface area is slightly smaller than that of powder synthesized at 1.8kW. It is important to prevent oxygen contamination during the high temperature steps of the reaction in order to ensure that only TiC is produced, since TiC is susceptible to oxidation and the formation of an oxygenated secondary phase.

ACKNOWLEDGEMENTS

The authors gratefully acknowledge the financial aid of the Brazilian research funding agencies FAPESP (Process # 07/59564-0) and CNPq (Process # 142911/2009-7), and the practical support of the Postgraduate Program in Materials Science and Engineering of the Federal University of São Carlos.

REFERENCES

[1] L. E. Toth, Transition metal carbides and nitrides. New York: Academic Press, 7, 275p (1971).

[2] V. Lipatnikov, A. Rempel and A. Gusev, Atomic ordering and hardness of nonstoichiometric titanium carbide, Int. J. Refrac. Met. Hard Mater.,15, 61-64(1997).

[3] F. Monteverde, V. Medri and A. Bellosi, Microstructure of hot-pressed Ti(C,N)-based cermets,J. Eur. Ceram. Soc., 22, 2587-2593 (2002).

[4] J. C. Caicedo, C. Amaya, L. Yate, M. E. Gómez, G. Zambrano, J. Alvarado-Rivera, J. Muñoz-Saldaña and P. Prieto, TiCN/TiNbCN multilayer coatings with enhanced mechanical properties,Appl. Surf. Sci., 256, 5898-5904 (2010).

[5] P. Ettmayer, H. Kolaska, W. Lengauer and K. Dreyert, Ti(C,N) cermets - metallurgy and properties,Int. J. Refrac. Met. Hard Mater.,13, 343-351 (1995).

[6] L. Chen, W. Lengauer, P. Ettmayer, K. Dreyer, H. H. Daub and D. Kassel,Fundamentals of liquid phase sintering for modern cermets and functionally graded cemented carbonitrides (FGCC),*Int. J. Refrac. Met. Hard Mater.*,**18**, 307-322(2000).

[7] J. Russias, S. Cardinal, Y. Aguni, G. Fantozzi, K. Bienvenu and J. Fontaine, Influence of titanium nitride addition on the microstructure and mechanical properties of TiC-based cermets,*Int. J. Refrac. Met. Hard Mater.*,**23**, 358-362 (2005).

[8] J. Chen, W. J. Li and W. Jiang, Characterization of sintered TiC–SiC composites,*Ceram.Int.*, **35**, 3125-3129 (2009).

[9] A. O. Kunrath, I. E. Reimanis and J. J. Moore, Microstructural evolution of itanium carbide – chromium carbide (TiC – Cr_3C_2) composites produced via combustion synthesis,*J. Am. Ceram. Soc.*, **85**, 1285-1290(2002).

[10] M. Masanta, P. Ganesh, R. Kaul and A. R. Choudhury, Microstructure and mechanical properties of TiB_2–TiC–Al_2O_3–SiC composite coatings developed by combined SHS, sol–gel and laser technology,*Surf. Coat.Tech.*, **204**, 3471-80 (2010).

[11] R. Bahl, A. Kuhmar, M. Vedawyas and D. Patel, Synthesis and characterization of TiC and TiCN coatings, *Appl. Phys. A*, **69**, S643-S646 (1999).

[12] Y. Luo and M. Matsuo Morphology of carbon/TiC composite films prepared by carbonization of polyimide/titania composites,*Polim.Bull.*, **64**, 939-951 (2010).

[13] R. Koc and J. Folmer, Synthesis of Submicrometer Titanium Carbide Powders,*J. Am. Ceram. Soc.*, **80**, 952-956 (1997).

[14] M. Ali and P. Basu, Mechanochemical synthesis of nano-structured TiC from TiO_2 powders, *J. Alloys and Compound*, **500**, 220-223(2010).

[15] Y. Leconte, H. Maskrot, L. Combemale, N. Herlin-Boime and C. Reynaud,Application of the laser pyrolysis to the synthesis of SiC, TiC and ZrC pre-ceramics nanopowders,*J. Anal. Appl. Pyrolysis*, **79**, 465-470(2007).

[16] L-. M. Berger, W. Gruner, E. Langholf and S. Stolle, On the mechanism of carbothermal reduction processes of TiO2 and ZrO2, *Int. J. Refrac. Met. Hard Mater.*,**13**, 353-358 (1995).

[17] Y. Luo, S. Ge, Z. Jin and J. Fisher, Formation of titanium carbide coating with micro-porous structure, *Appl. Phys. A*, **98**, 765-768(2010).

[18] C. K. Bang, Y. C. Hong and H. S. Uhm, Synthesis and characterization of nano-sized nitride particles by using an atmospheric microwave plasma technique, *Surface & Coatings Technol.*, **201**, 5007-5010(2007).

[19] T.P. Deksnys, R.R. Menezes, E. Fagury-Neto and R.H.G.A. Kiminami, Synthesizing Al_2O_3/SiC in a microwave oven: a study of process parameters,*Ceram. Int.*, **33**, 67-71 (2007).

[20] E. Fagury-Neto and R. H. G.A Kiminami, Al_2O_3/mullite/SiC powders synthesized by microwave-assisted carbothermal reduction of kaolin,*Ceram.Int.*, **27**, 815-819 (2001).

[21] J. D. Houmes and H.-C.Loye, Microwave Synthesis of Ternary Nitride Materials,*J. Sol. State Chem.*, **130**, 266-271 (1997).

[22] R. D. Peelamedu, M. Fleming, D.K. Agrawal andR. Roy, "Preparation of Titanium Nitride: Microwave-Induced Carbothermal Reaction of Titanium Dioxide", *J. Am. Ceram.Soc.*, **85**, 117-122 (2002).

[23] N. A. Hassine, J. G. P.Binner andT. E. Cross, Synthesis of refractory metal carbide powders via microwave carbothermal reduction,*Int. J. Refrac. Met. Hard Mater.*,**13**, 353-358 (1995).

[24] R. F. K Gunnewiek, P. M. Souto and R. H. G. A. Kiminami, Microwave-assisted synthesis of TiCN by carbothermal reduction, Microwave and RF power applications – 13th International Conference on Mirowave and RF Heating, J. Tao (org.), 205-208(2011).

EFFECT OF MICROWAVE PLASMA PROCESS CONDITIONS ON NANOCRYSTALLINE DIAMOND DEPOSITION ON AlGaN/GaN HEMT AND Si DEVICE METALLIZATIONS

N. Govindaraju and R. N. Singh

University of Cincinnati
Energy and Materials Engineering
Cincinnati, OH 45221-0012, U.S.A

ABSTRACT
    Thermal management plays a critical role in the continued development of modern microelectronics and high power electronics technology. This paper lays the foundation for the development of microwave plasma nanocrystalline diamond (NCD) based thermal interface materials for AlGaN/GaN High Electron Mobility Transistors (HEMTs). NCD is deposited, using Ar-rich microwave plasmas, on AlGaN/GaN HEMTs and Si devices with no visible damage to device metallization. Raman spectroscopy, optical and scanning electron microscopy are used to evaluate the quality of the NCD films. Results indicate that no visible damage occurs to the device metallization for temperatures below 290 °C for Si devices and below 320 °C for the AlGaN/GaN HEMTs. Possible mechanisms for metallization damage above the deposition temperature are enumerated. Electrical testing of the AlGaN/GaN HEMTs indicates that it is indeed possible to deposit NCD on GaN-based devices with no significant degradation in device performance.

INTRODUCTION
    High power electronic devices play a vital role in a variety of areas including power generation and transmission, electrical locomotion, avionics, oil drilling and exploration, industrial electronics, and control systems. However, device operation at high power levels results in excessive self-heating leading to premature failure. Therefore, it is important that effective thermal management strategies be developed for high-power and high-temperature electronics.
    A high-power electronics device which has received significant attention in recent years is the AlGaN/GaN High Electron Mobility Transistor (HEMT)[1]. AlGaN/GaN HEMTs designed for power electronics applications operate at high current densities (~ 1 A/mm) resulting in significant self-heating effects. Therefore, it is necessary to devise effective thermal management strategies for such devices in particular, and for high power electronics, in general. In this context, thin film diamond is an attractive material for thermal management of wide-bandgap semiconductor electronics for the following three reasons. Firstly, it has been established that thin film diamond has excellent thermal properties with thermal conductivity values well over 10 W/cmK even for low quality films, while for thick films they approach those of single crystal diamond[2]. Secondly, the lattice mismatch between diamond and other wide-bandgap semiconductors such as SiC, AlN, AlGaN and GaN is well below that of its mismatch with Si. Thirdly, the recent development of nanocrystalline diamond (NCD) films[3] opens up the possibility of developing diamond films with smooth surfaces. Smooth surfaces are a prerequisite for device applications, since they reduce thermal scattering, eliminate regions of stress concentration, and reduce the likelihood of dielectric failure. Smooth surfaces are important for reducing defect densities at interfaces. They also eliminate electric field

99

concentration due to jagged edges found on rough surfaces which can cause premature dielectric breakdown.

While a large body of literature exists on the use of diamond as a substrate material for thermal management applications [4-8], very little information is available on the deposition of diamond on top of wide-bandgap semiconductor devices, specifically for GaN-based devices. Such studies are challenging particularly because of the aggressive nature of the plasma environment used in diamond deposition.

The first step towards the realization of viable NCD-based thermal management technology for GaN HEMT devices is to ensure that the device metallization survives plasma processing conditions during diamond deposition. Once it has been successfully demonstrated that viable deposition conditions can indeed be realized, subsequent steps would involve improving the quality of the deposited NCD films and also ameliorating the effects of the plasma on device performance.

This paper addresses the first step in the realization of viable NCD-based thermal management technology for high power electronics applications by studying the effect of NCD deposition on device metallization. NCD films were deposited on devices fabricated on two different substrate materials – Si and GaN-on-SiC (HEMT devices). Test depositions on the Si samples led to the development of process conditions which did not result in any damage to the device metallization. Subsequently, NCD depositions on two AlGaN/GaN HEMTs samples demonstrated that it is indeed possible to realize process conditions, which resulted in no damage to device metallization. The results presented in the paper lay the groundwork for the development of NCD-based Thermal Interface Materials (TIMs) for thermal management of high-power and high-temperature electronics.

EXPERIMENTAL METHOD

The device wafers used in the experiments discussed below were fabricated by the Naval Research Laboratory (NRL) and consisted of a 3″ Si wafer and two samples of AlGaN/GaN HEMTs (GaN-on-SiC substrates). The two wide-bandgap semiconductor samples were approximately right-angled triangle in shape with ~3/8" sides. Since the number of samples available was limited, the following strategy was employed to optimize the process conditions and to maximize the probability of success when carrying out the final depositions on the wide-bandgap semiconductor samples. As stated above, the primary aim of the experiments discussed in this paper was to eliminate physical damage to the metallization on the devices. To this end, the Si device metallization was used as a test bed for developing the process conditions. Once the process conditions were developed on the Si device samples, these conditions were used for NCD deposition on the AlGaN/GaN HEMTs. It is to be emphasized here that both these materials systems had the same metallization and passivation schemes.

Diamond deposition was carried out in a 1.5 kW (ASTEX, 2.45 GHz) Microwave Plasma Chemical Vapor Deposition (MPCVD) system. It is to be noted here that all the depositions reported below used a process gas composition of $Ar:H_2:CH_4$ in the ratios 60:39:1 (flow rates: 60 sccm: 39 sccm: 1 sccm; please refer to Table I for deposition pressures), respectively. It was found from previous studies[9] that it was possible to deposit good quality diamond under these conditions. The temperature of the substrate during deposition was recorded using an infrared pyrometer (temperature range: 250 °C to 600 °C; spectral range: 2.0 μm to 2.6 μm), which was focused on the substrate surface through a quartz window at the top of the deposition system.

RESULTS AND DISCUSSION

The following steps were taken before carrying out diamond deposition on Si and HEMT devices.

i. Process conditions were optimized on bare Si wafers (p-type, (100) orientation, 1-20 Ω-cm) in order to locate the plasma position with respect to the substrate surface within 0.5" x 0.5" area. These samples were activated by ultrasonic agitation using 1-2 μm diamond suspension in ethanol.

ii. A proprietary spin coating process was developed which resulted in a uniform, dense nucleation layer and the plasma position was localized to an accuracy of 1/8"-1/4" within the 0.5" x 0.5" area.

Chemical Vapor Deposition of NCD on Silicon Devices

Since there was only one 3" silicon wafer available with devices fabricated on it, in order to maximize the utility of the wafer for optimizing the process conditions, the wafer was cleaved to produce multiple samples of approximately 0.5" x 0.5" size. The resulting pieces conveniently fit within the 0.5" x 0.5" area wherein the plasma was known to be located as a result of experiments described above. The metallization for the fabricated devices consisted of 250 nm of Au deposited on 20 nm of Ni. A passivation layer consisting of 100 nm of Si_3N_4 was deposited on top of the metallization. Table I lists the sample designations and the process conditions for the Si device samples.

Since Au and Si form an eutectic at 363 °C[10], an upper limit of 350 °C for NCD deposition on silicon device wafers chosen. Also, gold and nickel are completely soluble even at low temperatures[10]. Given that the thickness of Ni is an order of magnitude smaller than the thickness of Au, it is reasonable to assume that most, if not all, of the Ni will form an Au-rich, Au-Ni alloy. Therefore, for all practical purposes we can assume that essentially Au is in contact with Si at higher temperatures.

NCD was deposited on the first silicon device sample SiD1 under the conditions listed in Table I. The initial power for the NCD deposition on this sample was 1100 W, but was reduced to 980 W to keep the temperature close to 350 °C. The deposition time was confined to three hours in order to be able to quickly evaluate the results. It was found that at the end of the deposition run for SiD1, the metallization on the devices was damaged. It is to be noted here that a hydrogen plasma etch (30 Torr pressure; 600 W; 10 min duration) was carried out to clean the surface of the sample prior to the deposition of diamond. In order to limit damage to the Si_3N_4 passivation layer due the H_2 plasma etch, this process step was eliminated for all subsequent samples. The next sample, SiD2, was deposited under the same conditions as SiD1, except for the absence of the H_2 plasma etch.

Once again it was found that the device metallization was damaged under these process conditions. Figure 1 shows optical images (50X objective), of two specific devices on SiD2 before and after deposition. It can be seen from Fig. 1 that in the case of feature sizes of the order of a few microns, some of the metallization was etched away. On the other hand, it can be seen from Fig. 1 that for large metal features such as the approximately 70 μm x 70 μm contact pads, that not only was some of the metallization missing, but the metal layer suffered gross deformation wherein it crumpled and peeled off from the substrate surface. It is to be noted that at present we do not have information on the stresses developed in the films due to diamond deposition process conditions. This is a subject worthy of investigation, and further studies focused on this area are required. It is evident from the optical images that the damage to the metallization was significant and would compromise the device characteristics if the device were to be used as part of a circuit. Hence, the following experimental strategy was adopted to elucidate the underlying mechanisms which lead to the device degradation shown in Fig. 1.

The first approach was to reduce the deposition pressure, thereby reducing the ion density and hence effectively reducing the etch rate of the sample. NCD was deposited on two samples, SiD3 and SiD4 at pressures of 60 Torr and 70 Torr, respectively. Both these samples exhibited damage similar to the samples deposited at 80 Torr, with the extent of damage being reduced as the deposition pressure was reduced. Since there was an observed reduction in damage with the deposition pressure, the next logical step was to reduce the pressure further and to observe the effect on device metallization.

Table I. NCD deposition conditions for Si device experiments

| Sample | Deposition Conditions* | | | | | Remarks** |
	P (Torr)	T (°C)	Power (W)	t (h)	H_2O (°C)	
SiD1	80	365/350	1100/980	3	~17	H_2 plasma etch Metallization damaged
SiD2	80	358/332	1100/1065	3	17/12	No H_2 plasma etch Metallization damaged
SiD3	60	290	950	3	~11.5	No H_2 plasma etch Metallization damaged
SiD4	70	270	950	3	~ 11.5	No H_2 plasma etch Metallization damaged
SiD5	50	-	950	3 ⎫ 6	~11	No H_2 plasma etch No metallization damage
	50	-	950	3 ⎭	~11	
SiD6A	50	~245	950	3	~11	Annealed @ 400 °C (1 h); No H_2 plasma etch
	60	290	950	3	~12	Metallization damage after 60 Torr deposition
SiD7A6001	50	245/260	950	6	15.5	Annealed at 600 °C (2 h); No H_2 plasma etch Minimal damaged to metallization
SiD7A6002	50	260	940	6	~ 9	Annealed @ 600 °C (2 h); No H_2 plasma etch No metallization damage
SiD8	50	295/310	950	6	18/25	No H_2 plasma etch Metallization damaged
SiD9	50	290	950	6 ⎫ 12	10/11	No H_2 plasma etch Metallization damaged
	50	290	950	6 ⎭	~11	
SiD10	50	235/256	950/870	~ 6.5	~ 10	No H_2 plasma etch No metallization damage

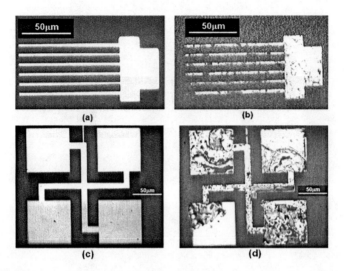

Figure 1. Optical images of individual devices on sample SiD2 before ((a) and (c)) and after ((b) and (d)) diamond deposition. Damage to device metallization is clearly evident, the granularity in the images after diamond deposition is due to the formation of a diamond film (50X objective, 50 μm scale bar).

Figure 2. SEM micrographs of sample SiD5 after six hours of diamond deposition (a) a single complete device (350X; 50 μm scale bar) (b) fine scale metallization feature on device shown in (a)- excellent diamond coverage is clearly evident. No damage to device metallization was observed on this sample (25 kX; 1μm scale bar).

In the case of the silicon device sample SiD5, there was no damage observed to any of the devices on the sample due to NCD deposition for three hours at a pressure of 50 Torr.

Therefore, an additional run for a period of three hours was carried out to ensure that no damage resulted from continued deposition under the same conditions. Figure 2 shows SEM micrographs of a representative device on SiD5 after a total deposition time of six hours – it can be readily seen that there is no damage to the metallization on this sample. Further, the coverage of diamond on this sample was excellent as is evident from Fig. 2 (b). In contrast, in sample SiD4 (deposition pressure: 60 Torr) the damage to the metallization was clearly visible.

The next step was to check for the effect of annealing and any other associated thermal effects. To this end, a series of Si samples were annealed in vacuum, prior to the diamond synthesis, at temperatures above the deposition temperature. Since the low temperature stage in the MPCVD system was not equipped with a substrate heater, sample SiD6A was annealed for 1 hour at 400 °C in a DC/RF magnetron sputter system (PVD 75, Kurt J. Lesker Co, PA, U.S.A) at a base pressure of 6.5 x 10^{-6} Torr. It is to be noted here that the substrate heater was not in direct contact with the sample and therefore the temperature of the sample may be lower than that of the heater. SiD6A did not exhibit any damage to the device metallization after annealing. In order to replicate the deposition conditions under which no metallization damage occurred, the first deposition run for sample SiD6A was carried out under the same conditions as that of sample SiD5 (i.e., at 50 Torr base pressure). At the end of the first deposition run, sample SiD6A showed no damage, thereby confirming the results of the sample SiD5. The next three hour deposition on sample SiD6A was carried out at a pressure of 60 Torr, resulting in device metallization damage. From these experiments it was very clear that for no metallization damage to occur, the pressure needs to be held at 50 Torr during deposition. Therefore, all subsequent depositions were carried out at a pressure of 50 Torr. Also, it was clear that annealing close to the eutectic temperature of the Au-Si binary alloy system did not damage the sample.

The next step was to check for the effect of higher annealing temperatures and also the effect of the difference in the coefficients of thermal expansion of different materials in the fabricated devices. To test for these effects, two samples were annealed simultaneously at 600 °C for 2 hours at a base pressure of 6.9 x 10^{-6} Torr in the PVD 75 system. Of these, one sample, SiD7A6001, was spin coated with diamond prior to the annealing. This would effectively test for the differences CTE values (Diamond: 1.18×10^{-6} K^{-1}, Au: 14.2×10^{-6} K^{-1}, Si: 2.6×10^{-6} K^{-1}, and Si_3N_4: 3.2×10^{-6} K^{-1}) between the different materials. The other sample, SiD7A6002, was spin coated *after* the annealing step, and therefore was tested for thermal effects on metallization due to prolonged exposure to high temperatures. Both samples did not exhibit any damage due to annealing at 600 °C for 2 hours. We may, therefore, conclude that the damage occurring to the metallization on Si devices was not due to a purely thermal effect, nor was it solely due to the differing CTE values of the fabricated devices and the deposited diamond thin film.

Since the sample SiD7A6001 was coated with diamond prior to annealing at 600 °C and did not exhibit any metallization damage, the sample was subject to a second, more direct approach to test for the effect of etching on the device metallization. SiD7A6001 was subject to similar deposition conditions as the sample which did not exhibit any damage (SiD5) but with the elimination of methane in the process gas flow for the entire length of a deposition run. The absence of methane would effectively eliminate the deposition of diamond and hence would clearly show the effect of etching on the device metallization. Etching of this sample for six hours resulted in very little damage to the device metallization, with only some slight etching of one of the fine metal lines on one of the devices. These results imply that very little damage results from etching the spin coated diamond sample without any diamond deposition. Further, it may be safe to assume that the small amount of etching evident on this sample may be ameliorated or even eliminated by increasing the thickness of the Si_3N_4 passivation layer.

As mentioned above, the only samples which exhibited little or no damage to the device metallization were those which were deposited at a pressure of 50 Torr and a substrate temperature of 260 °C or less. Therefore, the next series of experiments were designed to elucidate the effect of temperature on the device metallization at a *fixed pressure of 50 Torr*. In the case of samples SiD8 and SiD9, both of which were deposited with diamond at a temperature of 290 °C and above, damage to the device metallization was evident. However, in the case of samples SiD7A6002 and SiD10 there was no damage visible to both of these samples. From these experiments it was concluded that for diamond deposition to occur on Si device wafers with no damage to the device metallization, the deposition pressure needs to be maintained at 50 Torr and the deposition temperature should be below at least 290 °C, and below 260 °C as a conservative estimate.

Analysis of the Metallization Damage on Si Device Samples

Before discussing the results of the NCD deposition on the two wide-bandgap semiconductor samples, a preliminary analysis of the metallization damage observed on the silicon device samples is presented below. While the exact mechanism resulting in the observed damage is far from certain, instructive conclusions can be drawn from such an analysis.

As expected, EDX spectra (not shown here) taken at a point outside a damaged metallization pad on the sample SiD9 indicated negligible amounts of Au, with C and Si dominating the observed elemental concentration. When a similar spectrum was taken on a metallization on pad sample SiD9 a difference in contrast indicated the presence of Au, as expected, and the measured concentration of Au was higher (~5 atom %). It was observed that there were certain dark regions present on the metallization pads on the SEM micrograph. Upon measuring EDX spectra at one of these locations, it was found that the concentration of Au was negligible (Fig. 3). Further, close examination of such dark regions showed that while the entire region was covered with diamond, there was a noticeable height difference between the dark regions and the surrounding regions on the pad where the Au was present. The diamond in these dark regions was found to be a lower height as compared to the surrounding regions on the Au pad. The above observations indicate that in the case of devices fabricated on Si, diamond deposition at pressures greater than 60 Torr, and deposition temperatures above 300 °C, result in physical removal and transport of gold from the devices.

The pattern of damage on the Au pads indicated the preferential removal of Au in certain regions. Based on the experiments discussed above, it is clear that purely thermal annealing or etching effects do not result in damage to the device metallization. A combination of plasma and thermal effects seem to be required to result in metallization damage. There may be different mechanisms which may lead to the observed phenomenon. For instance, it is possible that a complex phenomenon involving the solid-state reaction of Au with the underlying Si leads to alloy formation which is preferentially etched due surface energy differences. Alternatively, local temperature differences in a plasma environment may lead to the formation of Au-Si alloy which may undergo grain growth and recrystallization leading to preferential removal of the metal in such regions as compared to the (unreacted) surrounding areas.

We would like to stress that these conclusions are preliminary and further in-depth study is required to elucidate the exact mechanism leading to the observed phenomenon of metallization damage in the Si device samples. In summary, the following inferences may be drawn from the study of the damage to device metallization on Si.

i. The damage to metallization on the Si device samples was not solely due to thermal effects, since annealing at two different temperatures, 400 °C and 600 °C, for prolonged periods did not result in any observable damage. It is to be noted that the above temperatures are well above the eutectic temperature of the Au-Si binary alloy system. Therefore, the formation of eutectic alone does not seem to result in observable damage to the device metallization.

ii. The damage to the metal layer does not seem to occur due to the differing CTE values for the multilayer thin films used in the device fabrication and diamond deposition.

iii. Etching the Si device samples in the absence of any diamond deposition does not result in any significant damage to the device metallization. The small amount of damage observed may be eliminated by increasing the thickness of the Si_3N_4 layer or by changing the passivation layer material.

iv. The above inferences lead us to suspect that the observed damage to the device metallization on the Si device samples was due to complex phenomena involving some form of solid-state reaction between the Au and the Si substrate followed by preferential removal of the alloy due to chemical or thermal effects, under the influence of the plasma, from areas where such a reaction has taken place.

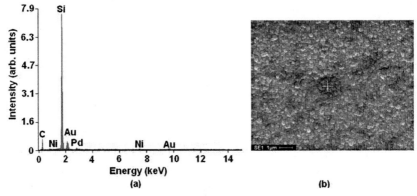

Figure 3. (a) EDX spectra at a point on the metallization pad shown in (b) on the sample SiD9 (the Au concentration was negligible < 1 atom %)

Chemical Vapor Deposition of NCD on Wide-Bandgap Semiconductor Samples

Table III lists the process conditions used for diamond deposition on the two wide-bandgap semiconductor device samples. The metallization for the fabricated devices consisted of 250 nm of Au deposited on 20 nm of Ni. A passivation layer consisting of 100 nm of Si_3N_4 was deposited on top of the metallization. Having established the deposition conditions under which no damage takes place to the device metallization in the case of the Si device samples, the same conditions were used for the initial NCD deposition of 6 hours on both the wide-bandgap semiconductor samples: 2112P and 2112D (sample designations provided by NRL). As expected, no device damage resulted from the first 6 hours of deposition. The samples were subsequently stored between sheets of paper under vacuum. When the samples were checked prior to the next set of deposition runs, superficial scratches were visible under an optical microscope under low magnification. Thorough inspection of both the samples under an optical microscope at high magnification and a SEM indicated that the observed scratches did not extend to the device metallization, leading us to conclude that the diamond deposited at lower temperatures might not be completely bonding with the underlying substrate material. In view of the fact that one of the primary aims of the experiments was to demonstrate the deposition of diamond on devices with no damage to the metallization, and the fact that number of samples available for experimentation were limited, the following experimental strategy was followed for subsequent deposition experiments on the two wide-bandgap semiconductor device samples.

Table III. Diamond deposition conditions for wide-bandgap semiconductor devices

Sample	Deposition Conditions[*]					Remarks[***]
	P (Torr)	T (°C)	Power (W)	t (h)	H$_2$O (°C)	
2112P	50	-	880/900	3		~ 10.5	Spin coat
	50	235/245	880/700	1		~ 10	
	50	< 240	800/700	2		~ 10	
	50	247/294	880/800	3		~21/~15	Spin coat
	50	< 235	800/815	3	21	10/18	
	50	247/264	800/900	3		21/~11	
	50	257/241	950/810	3		21/~11	
	50	295/255	950/830	3		11/17	
2112D	50	< 235	770	3		~ 10.5	Spin coat
	50	-	770	3	16	~ 10	
	50	311/266	1100/900	3		22/15	Spin coat
	50	310/296	1150/1100	1		~22/18	
	50	318/314	1250/1102	3		25	
	50	353/340	830/700	3		25/~9	

In the case of the sample 2112P, it was decided that the same process conditions would be maintained for all subsequent depositions. The experiments were carried out in increments of three hours in order to be able to carefully document the progress of the NCD deposition. Five such experiments were carried out (c.f. Table III) following the first 6 hour deposition of diamond resulting in a total deposition time of 21 hours. Prior to starting the second set of experiments a second spin coat of diamond was applied to seed the areas where the diamond had been removed (c.f. Table III). Figure 4 records the progression, using optical images, of diamond deposition on a single device on sample 2112P. The following process steps are recorded in Fig.4 : (a) before the deposition of diamond, (b) after the first 6 hour deposition, and the introduction of superficial scratches on the deposited diamond, (c) after the second spin coating step, and (d) after completion of deposition for 21 hours. It is clearly evident from the images that not only is there no visible damage to the device but also that there is indeed diamond deposition taking place on the samples as can be seen from the almost complete elimination of the superficial scratches in Fig. 4 (d). SEM micrographs showed excellent diamond coverage (Figs. 4(e) and (f)). EDX spectra for sample 2112P indicate that the percentage of carbon is of the order of 91 atom %. Figure 5 shows the normalized micro-Raman spectra (at 50X; 532 nm; 25 mW nominal power; Nicolet Almega, Thermo Fisher Scientific, Inc, MA, USA) of sample 2112P. The first order Raman peak (1332 cm^{-1}) due to diamond is clearly evident. It should, however, be noted here that this peak does not appear everywhere on the sample therefore indicating that the quality of the deposited diamond may not be uniform, even though SEM micrographs indicate conformal deposition.

Since diamond deposition with no damage to devices fabricated on a wide-bandgap semiconductor substrate was successfully demonstrated in the case of sample 2112 P, the next step was to systematically push the temperature of deposition upwards on the second sample, 2112D, in order to observe the evolution of the diamond deposition and its effect on the device metallization. As with sample 2112P, sample 2112D was subjected to a second spin coating of diamond (c.f. Table III). No damage to the metallization on this sample was visible after the first six hours of deposition. Table III lists the process conditions used for diamond deposition following the second spin coating. The temperature of deposition was progressively ramped up leading to a total deposition time of 16 hours. No damage was visible on any of the devices after 13 hours of deposition. After the final incremental deposition for three hours at ~ 350 °C, a peculiar type of damage was visible mostly on a particular type of device on this sample which we choose to designate as an "Interdigitated Device Structure" (IDS) for the purpose of this paper (Fig. 6).

It can be clearly seen that there is some form of damage to the contact pad metallization of the IDS. It is to be noted that the progression of diamond deposition was along similar lines as for 2112P up to the penultimate incremental deposition of diamond for three hours. In other words, as with sample 2112P, with each subsequent deposition the superficial scratches on the diamond were progressively filled with diamond deposit resulting in the removal of most of the superficial scratches on the sample. Also, a large majority of other devices were undamaged with excellent diamond coverage evident on the sample at the end of 16 hours of deposition.

EDX spectra on the damaged region of an IDS indicated that the Ni concentration was ~1.8 atom %, which was significantly higher as compared with earlier spectra measured when there was no damage to the metallization. This leads us to believe that some form of significant damage had occurred to the overlying Au layer leading to the Ni metallization being exposed. However, it was difficult to investigate this phenomenon in the absence of additional samples for further experimentation. Figure 7 shows the normalized micro-Raman spectra (50 X) for sample 2112D. The presence of the 1332 cm^{-1} diamond peak is evident. Unlike sample 2112P, this spectra was repeatable across the sample and therefore indicates that the diamond deposition may be more uniform in the case of this sample.

Electrical Characterization of AlGaN/GaN HEMTs

Following the demonstration of diamond deposition on GaN-based devices without any visible damage to the device metallization, the next step involved the electrical testing of the devices. The electrical characterization was performed by the Naval Research Laboratory (NRL) before and after the NCD deposition. Figure 8 shows some of the results of the electrical characterization at NRL. These plots show drain current vs. drain source voltage (I_{DS} vs. V_{DS}) data for both the wide-bandgap semiconductor samples.

It can be seen that there is a marginal lowering in the drain current after diamond deposition in the case of sample 2112 D. However, for sample 2112P there is a drastic deterioration in device characteristics. It is to be noted here that from the measured $I_{DS} - V_{DS}$ for sample 2112D, no significant enhancement in thermal performance was observed. Such an enhancement would be indicated by flat $I_{DS} - V_{DS}$ characteristics in the saturation region as opposed to the slight downward slope evident in the plots. It should be noted here that the transconductance characteristics followed a trend similar to that of the $I_{DS} - V_{DS}$ plots for both the samples, i.e., 2112D showed minimal change in the transconductance plots while 2112P did not. The gate leakage current values are comparable for both the samples.

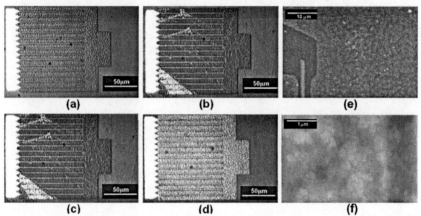

Figure 4 . Optical images (50X objective) of a device on sample 2112P: (a) before deposition, (b) after the introduction of superfical scratches in the diamond layer following 6 hours of deposition, (c) after the second spin coat of diamond - the scratches were not eliminated after the second spin coating step, and (d) after 21 hours of deposition - most of the surface scratches have been filled in with diamond deposit (e) SEM micrograph : 2 kX (10 μm scale bar), and (f) SEM micrograph: 20 kX (1 μm scale bar) - the diamond grains are not in focus due to surface charge buildup; the samples were not coated with a conducting material to facilitate further analysis at NRL.

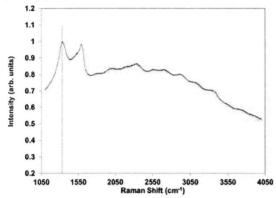

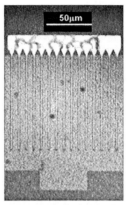

Figure 5. Normalized Raman spectra plot for the sample 2112 P.The vertical line indicates the location of 1332 cm⁻¹ first order Raman peak due to diamond.

Figure 6. Optical of IDS after the final incremental deposition of 3 hours on sample 2112D. Damage was visible to the contact pad metallization on the device.

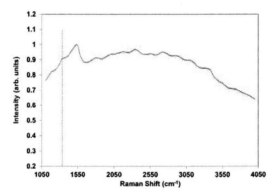

Figure 7. Normalized Raman spectra plot for the sample 2112 D. The vertical line indicates the location of 1332 cm⁻¹ first order Raman peak due to diamond.

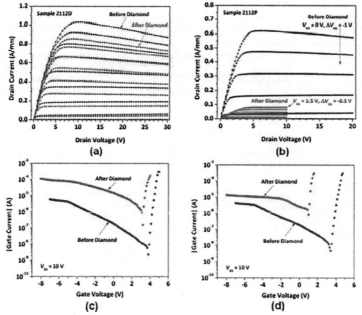

Figure 8. Drain current vs. drain-source voltage plots for: (a) sample 2112D and (b) sample 2112P. A marginal change in ($I_D - V_D$) characteristics was observed in the case of sample 2112D, while there was a significant reduction in the case of sample 2112P. Gate current vs. gate voltage plots for (c) sample 2112D and (d) sample 2112P. Both samples show increased I_G values after diamond deposition.

    The electrical characterization results in the case of the sample 2112P indicate that either the process steps carried out after diamond deposition (i.e., selective plasma etching of the diamond film to access the contact pads for device testing) or the diamond deposition process itself had resulted in the measured performance degradation. These results are particularly intriguing in the sense that while sample 2112D was processed at a higher temperature, it exhibits better device characteristics as compared to the sample 2112P. There could be a number of reasons for the observed change in the device characteristics – defects introduced into the device layer, surface metallization effects including increased resistivity due to diamond deposition. However, given the limited number of samples and devices tested, further replication and experimentation is required before definitive conclusions can be drawn from the observed data. Nevertheless, the results are interesting since they demonstrate that there is a possibility of realizing NCD deposition conditions which do not result in damage to device operation. Also, they are indicative that it may be possible to realize higher deposition temperatures (> 320 °C) without incurring damage to device performance. Higher deposition temperatures will result in higher quality diamond with better thermal properties and thermal performance of the HEMT devices.

CONCLUSIONS

Low temperature (<350 °C) diamond deposition was performed on a series of Si device samples and two wide-bandgap semiconductor HEMT device samples with the aim of elucidating the process conditions under which no damage occurs to the device metallization. In the case of the Si devices, no damage to the metallization was observed under deposition conditions of 50 Torr and average substrate surface temperatures of less than 260 °C. In the case of the GaN HEMT samples no damage was observed below an average substrate surface temperature of 320 °C. Further, Raman spectra measurements on both the wide-bandgap semiconductor samples indicated the presence of the first order Raman peak due to diamond at 1332 cm^{-1} indicating that diamond deposition was indeed taking place. Therefore, the deposition conditions under which no damage occurs to the device metallization have been delineated. In the case of the Si samples it was found that for average deposition temperatures greater than 290 °C (at 50 Torr deposition pressure) there was observable damage to the device metallization. Preliminary investigations indicate that the metallization damage may be attributed to a complex phenomenon involving a solid-state reaction between Au and Si followed by preferential removal of the alloy due to chemical or thermal effects, under the influence of the plasma, from areas where such a reaction has taken place. Further study is required to pinpoint the exact mechanisms which result in the observed metallization damage. In the case of the wide-bandgap semiconductor samples, it was found that diamond deposition on one sample above 350 °C, damage to the contact pad metallization on the "Interdigitated Device Structures" (IDSs) was observed. It is not clear as to why this type of damage seems to occur almost exclusively on the contact pads of the IDS type structure. Further investigation using multiple samples should help shed further light on the observed phenomenon. Also, it was found that the diamond deposited at low temperatures < 300 °C did not adhere completely to the substrate material. At higher temperatures there was a marginal improvement in film adhesion.

Preliminary electrical testing of the devices indicated that it is indeed possible for the devices to maintain their integrity even after being subject to harsh plasma processing conditions. However, since the major hurdle of depositing diamond thin films on devices without any damage to the metallization has been overcome, we anticipate that the problem of improving film adhesion may be addressed by further experimentation. Current research is focused on innovative ways to improve the diamond quality using multilayered films, and by performing further experimentation on GaN-based device structures. Finally, it may be emphasized that the deposition of diamond *over fabricated devices* is an exciting development, one which may have significant technological implications, and therefore merits further in-depth study and experimentation.

FOOTNOTES

[*]The gas composition was 60:39:1 (Ar:H$_2$:CH$_4$) for all deposition runs.

[**]All samples were seeded with 25 nm diamond by using the proprietary spin coating process.

[***]No H$_2$ plasma etch.

ACKNOWLEDGEMENTS

The authors are grateful for the support of the National Science Foundation (NSF; Grant No. ECCS0853789) and the Office Naval Research (ONR; Grant No.N00014-06-1-1147) for the present work. The authors would like to thank Dr. S.C. Binari and D. Meyer of the Naval Research Laboratory, Washington, D.C., for the fabrication and testing of the GaN-on-SiC devices. The authors would also like to thank Mr. Ratandeep Kukreja for help with the SEM micrographs. Any opinions, findings, conclusions or recommendations expressed in this material are those of the authors and do not necessarily reflect the views of the National Science Foundation.

REFERENCES

[1] U. K. Mishra, P. Parikh, and Y-F. Wu, AlGaN/GaN HEMTs—An overview of device operation and applications, *P. IEEE*, **90(6)**, 1022-31 (2002).

[2] J.E.Graebner, S. Jin, G.W. Kammlott, J.A. Herb, and C.F. Gardinier, Large anisotropic thermal conductivity in synthetic diamond films, *Nature*, **359**, 401-03 (1992).

[3] D.M. Gruen, Ultrananocrystalline diamond in the laboratory and the cosmos, *MRS Bull.*, **26(10)**, 771-76 (2001).

[4] R. C. Eden, Application of diamond substrates for advanced high density packaging, *Diam. Relat. Mater.*, **2**, 1051-58 (1993).

[5] J.E. Graebner and S. Jin, Chemical vapor deposited diamond for thermal management, *JOM – J. Min. Met. Mat. S.*, **50(6)**, 52-55 (1998).

[6] K. Jagannadham, T. R. Watkins, and R. B. Dinwiddie, Novel heat spreader coatings for high power electronic devices, *J. Mater. Sci.*, **37**, 1363-76 (2002).

[7] S. Jin and H. Mavoori, Processing and properties of CVD diamond for thermal management, *J. Electron. Mater.*, **27(11)**, 1148-53 (1998).

[8] M. Seal, Thermal and optical applications of thin film diamond, *Philos. T. R. Soc. S-A*, **342(1664)**, 313-22 (1993).

[9] N.Govindaraju, D. Das, R.N. Singh, and P.B. Kosel, High-temperature electrical behavior of nanocrystalline and microcrystalline diamond films, *J. Mater. Res.*, **23(10)**, 2774-86 (2008).

[10] *ASM Handbook, Vol. 3 - Alloy Phase Diagrams*, H. Baker and H. Okamoto (Eds.,), ASM International, Materials Park, 1992.

HIGH FREQUENCY MICROWAVE PROCESSING OF LITHIUM DISILICATE GLASS-CERAMIC

Morsi M. Mahmoud[1,2], Guido Link[1], Simone Miksch[1], Manfred Thumm[1,3]

1. Karlsruhe Institute of Technology (KIT), Institute for Pulsed Power and Microwave Technology (IHM), Karlsruhe, Germany.
2. Advanced Technology and New Materials Research Institute (ATNMRI), CSAT, Alexandria, Egypt.
3. Karlsruhe Institute of Technology (KIT), Institute of High Frequency Technology and Electronics (IHE), Karlsruhe, Germany.

ABSTRACT

A high frequency microwave processing technique was used to crystallize lithium disilicate (LS_2) stoichiometric glass. A gyrotron system with 30 GHz microwave frequency was used in the crystallization process of LS_2 glass. The glass was prepared by a regular melting and casting process using LS_2 glass frit. Thermal analysis of the prepared bulk glass was done to determine the nucleation and the crystallization temperatures. One stage heat-treatment crystallization regime was used to crystallize the studied LS_2 glass into glass-ceramic material using both 30 GHz microwave processing and conventional processing. The conventional crystallization was done in an electrical furnace. Characterization of the crystallized glass-ceramic samples was performed using X-ray diffraction (XRD), and scanning electron microscopy (SEM). High frequency (30 GHz) microwave processing was successfully used to crystallize LS_2 glass.

INTRODUCTION

Technological progress is difficult to achieve without new processing techniques for materials, such as microwave processing. Microwave materials processing technology is basically a powerful and different tool to process materials. Materials processing using microwave technology have several advantages when compared to the already known traditional materials processing techniques. These anticipated benefits include more precise and controlled volumetric heating, faster ramp-up to high temperature, lower energy consumption, and enhanced quality and properties of the processed materials[1-3].

The lithium disilicate (LS_2) glass system provides the basis for a large number of useful glass-ceramic products, such as cookware, radomes, ceramic composites, stovetops, and dental crowns. LS_2 glass system has been studied by many researchers [4-15] due to the fact that it was the first glass-ceramic material that was developed by Stookey (1959). LS_2 glass supplies a good glass forming system and crystallizes via homogenous nucleation. LS_2 glass-ceramic contains two solid phases with the same chemical composition: the amorphous phase and the crystalline phase[4].

The 30 GHz microwave processing is a millimeter wave microwave technique. The motivation to use the high frequency microwave technique is to take advantage of the benefits that this technique could offer. It provides a more homogenous field distribution as well as enhanced microwave absorption at room temperature which is critical for low loss materials such as glasses and ceramics [16, 17].

It has been reported that LS_2 glass could be crystallized using 2.45 GHz [18, 19]. Although it was possible to crystallize LS_2 glass at this frequency, a hybrid heating process was required. Because LS_2 glass has a low dielectric loss value at room temperature, it does not heat easily without the aid of hybrid heating at 2.45 GHz [20]. Variable frequency microwave (VFM) crystallization of LS_2 glass has been studied as well [20-22]. It has been demonstrated that LS_2 glass could be crystallized with the VFM processing without the aid of hybrid heating [20, 22]. The VFM crystallization concept and setup was based on the results of the cavity characterization experiment[20] that was done for the LS_2 glass sample in the VFM cavity as well as on the theoretical calculations[20] that was done for LS_2 glass in order to predict the microwave absorption of LS_2 glass using the molecular orbital model of microwave absorption[23, 24]. The calculations indicated the existence of strong microwave oscillations in LS_2 glass at higher frequency ranges such as 33 GHz and 59 GHz[20]. These microwave oscillations are expected to enhance the microwave absorption at this frequency range.

The aim of this work is to investigate experimentally the possibility to crystallize LS_2 glass system using one of such high frequencies (30 GHz) for microwave processing.

EXPERIMENTAL WORK

The stoichiometric composition (LiO_2-$2SiO_2$) of LS_2 glass has been provided as a glass frit by Ivoclar Vivadent AG, principality of Liechtenstein. The chemical composition of the glass frit was confirmed using X- Ray Fluorescence (XRF) technique (33.2 mol% Li_2O- 66.8 mol% SiO_2). The glass frit was melted in a platinum crucible in an electric furnace at 1450°C for 2 hours. The glass samples were then cast as rods of 2 cm in diameter in pre-heated graphite mold at 400°C. Annealing of these glass samples was performed at 400°C for 1 hour in an electric furnace, after which they were oven cooled to room temperature.

A simultaneous thermal analyzer instrument, NETZSCH model STA 449 C Jupiter®, was used to identify the crystal growth temperature peaks for the glass. The temperature range was from room temperature to 1450°C in a nitrogen atmosphere with a heating rate of 10 °C/min in Pt/Rh crucible. Glass specimens were cut into ~1 cm thick disks using a low-speed diamond saw. Based on the thermal analysis data, heat treatment was done to transform the glass samples into glass-ceramics using high frequency microwave processing in a one stage heat treatment process to 600°C using a 10 min. ramp then soaked for 1 min. to promote crystallization.

The microwave crystallization of the glass disks was done at the frequency of 30 GHz which has been generated by a compact gyrotron system with a maximum power level of 15 kW. The glass samples were placed in a microwave transparent thermal insulation setup in order to reduce the heat loss from the heated glass samples to the environment and also to minimize cracking during the microwave processing. The microwave experimental setup is shown in Fig.1. The temperature was controlled by a thermocouple (type S) that was in direct contact with the glass sample surface.

A conventionally crystallized LS_2 glass sample has been heat-treated in an electric furnace at 485°C (nucleation temperature) for 2 hours to ensure sufficient nucleation sites then heat-treated at 678°C (maximum crystallization temperature) for 24 hours in order to promote full crystallization. This conventionally crystallized sample will be used as a "control crystal".

The high frequency crystallized glass samples were characterized using X-Ray Diffraction (XRD) and Scanning Electron Microscopy (SEM). The X-ray patterns were obtained using an X-Ray Diffraction Spectrometer (SEIFERT) with a tube voltage of 40 kV and a current of 30 mA. A scanning speed of 0.05 degrees (θ)/min and a scan range 5-70 (2θ) were used for the analyses and the reference data for the interpretation of X-ray diffraction patterns were obtained from the ASTM X-ray diffraction file index.

The microstructures of the crystallized glass samples were examined using a SEM (PHILIPS XL 40). For SEM investigations, the surfaces of the glass-ceramic samples were first etched for 1 minute in 10% diluted hydrofluoric acid, rinsed with distilled water and acetone, dried and then coated with a thin film of evaporated gold.

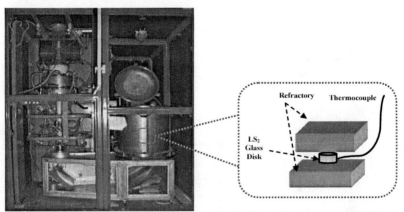

Figure 1. Gyrotron system for materials processing, 30 GHz, with a schematic of the LS_2 glass microwave processing setup.

RESULTS AND DISCUSSION

Fig. 2 shows the differential scanning calorimetry (DSC) and the thermogravimetric (TG) data of the prepared bulk LS_2 glass. An endothermic nucleation temperature was detected around 486°C. The exothermic crystallization peak was detected around 678°C (T_c) that represents the heat released due to the crystallization reaction. An endothermic melting peak was observed around 1052°C. Very close and similar values of the nucleation and crystallization temperatures have been also reported by others workers [20, 25] for the bulk stoichiometric composition of LS_2 glass. At 960°C and 1052°C two endothermic DSC peaks were detected. Those two endothermic peaks are characteristics for the polymorphous transformation of the stoichiometric LS_2 crystal phase and for the congruent melting of the same crystalline phase[4]. These DSC data was used to design the microwave crystallization process of the LS_2 glass in a one stage heat treatment regime. The TG data of the prepared glass indicates that there is no significant weight loss in the studied temperature range.

Fig. 3 shows the XRD patterns of the 30 GHz microwave crystallized LS_2 glass sample at 600°C for 1 minute and the conventionally crystallized LS_2 glass sample "control crystal". The XRD patterns indicate that the 30 GHz microwave crystallized sample was successfully crystallized into glass-ceramic material and has developed the same crystal phase with the characteristic XRD peaks of lithium disilicate phase. The crystal phase in both samples was identified as "Orthorhombic Ccc2" lithium disilicate crystal phase and matched the ICCD-XRD pattern [26, 27]. The XRD patterns of both samples indicate that the 30 GHz microwave crystallization occurs in a significantly shorter time, lower temperature and without hybrid heating.

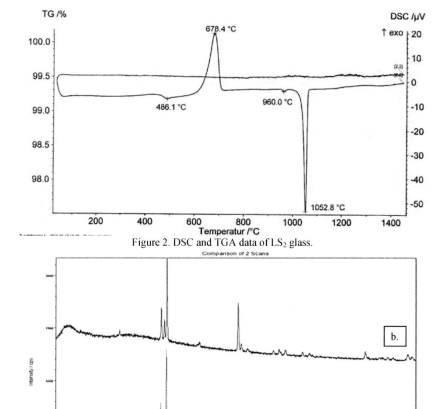

Figure 2. DSC and TGA data of LS_2 glass.

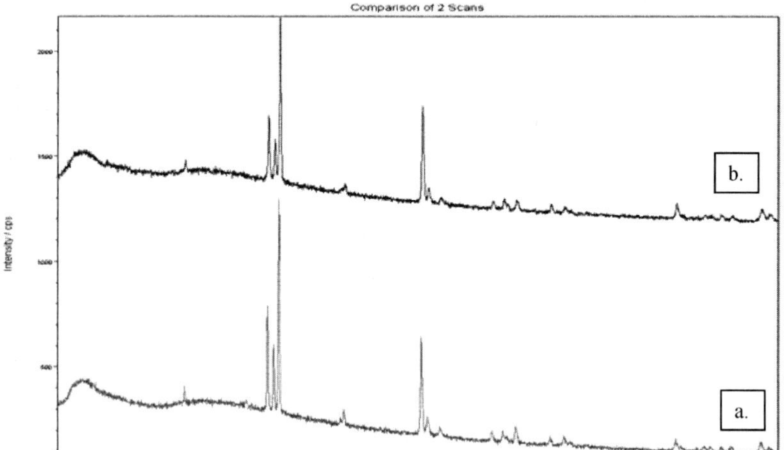

Figure 3. XRD patterns of LS_2 glass-ceramic samples crystallized using a) Conventional heating - 680°C for 24 hours "control crystal" and b) 30 GHz microwave heating- 600°C for 1 minute.

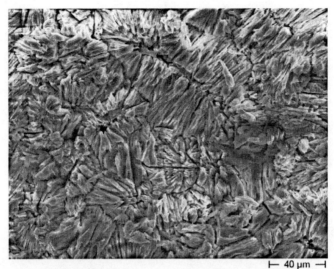

Figure 4. SEM micrograph of the 30GHz microwave crystallized LS$_2$ glass sample 600°C- 1 min.

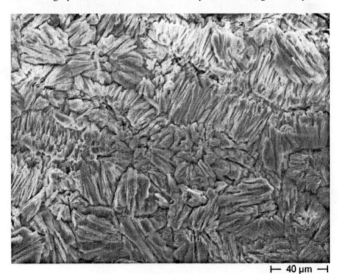

Figure 5. SEM micrograph of the conventionally crystallized LS$_2$ glass sample "control crystal" 680°C- 24 hours.

Fig.4 and Fig.5 show the SEM micrographs of the 30 GHz microwave crystallized sample at 600°C for 1 minute and the conventionally crystallized control crystal, respectively. It is clear from the graphs that both samples have developed the typical characteristic and the microstructural features of polycrystalline lithium disilicate glass-ceramic[4]. Both glass-ceramic samples exhibit the tightly interlocking crystal form that is characteristic of the corrugated sheets or layers of LS_2 glass-ceramics[4].

From the XRD patterns and the SEM micrographs of the 30 GHz microwave crystallized LS_2 glass and the control crystal, it is clear that 30 GHz microwave frequency was able to crystallize LS_2 glass into glass-ceramic materials. This experimental finding is on good agreement with the theoretical calculations[20] that was done for LS_2 glass in order to predict the microwave absorption of LS_2 glass using the molecular orbital model of microwave absorption. The development of the lithium disilicate crystal phase in the 30 GHz microwave crystallized sample in a relatively short time, lower temperature and without the aid of hybrid heating makes the high frequency microwave processing technology an attractive processing technique and also opens new opportunities and promising alternative ways in the production of the lithium disilicate based glass-ceramics products.

CONCLUSION

High frequency, 30 GHz, microwave processing was used successfully to crystallize lithium disilicate stoichiometric glass composition into glass-ceramic material in a relatively short time, low temperature and without the aid of hybrid heating using one stage heat treatment. This experimental investigation of the 30 GHz microwave crystallization of LS_2 glass was on good agreements with the theoretical calculations that were done using the molecular orbital model of microwave absorption. The microwave absorption in LS_2 glass is highly dependent on the microwave frequency.

ACKNOWLEDGMENT

The support of Ivoclar Vivadent AG, Principality of Liechtenstein is kindly acknowledged

REFERENCES
1. Clark, E.D., C.D. Folz, E.C. Folgar, and M.M. Mahmoud, *Microwave Solutions for Ceramic Engineers.* Westerville,Ohio: The American Ceramic Society. 494. 2005
2. Clark, D.E. and W.H. Sutton, *Microwave processing of materials.* Annual Review of Materials Science, 26: p. 299-331. 1996.
3. *Materials Research Advisory Board,Microwave Processing of Materials*: National Research Council, National Academy Press. 1994
4. Holand, W. and G. Beall, *Glass-ceramic Technology*: The American Ceramic Society. 2002
5. James, P.F., A. Paul, R.M. Singru, C. Dauwe, L. Dorikens-Vanpraet, and M. Dorikens, *Positron lifetimes in lithium disilicate glass with different degrees of crystallization.* Journal of Physics C: Solid State Physics, 8(4): p. 393-8. 1975.
6. Zanotto, E.D. and P.F. James, *Experimental tests of the classical nucleation theory for glasses.* Journal of Non Crystalline Solids, 74: p. 2-3. 1985.
7. Joseph, I., Nucleation and related phenomena in lithium disilicate glass, 152 pp, 1985

8. Goharian, P., A. Nemati, M. Shabanian, and A. Afshar, *Properties, crystallization mechanism and microstructure of lithium disilicate glass-ceramic.* Journal of Non-Crystalline Solids, 356(4-5): p. 208-214. 2010.
9. Burgner, L.L., Crystallization kinetics of lithium disilicate and sodium silicate glasses, 291 pp, 2000
10. Iqbal, Y., W.E. Lee, D. Holland, and P.F. James, *Metastable phase formation in the early stage crystallisation of lithium disilicate glass.* Journal of Non-Crystalline Solids, 224(1): p. 1-16. 1998.
11. Deubener, J., R. Brueckner, and M. Sternitzke, *Induction time analysis of nucleation and crystal growth in di- and metasilicate glasses.* Journal of Non-Crystalline Solids, 163(1): p. 1-12. 1993.
12. Barker, M.F., T.H. Wang, and P.F. James, *Nucleation and growth kinetics of lithium disilicate and lithium metasilicate in lithia-silica glasses.* Physics and Chemistry of Glasses, 29(6): p. 240-8. 1988.
13. McCracken, W.J., D.E. Clark, and L.L. Hench, *Aqueous durability of lithium disilicate glass-ceramics.* American Ceramic Society Bulletin, 61(11): p. 1218-23. 1982.
14. Burgner, L.L. and M.C. Weinberg, *Assessment of crystal growth behavior in lithium disilicate glass.* Journal of Non Crystalline Solids, 279: p. 28-43. 2001.
15. Davis, M.J., P.D. Ihinger, and A.C. Lasaga, *Influence of water on nucleation kinetics in silicate melt.* Journal of Non Crystalline Solids, 219: p. 62-69. 1997.
16. Link, G., L. Feher, M. Thumm, H.J. Ritzhaupt-Kleissl, R. Bohme, and A. Weisenburger, *Sintering of advanced ceramics using a 30-GHz, 10-kW, CW industrial gyrotron.* Plasma Science, IEEE Transactions on, 27(2): p. 547-554. 1999.
17. Thumm, M., L. Feher, and G. Link, *Micro- and millimeter-wave processing of advanced materials at Karlsruhe Research Center,* in *Novel Materials Processing by Advanced Electromagnetic Energy Sources.* Elsevier Science Ltd: Oxford. p. 93-98. 2005
18. Boonyapiwat, A., D.C. Folz, and D.E. Clark, *Microwave crystallization of glass.* Ceramic Transactions, 101(Surface-Active Processes in Materials): p. 87-96. 2000.
19. A.D.Cozzi, Z.Fathi, R.L.Schulz , and D.E.Clark. *Nucleation and crystallization of Li2O-2SiO2 in a 2.45 GHz microwave field.* in *17th Annual Conference on Composite and Advanced ceramic Materials.* Cocoa Beach, FL, Jan. 10-15, 1993. .p. 856-862.1993
20. Mahmoud, M.M., Crystallization of lithium disilicate glass using variable frequency microwave processing, Materials Science and Engineering Department, Virginia Tech, PhD Dissertation, 215, 2007
21. Mahmoud, M.M., D. Folz, C. Suchicital, and D.E. Clark, *Crystallization of Lithium Disilicate Glass By Variable Frequency Microwave (VFM) Processing,* in *Proceedings of the Fourth World Congress on Microwave and Radio Frequency Applications, November 2004,* R.L.S.a.D.C. Folz, Editor. Microwave Working Group, Ltd., Arnold, MD: Austin, TX. p. 271-77. 2005.
22. Mahmoud, M.M., D. Folz, C. Suchicital, D.E. Clark, and Z. Fathi, *Variable Frequency Microwave (VFM) Processing: A New Tool to Crystallize Lithium Disilicate Glass,* in *Ceramic Engineering and Science Proceedings: Advances in Bioceramics and Biocomposites II,* M. Mizuno, Editor. The American Ceramic Society,Westerville,Ohio. p. 143-153. November 2006.

23.   West, J.K. and D.E. Clark, *A Molecular Orbital Model for Absorption of Microwave Energy by Materials.* Proceedings, 2nd World Congress on Microwave and Radio Frequency Processing, Orlando, Florida, Ceramic Transactions, 111: p. 43-56. 2000.

24.   West, J.K. and D.E. Clark, *Microwave absorption by materials: Theory and application.* Ceramic Transactions, 101(Surface-Active Processes in Materials): p. 53-73. 2000.

25.   Soares Jr, P.C., E.D. Zanotto, V.M. Fokin, and H. Jain, *TEM and XRD study of early crystallization of lithium disilicate glasses.* Journal of Non Crystalline Solids, 331: p. 1-3. 2003.

26.   De Jong, B.H.W.S., P.G.G. Slaats, H.T.J. Supèr, N. Veldman, and A.L. Spek, *Extended structures in crystalline phyllosilicates: silica ring systems in lithium, rubidium, cesium, and cesium/lithium phyllosilicate.* Journal of Non-Crystalline Solids, 176(2-3): p. 164-171. 1994.

27.   Fuss, T., A. Mogus-Milankovic, C.S. Ray, C.E. Lesher, R. Youngman, and D.E. Day, *Ex situ XRD, TEM, IR, Raman and NMR spectroscopy of crystallization of lithium disilicate glass at high pressure.* Journal of Non-Crystalline Solids, 352(38-39): p. 4101-4111. 2006.

MICROWAVE SINTERING OF A PZT/Fe-Co NANOCOMPOSITE OBTAINED
BY IN SITU SOL-GEL SYNTHESIS

Claudia P. Fernández[1], Ducinei Garcia[2], Ruth H. G. A. Kiminami[1]

[1]Federal University of São Carlos, Materials Engineering Department São Carlos, SP, Brazil, 13565-905 São Carlos, SP - Brazil

[2]Federal University of São Carlos, Department of Physic, 13565-905 São Carlos, SP - Brazil

ABSTRACT

Microwave sintering is a technique with an enormous potential for the manufacture of ceramic materials with controlled microstructures, providing uniform heating and more homogeneous and controlled grain size distribution than conventional sintering. This paper discusses the effect of ultra rapid microwave-assisted sintering versus conventional sintering on the microstructural and physical characteristics of the particulate composite PZT/Fe-Co in a 1:1 ratio, prepared *in situ* via sol-gel synthesis to obtain a nanostructured system with a highly homogeneous two-phase distribution. The samples sintered by the two sintering processes were characterized microstructurally by scanning electron microscopy (SEM) and X-ray diffraction (XRD) and their apparent density and porosity were measured by the Archimedes method. These characterizations revealed that the two methods of synthesis yielded the particulate composite, and that the grains of the ferromagnetic phase were distributed evenly within the ferroelectric matrix. However, the microwave-sintered samples retained a more homogeneous average grain size than the conventionally sintered samples.

Keywords: Microwave sintering, PZT/Fe-Co, Sol-gel, Synthesis *In Situ*.

INTRODUCTION

Multiferroic magnetoelectric (ME) composite materials such as ferromagnetic and ferroelectric heterostructures, have properties of both electric and magnetic ordering. These materials have been intensely investigated recently due to their attractive physical properties [1,2], as well as their potential new applications in multifunctional materials for next generation electronic devices that exploit control of the magnetic (spin) state via electric fields and/or vice versa, with potential for spintronic devices, solid-state transformers, high sensitivity magnetic field sensors, and electromagnetooptic actuators [3,4]. Within this context, magnetoelectric multiferroic composite materials now pose challenges in terms of devising new synthesis methods, as well as in the use of unconventional methods of sintering to enable optimization of the properties derived from the coupling of ferroelectric and ferromagnetic systems [3, 5, 6, 7].

Wu et al. and Iordan et al. [8,9] have shown that *in situ* synthesis is a successful approach for the synthesis of ceramic composites, which enhances the mixture of the two phases, resulting in more homogeneous microstructures. Microwave sintering, a new technology developed in the mid to late 1980s, has proved to be a successful processing technique [10, 11] characterized by fast densification of ceramic materials [11,12]. It makes use of dielectric loss of materials in microwave electromagnetic fields to heat the sample rapidly to the sintering temperature. A wide variety of oxides and non-oxides have been sintered successfully by microwaves [11,12]. Compared with traditional heating, the effect of microwave heating is produced mainly by continuous changes in the polarization of the material in an alternating electromagnetic field. The continuous changes in the dipoles in materials can produce strong vibrations and friction, generating heat to achieve sintering [12,13].

This method of heating differs from conventional processes in that microwaves facilitate the transfer of energy directly into the material, providing volumetric heating [10-13]. Due to the penetration of microwaves into microwave-absorbing materials, microwave heating is volumetric and material-dependent.

The direct interaction of energy in the volume of the material eliminates energy wastage by heating the walls of the furnace and refractory supports simultaneously [14-16], enabling the application of high heating rates allied to shorter processing times. This allows for significantly reduced energy consumption, particularly in processes that require high temperatures, since heat losses increase dramatically at higher temperatures [15].

Heating with microwave energy offers several advantages over conventional heating techniques, such as energy savings, considerably shorter processing times, and high heating rates (limited, in theory, only by the power used), yielding fine and uniform microstructures [16].

This paper reports on the synthesis of $Pb(Zr_{0.52}Ti_{0.48})O_3$-$Fe_2CoO_4$ (PZT/Fe-Co) (1:1) composite ceramics produced via *in situ* processing based on a sol-gel method. The main purpose of the present study is to investigate the effect of microwave sintering on the structural properties of the ceramics.

EXPERIMENTAL PROCEDURES

Processing of PZT/Fe-Co

Ceramic precursors of $x(Pb(Zr_{0.53}Ti_{0.47})O_3)$ $(1-x)(Fe_2CoO_4)$ with ($x=0.5$) were synthesized by *in situ* processing based on a sol-gel method.

The solution was prepared with a molar ratio of Pb: Zr: Ti= 1: 0.53 :0.47, and lead nitrate $Pb(NO_3)_2$ was dissolved in aqueous HNO_3 0.5N, which was then mixed with the ethanol solution containing the required contents of titanium tetrabutoxide ($C_{16}H_{36}O_4Ti$) and zirconium butoxide, ($C_{16}H_{36}O_4Zr$- 80% in butanol). Appropriate portions of (Fe $(NO_3)_3.9H_2O$) and (Co $(NO_3)_2.6H_2O$) were dissolved in an aqueous solution of HNO_3 0.5N to reach a molar ratio of Fe:Co= 2:1. Both Ti-Zr-Pb and Fe-Co solutions were subjected to hydrolysis and polycondensation reactions. The two solutions were then mixed and treated by adding ammonium hydroxide (NH_4OH) to promote the formation of nanosized particles. The sol was aged for 12 hours to produce the gel. Finally, the gel was dried at 80°C and calcined.

The calcined powder was pressed into disc-shaped compacts with a diameter of 0.5 mm and thickness of about 2mm under uniaxial pressure of approximately 120 MPa and isostatic pressure of 200MPa, followed by sintering. Microwave sintering (MS) was carried out at 2.45 GHz, which was generated by a 6 kW magnetron (Cober Electronics, model MS6K). The samples were heated at a rate of 100°C/min to 1050–1150°C, soaked for 15 min (in air), and then cooled. Conventional sintering (CS) involved sintering the samples at 1050–1150°C for 3 h (in air), following by cooling at a rate of 5°C/min. The PZT/Fe-Co samples were placed on a plate on a thin layer of PZT powder (to prevent the reaction of PZT with alumina), which was then covered with an AlO_3/mullite crucible to enclose the PZT powders and thus minimize PbO volatilization.

Characterization

The gels were characterized by X-ray diffraction (XRD), termogravimetry (TG) and differential thermal analysis (DTA), N^{+2} physisorption (BET) and scanning electron microscopy (SEM). An X-ray diffraction (XRD) analysis of the sample was performed to assess the degree of crystallization and identify the phases of the powder. XRD patterns of PZT/Fe-Co were recorded on a Rigaku D/II-B X-ray diffractometer using CuKa radiation. Termogravimetry (TG), and differential thermal analysis (DTA), N^{+2} physisorption (BET) and scanning electron microscopy (SEM) were performed with a Netzsch STA 429 CD thermal analyzer, BET (Micrometrics, Gemini-2370) and SEM-FEG (Phillips, XL30 FEG), respectively. The apparent densities of the sintered pellets were measured by the Archimedes method with water, and the microstructure was examined by scanning electron microscopy (SEM) used a SEM-FEG (Phillips XL30 FEG-SEM). The PZT/Fe-Co samples were subjected to conventional dilatometric measurements in a Netzsch TASC 414/2 Dilatometer model 402 EP to evaluate the relative linear shrinkage at temperatures near 1200°C, applying a constant heating rate of 10°C/min.

RESULTS AND DISCUSSION

Figure 1 illustrates the results of the TG/ DTA analysis of the dried gel of the PZT/Fe-Co composite obtained by *in situ* sol-gel processing. In this figure, note the endothermic peak in the DTA curve at 100°C, which corresponds to the loss of chemisorbed water and of water absorbed from the volatilization of ammonia contained in the sample, which originates from the synthesis process. The endothermic peak that appears in the temperature range of 180 to 300°C can be attributed to the melting process of NH_4NO_3 composed of nitric acid and ammonium hydroxide and its subsequent volatilization, as indicated by the large mass loss between 100-300°C shown on the TG curve. Similar effects were observed by Raju et al. [12] in the synthesis of PZT by the citrate gel route. The endothermic peak between 300-400°C corresponds to the decomposition of Co, Fe and Pb hydroxides formed by (OH-) added to the base system. Lastly, the two endothermic peaks appearing in the temperature range of 450and 500°C correspond to the nucleation of ions which result in the formation of PZT phase and Fe-Co. However, the formation of the phases of interest is expected to occur above 600°C.

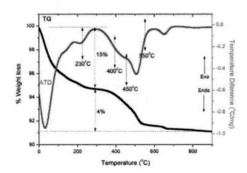

Figure 1 – TG/DTA curves of dry gel of 1:1 PZT/Fe-Co composite

Figure 2 depicts the XRD patterns of dry gel calcined at 600°C and 800°C for 2 h. Note that the sample calcined at 600°C Figure 2 (a) shows crystallization of Pb(ZrTi) O_3 Fe_2CoO_4 phases and the presence of unidentified secondary phases. This confirms the result and analysis of the endothermic peaks in the temperature range of 450-500°C on the ATD curve in Figure 1. Raising the calcination temperature to 800°C for 2 h (Fig. 2 (b)) increased the degree of crystallinity of both phases, and the identified peaks corresponded mostly to the cubic Fe_2CoO_4 spinel structure, according to the JCPDS database (22 - 1086), and to tetragonal Pb $(Zr_{0.52}$ $Ti_{0.48})$ O_3perovskite (33-0784). These results indicate that the synthesis method yielded a material with the simultaneous formation of two phases, PZT and Fe-Co, but without the formation of secondary phases, starting from an amorphous dry gel, which is important for the PZT/Fe-Co composite presenting a good ME coupling effect.

The BET measurement of the surface area of the powder calcined at 800°C was 3.24 m²/g, and the crystallite size calculated by the Sherry equation was 21.98nm for the ferroelectric phase (PZT) and 38.69 nm for the ferromagnetic phase (Fe-Co). These crystallite sizes are smaller than those obtained by Iordam et al. [9], who reported sizes of 150-180nm for the PZT phase and 60-80nm for the Fe-Co phase of the composite.

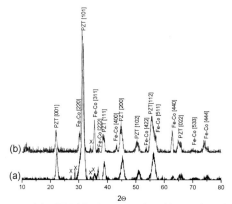

Figure 2 – XRD patterns of the PZT / Co-Fe composite with a molar ratio of 1:1, prepared in situ and calcined at: (a) 600°C, and (b) 800°C for 2 h; x secondary phases

As Figure 3 (a) shows, the above discussed values are confirmed by the SEM results of the powder of the PZT/Fe-Co composite prepared *in situ* and calcined at 800°C/2h. Note that the soft clusters are composed of primary particles with a size of about 50nm and a spherical morphology, and the sample is devoid of impurities, as indicated by the electron diffraction pattern (EDS) in Figure 3 (b).

(a)   (b)

Figure 3.SEM images of in situ composite PZT/Co-Fe calcined at 800°C/2h (a) EDS the powder (b).

Figure 4 presents the results of the dilatometric evaluation of the material sintered conventionally at a constant heating rate of 10°C/min. An analysis of linear shrinkage as a function of temperature (at a constant heating rate of 10°C/min) (see Fig. 4) indicated that shrinkage of the sample started at 897°C and continued up to 1038°C, with an initial shrinkage of 8%. This shrinkage is attributed to the onset of densification of the compact and the marked reduction of pores in the intermediate stage up to 1120°C, with shrinkage of 12%. A low shrinkage rate was observed in the final stage of sintering [17], and total shrinkage above 1150°C was about 26%, which is also related with liquid phase formation due to melting of PbO, which occurs at temperatures above 888°C. Based on these dilatometric results obtained by conventional heating, both conventional and microwave sintering were performed at temperatures ranging from 1050°C to 1150°C.

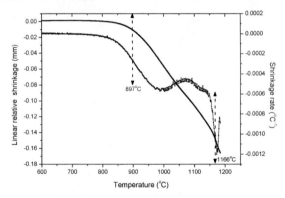

Figure 4 – Relative linear shrinkage curves and shrinkage rates of the PZT/Fe-Co composite as a function of temperature

Figure 5 shows the densification curves of the microwave-sintered (Fig. 5 (a)) and conventionally sintered (Fig. 5 (b)) samples. In both cases, maximum density was reached at 1125°C, in 15 min by microwave-assisted sintering at a heating rate of 100°C/min and a firing cycle of only 3 hours, and by conventional sintering at a heating rate of 5°C/min and a firing cycle of 11 hours. Microwave sintering led to 90.9% and conventional sintering to 96.7% of theoretical density. The density was found to decrease above 1125°C due to volatilization and loss of PbO, leading to the formation of nonstoichiometric ZrO_{2-x} in the structure.

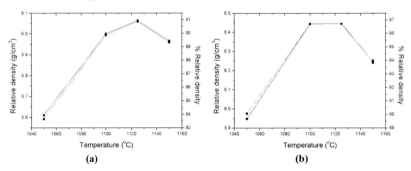

(a) **(b)**

Figure 5 – Relative density curves of composites sintered by: (a) microwaves, and (b) conventional processing

Note the presence of Fe_2CoO_4 (22-1086) and Pb $(Zr_{0.52}Ti_{0.48})O_3$ (33-0784) crystalline phases, without secondary phases in the case of conventionally sintered samples, in the XRD patterns of the samples sintered at 1125°C, which showed much higher densification in 15 min by microwave sintering than in 3 hours by conventional sintering (Fig. 6). Figure 6 (a) indicates the absence of a chemical reaction or interdiffusion between two phases. The microwave-sintered samples showed the presence of ZrO_2 (37-1484) due to the elimination of lead from the system, since, unlike conventional sintering (Fig. 6 (b)), the atmosphere cannot be controlled in microwave processing.

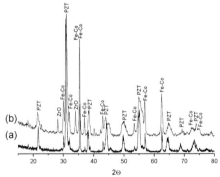

Figure 6 – XRD patterns of the PZT/Fe-Co composite prepared in situ and sintered at 1125°C by the conventional route for 3h (a) and by microwaves for 15min (b)

Figures 7 and 8 show SEM micrographs of the microstructure of the PZT/Fe-Co composite sintered at 1125°C by microwaves and by conventional processing, respectively.

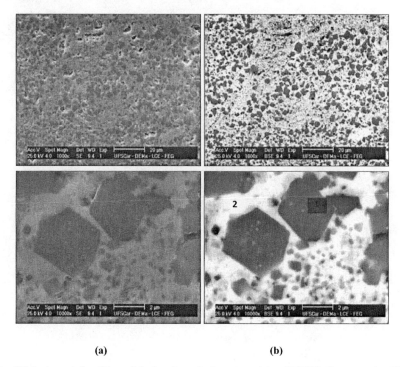

(a)   (b)

Figure 7 – SEM micrographs of the polished surface of microwave-sintered PZT/Fe-Co composite: SE image (a) BSE image (b). (black (1): Fe-Co; white(2): PZT; gray(3): ZrO2)

The SEM images in BSE show two different phases, with the white region (2) corresponding to PZT phase and the dark region (1) to ferrite phase, as indicated by the XRD patterns in Figure 6 for the two sintering processes. The micrographs of microwave-sintered samples in Figure 7 show the presence of two main phases (ferrite and PZT). The small gray regions shown in Figure 7 (b) and identified by the number 3 correspond to ZrO_2, as observed in the XRD pattern in Figure 6. The micrograph in Figure 7 shows a very homogeneous distribution of ferrite phase, with regions of ferrite with an average grain size of 3μm surrounded by PZT with an average grain size of approximately 600nm.

These grains did not grow due to the high heating rate and the short exposure time of the sample at the sintering temperature. In Figure 8, the samples sintered conventionally show a homogeneous distribution of ferrite phase, without the presence of secondary phase (primary phase ferrite and PZT), with average ferrite grains of 6μm mixed in the PZT matrix, whose grain size is about 1μm. The grain sizes of both the ferrite and PZT phases of composites sintered conventionally *in situ* are smaller than those reported by Iordam et al. [9] and Weng et al. [18], who obtained ferrite phase with grain sizes of 10-15μm and PZT phase with grain sizes of 1-4μm, as well a ZrO_2.

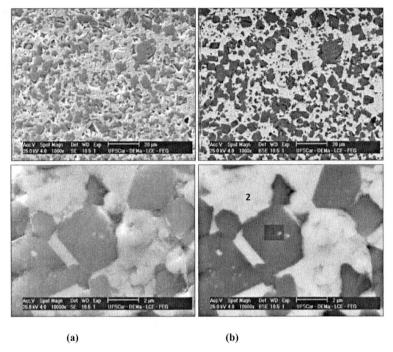

(a)   **(b)**

Figure 8 – SEM micrographs of the polished surface of conventionally sintered PZT/Fe-Co composite: SE image (a) BSE image (b). (black (1):Fe-Co; white(2): PZT)

It is worth noting that the distribution of the ferrite phase within the matrix after conventional sintering was far more homogeneous than that of ME composites produced *in situ* by simple mixing of phases obtained separately, as reported by [19-22].

Moreover, in the case of microwave-sintered composites (Fig. 7), the grains of the ferrite phase were surrounded by PZT phase, with percolation occurring on a lesser scale than in conventional sintering (Fig. 8), demonstrating the efficiency of microwave-assisted sintering. Therefore, microwave sintering produced a homogeneous microstructure with smaller grain sizes than those reported in the literature. Moreover, compared to the results reported by Agrawal et al. [24], it reduced the level of ferrite phase percolation when compared to conventional sintering. The powder of the composite PZT/Fe-Co, with molar ratio 1:1, synthesized by sol-gel method in situ after sintering by both microwave and conventional microstructures without resulting in aggregation and formation of ferrite blocks, as is usually found in the simple mixture of ferrite and PZT, when synthesized separately. Conventional sintering yielded a microstructure with ferrite phase distributed homogeneously in the PZT phase, without the presence of secondary phases, and a smaller average grain size of ferrite phase than that reported in the literature.

CONCLUSIONS

A PZT/Fe-Co nanocomposite produced by *in situ* sol-gel synthesis was microwave-sintered. The PZT/Fe-Co powder synthesized by both microwave and conventional sintering procedures showed microstructures with no formation of aggregates or ferrite blocks, which are commonly found in simple mixtures of ferrite and PZT that are synthesized separately. The *in situ* sol-gel synthesis yielded more homogenous powder with different sintering kinetics. A comparison of the microwave-sintered microstructure and the one sintered conventionally showed differences in grain size and distribution. The microwave-sintered nanocomposites showed ferrite grains surrounded by PZT phase, with percolation occurring on a lesser scale than in conventional sintering, demonstrating the efficiency of microwave-assisted sintering in only 15 minutes at a heating rate of 100°C/min.

ACKNOWLEDGEMENTS

The authors gratefully acknowledge the financial aid of the Brazilian research funding agencies FAPESP (Process # 07/59564-0 and 08/04025-0) and CAPES, and the practical support of the Postgraduate Program in Materials Science and Engineering of the Federal University of São Carlos.

REFERENCES

[1] W. Eerenstein, N.D. Mathur, and J.F Scott, Multiferroic and Magnetoelectric Materials, *Nature*, **442**, 759-65 (2006).

[2] M. Fiebig, Revival of the Magnetoelectric Effect. *J. Phys.* D, 38, R123–52 (2005).

[3] C.A.F. Vaz, J. Hoffman, C.H. Ahn and R. Ramesh., Magnetoelectric Coupling Effects in Multiferroic Complex Oxide Composite Structures, **22**, 2900-18 (2010).

[4] J. Ma , J. Hu, Z. Li, and C. Nan. Recent Progress in Multiferroic Magnetoelectric Composites: from Bulk to Thin Films, *Adv. Mater,* **23**, 1062–87 (2011).

[5] J. Ryu, S. Priya, K. Uchino and H. Kim,, Magnetoelectric Effect in Composites of Magnetostrictive and Piezoelectric Materials, *J. Electroceram,* **8**, 107-19 (2002).

[6] C.W Nan, M.I Bichurin, S. Dong, D. Viehland and G. Srinivasan, Multiferroic Magnetoelectric Composites: Historical Perspective Status, and Future Directions", *J. Appl. Phys.*, **103**, 031101-35 (2008).

[7] V. Corral-Flores, D. Bueno-Baques, and R.F. Ziolo, Synthesis and Characterization of Novel $CoFe_2O_4$–$BaTiO_3$ Multiferroic Core–Shell-Type Nanostructures, *Acta Mater.*, **58**, 764–69 (2010).

[8] D. Wu, W. Gong, H. Deng and M. Li, Magnetoelectric Composite Ceramics of Nickel Ferrite and Lead Zirconate Titanate Via *In Situ* Processing, *J. Phys.* D, **40**, 5002–5 (2007).

[9] A.R Iordan, M. Airimioaiei, M.N Palamaru, C. Galassi, A.V Sandu, C.E. Ciomaga, F. Prihor, L. Mitoseriu and A. Ianculescu, In Situ Preparation of $CoFe2O4$-$Pb(ZrTi)O_3$ Multiferroic Composites by Gel-combustion Technique, *J. Eur. Ceram. Soc.*, **29**, 2807-13 (2009).

[10] H. Hao, L. Xu, Y. Huang, X. Zhang and Z. Xie, Kinetics Mechanism of Microwave Sintering in Ceramic Materials, *Sci. China Ser. E-Tech. Sci.*, **52**, 2727-31 (2009).

[11] C.Y. Fang, C.A. Randal, M.T. lanagan and D.K. Agrawal, Microwave Processing of Electroceramic Materials and Devices, *J. Electroceram.*,, **22**, 125-30 (2009).

[12] K. Raju, and P. V. Reddy, Synthesis and Characterization of Microwave Processed PZT Material, *J Curr Appl Phys*, **10**, 31–5 (2010).

[13] M. Panneerselvam, K.J. Rao, A Microwave Method for the Preparation and Sintering of β'-SiAlON. *Mater Res Bull*, **38**, 663—74 (2003).

[12] R. R. Menezes, P. M. Souto and R. H. G. A. Kiminami, "Microwave Sintering of Ceramics. Part III: Sintering of Zirconia, Mullite and Alumina", *Cerâmica*, **53**, 218-226 (2007).

[14] R. R. Menezes, P. M. Souto and R. H. G. A. Kiminami, "Microwave Hybrid Fast Sintering of Porcelain Bodies", *J Mater Process Tech*, **190**, 223–29 (2007).

[15] S.A. Freeman, J.H. Booske, and R. F. Cooper, "Microwave Field Enhancement of Charge Transport in Sodium Chloride", *Phys Rev Lett*, **74**, 2042-44(1994).

[16] R. R. Menezes, P. M. Souto and R. H. G. A. Kiminami, "Microwave Sintering of Ceramics. Part II: Sintering of ZnO-CuO Varistors, Ferrite and Porcelain Bodies", *Cerâmica*, **53**, 108-15 (2007).

[17] W. Kingery, Introduction to Ceramics.Vol.1. (New York, NY: John Willy & Sons, (1996).

[18] L. Weng, Y. Fu, S. Song, J. Tang, and J.Li, Synthesis of Lead Zirconate Titanate-Cobalt Ferrite Magnetoelectric Particulate Composites Via an Ethylenediamenetetraacetic Acid Citrate Gel Process, *Scripta Mater*, **56**, 465-68 (2007).

[19] B. K. Bammannavar, L. R. Naik, and B. K. Chougule, Studies on Dielectric and Magnetic Properties of (x)$Ni_{0.2}Co_{0.8}$ Fe_2O_4 +(1−x) Barium Lead Zirconate Titanate Magnetoelectric Composites, *J Appl Phys*, **104**, 064123-8 (2008).

[20] J. Ryu, A. V. Carazo, K. Uchino and H.E. Kim, Piezoelectric and Magnetoelectric Properties of Lead Zirconate Titanate/Ni-Ferrite Particulate Composites", *J Electroceram*, **7**, 17–24 (2001).

[21] Arif D Sheikh and V L Mathe. Effect of the Piezomagnetic $NiFe_2O_4$ phase on the Piezoelectric $Pb(Mg_{1/3} Nb_{2/3})_{0.67}Ti_{0.33}O_3$ Phase in Magnetoelectric Composites. *Smart Mater Struct*, **18**, 065014-9 (2009).

[22] Q.H. Jiang, Z.J. Shen, J.P. Zhou, Z. Shi and C.W Nan, Magnetoelectric Composites of Nickel Ferrite and Lead Zirconnate Titanate Prepared by Sparkplasma Sintering, *J Europ Ceram Soc*, **27**, 279–84 (2007).

[23] S. Agrawal, J. Cheng, R. Guo, A.S. Bralla, R.A. Islam and S.Priya, Magnetoelectric Properties of Microwave Sintered Particulate Composites, *Mater lett*, **63**, 2198-200 (2009).

INVESTIGATION ON MICROSTRUCTURAL CHARACTERIZATION OF MICROWAVE CLADDING

Dheeraj Gupta[a1], Apurbba Kumar Sharma[b2]*, Guido Link[c3], Manfred Thumm[c4]

[a]Department of Mechanical Engineering, Graphic Era University, Dehradun, India
[b]Department of Mechanical and Industrial Engineering, Indian Institute of Technology Roorkee, 247667, India
[c]Karlshrue Institute of Technology, Karlsruhe 76021, Germany

E-mail: [1]guptadheeraj2001@gmail.com ,[2]akshafme@iitr.ernet.in, [3]guido.link@kit.edu, [4]manfred.thumm@kit.edu

*Corresponding author

ABSTRACT

In the present work cladding of nickel based and cermet (WC10Co2Ni) powder on austenitic stainless steel SS-316 substrate has been developed in a home microwave oven at frequency 2.45 GHz and power 900 W. The nickel-based and WC10Co2Ni clads were developed by exposure of microwave radiation for the duration of 360 s. Clads were metallurgically bonded with the substrate by partial mutual diffusion of elements. The microstructure along the cross-section of Ni-based clad (thickness, ≈1 mm and ≈2 mm) of WC10Co2Ni shows significantly less porosity (<1%) and no visible interfacial cracks. The Ni-based clad has cellular microstructure where nickel are segregated inside the cells, while chromium was segregated at the cell boundaries. Typical average microhardness of nickel based clad was ≈300 Hv. The developed cermet clad shows composite characteristics. The carbides formed during cladding (iron-tungsten) show skeleton like morphology which is uniformly distributed in the soft matrix of iron and nickel. The microhardness of cermet clad was ≈1064±99 Hv.

INTRODUCTION

Microwaves are primarily coherent, polarized electromagnetic waves; their frequency lies between 0.3 to 300 GHz and wavelength varies from 1 mm to 1 m. Since the early 1950, microwaves have been in used as a heat generation source [1] for food processing and rubber vulcanizations. Microwave processing of materials is different from the conventional thermal processing methods. Microwave energy heats the material at molecular level, which leads to uniform bulk heating. In the conventional heating systems, however, the material gets heated from the surface to interior with thermal gradient [2, 3]. This characteristic (volumetric heating) of microwave processing has attracted researchers and industrialist who have identified potentials of microwave energy in processing of engineering materials at high temperature too. However, application of microwaves for processing engineering materials, are limited to a range of ceramics and dielectrically lossy materials, both in bulk as well as in particulate form. These materials readily couple with microwaves at room temperature. Till the year 1998, microwave energy was effectively used in processing ceramic and ceramic composites and very few studies were reported on heating of metallic materials through microwave radiation. Microwave radiation gets reflected by bulk metals at ordinary conditions, owing to which bulk metals cannot directly interact with microwave radiation [4, 5].

Application of microwave energy in processing metallic materials is quite challenging. The problems associated with microwave heating of metals arise basically due to two factors- (i) microwave absorption coefficient for metals at 2.45 GHz radiation is significantly less at room temperature [5], and (ii) thermal instabilities which can potentially lead to the phenomenon of thermal

133

runaway [6]. This makes it extremely difficult to achieve heating in metallic materials through microwave without using hybrid heating technique [2, 7]. A unique characteristic of microwave heating is inverse thermal gradient in which heat flows from interior (core) to the surface of the target material body which may lead to poor microstructure at the surface [6]. The problem can be resolved to a large extent using hybrid heating technique, where the resultant temperature profile will be almost uniform from the surface to the material interior and it is possible to obtain unidirectional microstructure. Thus, microwave hybrid heating (MHH) technique becomes popular in which a combination of conventional as well as microwave heating is used to advantage.

In 1999, a US research group first reported sintering of metallic materials through microwave radiation of frequency 2.45 GHz [8]. Later, several authors have reported sintering of metallic materials through microwave heating [8-12]. Gupta and Wong (2005) [13] reported sintering of aluminum, magnesium and lead. Cho and Lee (2008) [14] used microwave heating for metal recovery from stainless steel mill scale. Takayama et al. (2008) [15] have reported production of pig iron by microwave processing of mixed magnetite and carbon powder at 2.45G Hz and 30 GHz microwave frequencies. Borneman and Saylor (2008) [16] reported development of coating of friction reducing alloys using CuNi powder on Ti-6Al-4V substrate through microwave radiation. Sharma et al. (2009) and Srinath et al. (2011) [17] have reported joining of metallic materials by using domestic multimode applicator with 2.45 GHz frequency at 900 W power. Further, an Indian research group extended the application of microwaves as an alternative source of energy for producing cladding of metallic and non metallic materials on metallic substrate using domestic multimode applicator [Sharma and Gupta, 2010]. Gupta and Sharma (2011) [18] have developed WC10Co2Ni cladding on austenitic stainless steel SS-316 through microwave hybrid heating using a domestic microwave oven and reported significantly high dry sliding wear resistance of microwave clad.

In the present work, the MHH technique was employed to develop cladding of Ni-based and WC10Co2N powder on austenitic stainless steel substrate using a multimode microwave applicator. The clads were characterized using various techniques.

EXPERIMENTAL PROCEDURE

In the present work, clads of Ni-based (EWAC) and WC10Co2Ni were been developed on austenitic stainless steel (SS-316) using a multimode domestic microwave oven (f=2.45 GHz, 1 kW). The following sections briefly describe the development and characterisation of the clads.

Materials Details

In order to develop clads on austenitic stainless steel, Ni-based powder (EWAC) and cermet (WC10Co2Ni) powder of average particle size 40 μm were used. The chemical composition of substrate SS-316 and powders are shown in Table 1.

Table 1 Chemical composition (wt %) of materials used.

Element	Fe	Cr	Ni	Mo	W	Co	C	Si	Mn	Others
Substrate (SS-316)	balance	17.3	13.1	2.66	—	—	0.02	0.73	1.71	0.5
EWAC powder	—	0.17	balance	—	—	—	0.2	2.8	—	—
WC based powder	—	—	2.0	—	balance	10.0	5.0	—	—	—

Typical morphology of raw EWAC powder and WC10Co2Ni powder used for deposition is illustrated in Figure 1(a) and (b) respectively. Spherical morphology of the raw EWAC and WC10Co2Ni powder is clearly seen in Figure 1. Austenitic stainless steel plates having dimensions 35 mm×12 mm×6 mm were used as substrate material.

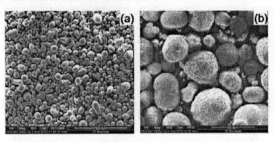

Figure 1 Morphology of raw powder: (a) EWAC, (b) WC10Co2Ni powder.

Development of Clads

Preparation of raw powder and substrate is critical in development of cladding. In the present work, the substrates were cleaned in alcohol in an ultrasonic bath prior to deposition. The powder particles were pre heated at 100 °C for 24 hours in a conventional muffle furnace. Preheating removes possible moisture content in the powder. The powder was preplaced manually on SS-316 substrate maintaining an approximately uniform thickness.

Metals are opaque to microwaves at 2.45 GHz and ordinary conditions owing to which bulk metals cannot couple with microwave at room temperature. Bulk metallic materials (such as solid steel) reflect at room temperature due to low penetration depth, setting up a high voltage between the metal and the magnetron. When this voltage surpasses a threshold, it discharges as a visible spark. The cermet tungsten carbide too exhibit almost similar response to microwave radiations. However, metal powder at room temperature absorbs microwave radiation and gets heated and melted very effectively if the size of powder particles is comparable with its skin depth. The values of the skin depths for nickel and tungsten carbide are ≈ 0.12 μm [19] and ≈ 4.7 μm respectively at 2.45 GHz [20]. The skin depth of powders used is less than the particles sizes (≈ 40 μm), and hence the particles cannot directly interact with microwave radiation at room temperature. In order to overcome the problem, clads were developed by microwave hybrid heating technique using charcoal as susceptor. In order to avoid possible contamination of cladding by susceptor powder used in the MHH, a 99% pure graphite sheet was used as a separator between the susceptor and powder as shown in the Figure 2.

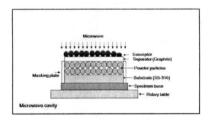

Figure 2 Schematic of the microwave hybrid heating configuration used for cladding.

The figure shows a schematic view of the MHH arrangement adopted during the trials. Details of development of cladding using MHH with suitable examples have been discussed elsewhere [21, 22].

In the present investigation, experimental trials were carried out in a domestic microwave oven while varying interaction time from 180 s to 420 s in a step of 60 s for a constant thickness of preplaced powder layer. Clads of EWAC and cermet (WC10Co2Ni) powder systems were achieved at the microwave exposure time of 360 s.

Characterization of the Clads

The clads were subsequently washed thoroughly with acetone in an ultrasonic cleaner prior to proceeding for characterization. The clads were cut along the clad thickness at the center using a low speed diamond cutter (Model: BAINCUT - LSS, Make: Chennai Metco, India). The cut sections of clad specimens were cold mounted using resins for polishing. The mounted specimens were polished first by emery paper of 320 grit, followed by further polishing using emery papers of grades 600, 800, 1X, 2X, 4X, and finally with 1μm diamond paste in a cloth wheel machine. Specimens were then water cleaned and dried in hot air.

The analysis of clad microstructures and chemical composition of the clads were carried out in a field emission scanning electron microscope at an acceleration voltage of 20 kV equipped with energy dispersive x-ray detector (Model: Quanta 200 FEG). Vicker's microhardness measurement along the thickness of the clad and substrate was accomplished by Vicker's microhardness tester (Mini load, Leitz, Germany) at the load of 50 g applied for 30 s. The indentations for Vicker's hardness measurements were made at an interval of 80 μm starting from the substrate material towards the top of the clad surface. Three indentations were carried out laterally in each location. Averages of these measurements were considered.

RESULTS AND DISCUSSION

Ni based (EWAC) Cladding

Nickel is a characteristically tough metal and has high oxidation and corrosion resistance at room and elevated temperature; also nickel-based materials exhibit superior wear (abrasion and erosion) and corrosion resistance. Accordingly, EWAC ($\approx$ 97% Ni) powder was selected for development of cladding on SS-316 substrate. The experiments were carried out at different exposure time of microwave radiation from 120 s to 360 s. The clad obtained at exposure 360 s was found metallurgically bonded well with the substrate.

Microstructure Study

A typical cross section of the developed EWAC clad as observed through field emission scanning electron microscope is shown in Figure 3 (a). The clad of $\approx$ 1mm thickness was developed, which is free from any visible solidification interfacial crack as shown in Figure 3. The porosity of developed clad was measured by linear point count method which showed average porosity of 1.01%. The clad is metallurgically bonded with the substrate by partial mutual diffusion of elements.

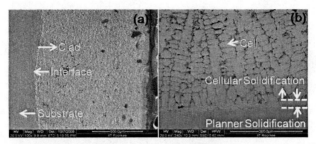

Figure 3; (a) Microstructure of clad cross section, (b) back scattered electron image of clad cross section showing cellular growth.

Microwave heating is known for molecular heating whereby entire volume of exposed material is heated simultaneously (volume heating) which make more alternative attractive heating source for realization of cladding for engineering materials. The volumetric nature of heating causes the exposed materials to get rapidly heated to elevated temperature with less thermal gradient. The associated less thermal gradient during microwave cladding causes unidirectional thermal gradient owing to which clad cross section shows the growth of clad from planner for a distance of approximately 20 micron; following this; the entire clad has uniform cellular structure as shown in back scattered electron micrograph (Figure 3 (b)). The microstructure of clad cross section shows no transition from cellular to dendrite which is one of the major significances of the process. The unidirectional microstructure in microwave clad provides uniform mechanical properties in the entire clad. This is attributed to volumetric nature of heating in which the resultant mean temperature profile is relatively uniform throughout the volume of the target material. In MHH, however, the relatively uniform temperature reduces the possibility of non-uniformity in the resulting microstructure. As microwave cladding process is relatively faster (maximum exposure time: 360 s only), therefore, more homogenous microstructure can be expected.

The back scattered electron image of clad cross section shows the segregation of light atomic weight element (dark phase) at the cell boundaries, while heavy atomic weight (white phase) inside the cell. The distribution of elements in clad was also analyzed through X-ray elemental mapping which is illustrated in Figure 4. Figure 4 (a) shows the back scattered electron image (BSE) area considered for scanning. Figure 4 (b) clearly shows that chromium (yellow) is distributed around the cells boundaries, while iron and nickel are homogeneously distributed inside the cells as clearly seen in Figure 4 (c, d) respectively. It is evident from the Figure 4 (b and c) that chromium and iron from the stainless steel substrate got partially diluted to the clad during microwave irradiation by way of convection current of the melt-pool which helps in formation of metallurgical bonds. The diluted chromium had further reacted with carbon (present in EWAC powder, Table 1) and formed chromium carbide which got precipitated at the cell boundaries. Carbon has more affinity towards chromium than iron [23] which causes chromium carbide to form during solidification. This microsegregation effect of solute (chromium) around the cell wall is due to the value of partition coefficient (K) less than unity (K <1) as obtained from the phase diagram of binary Ni-Cr alloy. The solute may react with free carbon which is present in raw powder as well as may get diffused from the separator (graphite) used during microwave processing.

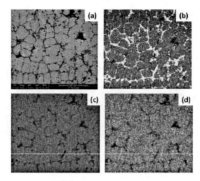

Figure 4 Elemental X-ray mapping of EWAC clad: (a) back scattered micrograph (scanned area); elemental distribution of (b) chromium, (c) iron and (d) nickel.

Microhardness Study

The microhardness of the clad across the clad cross-section was also evaluated. The distance between two successive indentations (Vicker's) was maintained at 80 μm with additional indentations on the interface and on the substrate. The distribution of microhardness is illustrated in Figure 5. As observed from the Figure 5, the distribution of microhardness in the clad section is not uniform, which is attributed to the presence of different phases of varying microhardness. The average microhardness of the clad is 304±48 Hv while that of the SS-316 substrate is 175 Hv. The microhardness measured at the interface is 240 Hv. Typical indentation geometries of the clad and SS-316 substrate are shown in Figure 5 (b, c). The indentation on the clad is seen restricted by the harder carbide phases in the deformed geometry Figure 5 (b) while an ideal pyramidal geometry on the metallic substrate (SS-316) could be observed Figure 5 (c).

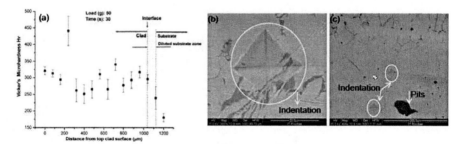

Figure 5   Assessment of Vicker's microhardness: (a) Microhardness profile, (b) SEM micrograph showing indentation geometry in microwave clad, (c) indentation geometry on the austenitic stainless steel.

WC10Co2Ni Cladding

Hardness as well as toughness both are important for higher wear performance of a mating surface in sliding type of tribocontact. Tungsten carbide has good hardness and wear resistance, but poor toughness. However, tungsten carbide provides better wear resistance while dispersed in a tough phase. Thus, WC–Co–Ni can be one of the best systems for a combination of high hardness and toughness. In this system, cobalt (Co) acts as a binder, which is responsible for densification through wetting, spreading and agglomerations during liquid phase sintering [24]. Therefore, WC10Co2Ni system has been chosen as the clad material for the present study. Microwave cladding of WC10Co2Ni powder was successfully carried out on austenitic stainless steel substrate using a multimode domestic microwave applicator, operated at 900 W and 2.45 GHz. The clads were obtained at the exposure duration of 360 s.

Observations on Clad Microstructure

The back scattered electron image of a typical transverse section of the WC10Co2Ni cladding is illustrated in Figure 6. The cladding with an average thickness of ≈ 2 mm shows good metallurgical bonding with substrate through partial mutual diffusion of elements like iron (Fe), chromium (Cr) from the substrate to clad and tungsten from clad to substrate. The porosity of the developed clad was measured by linear point count method; results show significantly less porosity of 0.89 %. The microstructure presented in Figure 6 further reveals that the developed clad is free from interfacial cracking despite of the significant mismatch of coefficient of thermal expansion of clad material and substrate SS-316. The crack free clad microstructure is also indicative of uniform heating associated with microwave processing.

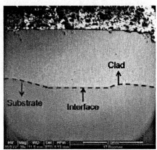

Figure 6. BSE image of transverse WC10Co2Ni clad surface.

The developed clad microstructure show dominating skeleton like (Figure 7) structure of binary carbides (tungsten and iron)$_6$C in a metallic matrix of iron, nickel and chromium. The phenomenon of formation of skeleton like binary carbide phase is attributed to the nucleation of austenite, which grows at the boundary of the liquid solution along with impoverished tungsten and carbon which may precipitate from the liquid solution during solidification of Fe-W-C alloys [25]. The hard carbide phase of skeleton are uniformly precipitated in a tough metallic matrix. Thus, it is expected that when developed clads will be exposed to severe dry sliding conditions, the composite like skeleton –studded structure of developed clad could bear the sliding environment better owing to the presence of hard carbide phase in the tough and ductile metallic matrix.

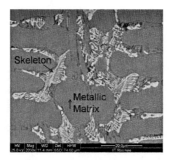

Figure 7. A typical back scattered electron image of clad cross section.

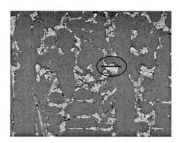

Figure 8. Location for X-ray elemental line scanning on clad cross section.

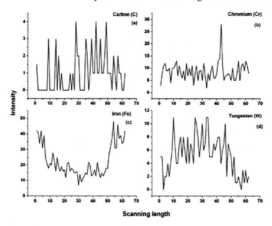

Figure 9. X-ray elemental distribution in WC10Co2Ni cladding; (a) carbon, (b) chromium, (c) Iron, (d) tungsten.

The compositions of matrix and skeleton were analyzed through X-ray elemental line scanning as shown in Figure 8. The corresponding elemental distributions are presented Figure 9. The X-ray elemental study exhibits that, the matrix consists of elements iron (Fe) and chromium (Cr), where Fe is dominating. The elemental distribution of the skeleton structure, on the other hand, consists of dominating tungsten (W), iron (Fe) and carbon (C). Thus, it is likely that the matrix phase consists of the relatively soft metallic materials (Fe and Cr) which get diluted from the metallic substrate. Thus, there exists a strong possibility of exhibiting higher wear resistance by the clads with well distributed hard metallic carbides (reinforcements) and a tough metallic matrix.

Microhardness Study

Hardness of a material is one of the most important factors, which influence wear performance of the material. Generally, increasing the hardness of components can enhance the wear resistance ability although the effect of hardness is not straightforward. The Vicker's microhardness of clad layer over the cross-section was evaluated. The distance between two indentations was kept 120 μm with an additional indentation at the fusion line or interface and the substrate. The distribution of microhardness is shown in Figure 10. The average microhardness of the clad section is $\approx$ 1064±99 Hv. The distribution of microhardness in the clad section is not observed to be uniform.

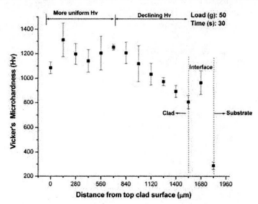

Figure 10. Vicker's microhardness distribution through a typical section of WC10Co2Ni clad.

An increasing tendency in hardness away from the substrate is seen. The microhardness from top of the clad to the half of clad is approximately uniform ($\approx$ 1100 Hv), which however reduces and approaches $\approx$ 800 Hv at the clad–substrate interface. The higher standard deviation in the measurements can be attributed to the indentations being carried out at the hard skeleton structure as well as at the tough matrix phase.

The typical indentation geometries in microwave clad is shown in Figure 11. The indentation (A) as shown in Figure 11, is located in between the soft matrix and skeleton. The geometry of indentation shows, the diagonals have been depressed and extended towards apex of the pyramid by virtue of which it has grown bigger. The indentations at (B) and (C) on the skeleton of the mainly tungsten based element are lower in size compared to the indentation (A) and subsequently, high microhardness results.

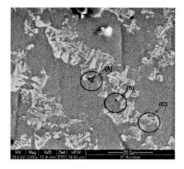

Figure 11 Micrograph showing the Vicker's indentations in a WC10Co2Ni clad section.

CONCLUSIONS

A novel technique for the development of cladding has been illustrated; microstructural characterization of the developed Ni-based (EWAC) and WC10Co2Ni clads on SS-316 substrate has been carried out. The major conclusions drawn from the above studies are:

(1)   The metallic EWAC powder could be melted by using domestic microwave oven with power 900 W at frequency 2.45 GHz to cause the dilution and form the clad.

(2)   The controlled dilution of elements (substrate or/and clad) results in good metallurgical bonding of the molten particles with the substrate.

(3)   The developed clads reveal cellular solidification; which demonstrated the significance of microwave heating (volumetric heating).

(4)   Chromium carbide is precipitated around the cell boundaries while iron and nickel is uniformly distributed around the cells.

(5)   The average microhardness of the clad section is 304±48 Hv and mean observed porosity of the clad is significantly less (1.01%).

(6)   The WC10Co2Ni clad of thickness ≈ 2 mm could be developed by the exposure of microwave radiation at 2.45 GHz and power 900 W for the duration of 360 s.

(7)   The cermet clad is free from interfacial cracking and exhibit significantly less porosity (0.89%) while compared to other cladding technique (2-3% in laser cladding [26]).

(8)   The clad is metallurgical bonded well with the substrate by partial mutual diffusion of the elements.

(9)   Skeleton like structure of carbides of Fe and W are formed in clads during cladding.

(10)   The average microhardness of developed clad is $\approx 1064\pm99$ Hv.

REFERENCES

[1]Guido Link, Lambert Feher, Manfred Thumm, Hans-Joachim Ritzaupt-Kleissl, Roland Bohme and Alfons Weisenburger, Sintering of Advanced Ceramic Using a 30-GHz, 10-kW, CW industrial Gyrotron, *IEEE Transactions on Plasma Science*, **27**-2, 547-554 (1999).

[2]A. K. Sharma, S. Aravindan and R. Krishnamurthy, Microwave glazing of alumina–titania ceramic composite coatings, *Mater. Lett.*, **50**, 295–301 (2001).

[3]Apurbba Kr. Sharma and R. Krishnamurthy, Microwave processing of sprayed alumina composite for enhanced performance. *J. Eur. Ceram. Soc.* **22** , 2849–2860 (2002).

[4]Dheeraj Gupta and A K Sharma, On Development and Performance of Microwave Induced Metal-Ceramic Composite Cladding, *In Proceedings of an International Conference on Processing and Fabrication of Advanced Materials XIX*, New Zealand 90-101 (2011).

[5]C. Leonelli, P. Veronesi, L. Denti, A. Gatto and L. Iuliano, Microwave assisted sintering of green metal parts, *J. Mater. Process. Technol.* **205**, 489–496 (2008).

[6]Morteza Oghbaei and Omid Mirzaee, Microwave versus conventional sintering: A review of fundamentals, advantages and applications *J. Alloys Compd.* , **494**, 175-189 (2010).

[7]Yu. V. Bykov, K. I. Rybakov and V. E. Semenov, High-temperature microwave processing of materials, *J. Phys. D: Appl. Phys.*, **34**, R55 –75 (2001).

[8]Rustum Roy, Dinesh Agrawal, Jiping Cheng and Shalva Gedevanishvili, Full sintering of powdered-metal bodies in a microwave field, *Nature*, **399**, 668-670 (1999).

[9]S. S. Panda, V. Singh, A. Upadhyaya and D. Agrawal, Sintering response of austenitic (316L) and ferritic (434L) stainless steel consolidated in conventional and microwave furnaces, *Scr. Mater.*, **54**, 2179-2183 (2006).

[10]K. Saitou, Microwave sintering of iron, cobalt, nickel, copper and stainless steel powders. *Scr. Mater.* **54**, 875–879 (2006).

[11]P. Chhillar, D Agrawal, and J. H. Adair, Sintering of molybdenum metal powder using microwave energy, *Powder Metallurgy*, **51-2**, 182-187 (2008).

[12]A. Mondal, A. Upadhyaya and D. Agrawal, Microwave Sintering of W-18Cu and W-7Ni3Cu Alloys, *J. Microwave Power Electromagn. Energy,* **43-1** ,11-16 (2009).

[13]M. Gupta and W.L.E. Wong, Enhancing overall mechanical performance of metallic materials using two-directional microwave assisted rapid sintering, *Scr. Mater.* , **52**, 479– 483 (2005).

[14]Seungyoun Cho and Joonho Lee, Metal recovery from stainless steel mill scale by microwave heating, *Met. Mater. Int.,* **14- 2**, 193-196 (2008).

[15]Sadatsugu Takayama, Guido Link, Akihiro Matsubara, Saburo Sano, Motoyasu Sato and Manfred Thumm, Microwave Frequency Effect for Reduction of Magnetite, *Plasma and Fusion Research: Regular Articles* , **3**, S1036-1-4 (2008).

[16]Borneman Karl Lee and Matthew David Saylor, Microwave Process for Forming a Coating. *United State Patent*, US 0138533A1 (2008).

[17]A. K. Sharma, M. S. Srinath and Pradeep Kumar, Microwave Joining of Metallic Materials, *Indian Patent*, 1994/Del/2009 (2009).

[18]D. Gupta and A.K. Sharma, Investigation on sliding wear performance of WC10Co2Ni cladding developed through microwave irradiation, *Wear* , **271** , 1642– 1650 (2011).

[19]D. Gupta and A.K. Sharma, Development and Microstructural Characterization of Microwave Cladding on Austenitic Stainless Steel, *Surf. Coat. Tech.* , **205**, 5147–5155 (2011).

[20]K. Rodiger, K. Dreyer, T. Gerdes and M Willert Porada, Microwave Sintering of Hardmetals, *Int. J. Refract. Met. Hard Mater,* **16**, 409-416 (1998).

[21]A.K. Sharma, and Dheeraj Gupta, A method of cladding/coating of metallic and non- metallic powders on metallic substrates by microwave irradiation, *Indian Patent,* Application No. 527/Del/2010 (2010).

[22]Dheeraj Gupta and A K Sharma, Development of Erosion Resistant Cladding on Austenitic Stainless Steel through Microwave Heating. *In Proceedings of the Third International and Twenty Fourth All India Manufacturing Technology, Design and Research Conference (AIMTDR) India,* 1029-1034 (2010).

[23]S.D. Carpenter, D. Carpenter, J.T.H. Pearce, XRD and electron microscope study of an as-cast 26.6% chromium white iron microstructure, *Materials Chemistry and Physics,* **85**, 32–40 (2004).

[24]Paul C. P., Alemohammad H., Toyserkani E., Khajepour A. and Corbin S., Cladding of WC– 12Co on low carbon steel using a pulsed Nd:YAG laser, *Mater. Sci. Eng. A,* **464**, 170–176 (2007).

[25]Taran Yu N., Ivanov L. I. and Moshkevich L. D,. Morphology of the eutectic in Fe-W-C alloys, *Metal science and heat treatment,* **14**, 3-6 (1972).

[26]Balla Vamsi Krishna, Bose Susmita, Bandyopadhyay Amit, Microstructure and wear properties of laser deposited WC–12%Co composites, *Materials Science and Engineering A,* **527**, 6677–6682 (2010).

DILATOMETRIC STUDY AND IN SITU RESISTIVITY MEASUREMENTS DURING MILLIMETER WAVE SINTERING OF METAL POWDER COMPACTS

G. Link[1], M. M. Mahmoud[1,2], M. Thumm[1]
1 Karlsruhe Institute of Technology, IHM, Germany
2 Advanced Technology and New Materials Research Institute (ATNMRI), MuCSAT, Alexandria, Egypt.

ABSTRACT

Experimental investigations on millimeter-wave sintering of iron and steel powder compacts have been performed in a compact 15 kW gyrotron system operating at 30GHz in reducing atmosphere. Linear shrinkage, thermal expansion and changes in ohmic resistance were measured as a function of temperature using a modified dilatometer system. This data contains useful information about microstructural changes with respect to the inter-particle electrical contacts during the sintering process.

INTRODUCTION

Microwave sintering technology has been developed in the field of ceramics and composites in the last decades motivated by the potential of a direct and instantaneous volumetric heating. This allows more rapid heating in comparison to the conventional heating process and very often an enhanced sintering has been reported. This leads to a reduction of process temperature and/or processing time and may result in an increased productivity and a reduction of energy consumption.

Microwave sintering of metal powders is a new activity with growing interest since Roy, Agrawal and co-workers published their results some years ago [1,2]. Although metallic materials are known to reflect microwaves rather than absorbing them, it has been demonstrated that sintering of metal powder compacts using microwave heating is possible. In former investigations, a volumetric heating by millimeter waves could be demonstrated by measurement of temperature gradients within the metal powder compact during the heating cycle [3].

The characterization of the sintering progress by density measurement or by using a dilatometer is a standard procedure in case of ceramic materials where the relative density is a unique macroscopic variable that generally used in classical sintering models. In case of compacts obtained from ductile metal powders with green densities up to 90%, the classical sintering models can hardly prescribe macroscopic properties in relation to the density. . In that case, the classical sintering models can hardly prescribe macroscopic properties in relation to the relative density. Therefore different measurement techniques have to be developed.

At the Bayreuth University, Germany, the DC electrical conductivity was measured on steel powder compacts at ambient temperatures after different steps of sintering [4]. A strong increase in electrical conductivity was measured, although the measured microwave power dissipation was almost not changing, what was found to be contradictory.

In order to investigate the progress of sintering in die-pressed steel powder compacts, the electrical resistivity was measured simultaneously during the millimeter-wave sintering cycle in a modified dilatometer set-up.

EXPERIMENTAL WORK

The millimeter-wave sintering was performed in a compact 30 GHz gyrotron system [5]. The millimeter-wave power generated in a so called gyrotron oscillator can be controlled from 0 up to 15 kW. The millimeter-wave radiation is launched via a quasi-optical transmission line through a vacuum-sealed boron nitride window into the applicator. The applicator has a hexagonal geometry and is equipped with a rotating mode stirrer, which results in a very homogeneous field distribution. As schematically shown in Fig. 1, this system can be used in combination with a special dilatometer set-up so that information on sintering kinetics of the studied samples as a function of temperature can be obtained. It is a commercial dilatometer (type L75 from Linseis Company, Germany), adapted to the millimeter-wave applicator so that the millimeter-wave leakage could be excluded. The dilatometer was modified, so that the dc electric resistance could be measured during the sintering process, parallel to the measurement of changes in sample length. The resistance was measured using a Keithley digital multimeter Model 2002 in combination with the four-wire method. Therefore, two platinum wires were in contact with the top and the bottom surfaces of the cylindrical shape sample, respectively. This sample electrode arrangement is assembled into the dilatometer set-up whereas the spring force of the dilatometer, which clamps the sample in-between the sample holder end-plate and the dilatometer sensor rod, makes sufficient contact to the sample. The benefit of this arrangement with respect to the improved electrode contact has to be paid by an additional source of errors in the dilatometer signal, since any length changes of the platinum wire electrodes used are superimposed to the intrinsic sample signals. So, the dilatometer results will give qualitative information rather than precise quantitative data about the sintering process. The electrical resistivity ρ is calculated from the measured resistance R by equation (1):

$$\rho = R \frac{(d/2)^2 \pi}{l} \qquad (1)$$

where d is the diameter of the cylindrical sample and l its length.

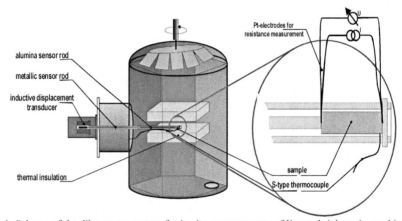

alumina sensor rod

metallic sensor rod

inductive displacement transducer

thermal insulation

Pt-electrodes for resistance measurement

sample

S-type thermocouple

Fig. 1: Scheme of the dilatometer system for in-situ measurements of linear shrinkage in combination with resistance measurement.

The heating process was controlled along a preset temperature-time program using the temperature signal of an S-type sheath thermocouple placed at the sample surface. All experiments were performed in nitrogen atmosphere at ambient pressure. Samples with 5 mm in diameter and about 20 mm in length with green densities of about 7 g/cm^3 were used in two different qualities. One type of samples was consisted of pure atomized steel powder from KOBE Steel, Ltd. with a purity of more than 99.6% while the other one was mixed with 0.7 weight% graphite powder from Asubery with purity higher than 95%. Samples were heated with a rate of 10 °C/min up to 1140 °C and soaked for 10 min. at this temperature. After free cooling, a subsequent identical run was performed, using the same sample.

RESULTS AND DISCUSSION

Figures 2 and 3 show the experimental data of both subsequent temperature cycles for a pure iron sample and a sample including some graphite powder, respectively. The left graphs show the absolute changes in length from the dilatometer, the measured resistivity and the corresponding temperature as a function of time. If the data is plotted as function of temperature as shown in the right graphs, the data from the first and second heating cycle can be compared more easily. Thus, extrinsic effects from the sintering process can be distinguished from intrinsic temperature dependent effects.

A positive slope at the beginning of the dilatometer curves due to thermal expansion until a temperature of about 700 °C is common in both materials. Thereafter, the slope gets slightly negative indication the onset of sintering. In the pure iron sample, at a temperature of about 900 °C a sudden change in length can be observed in the pure iron sample due to the □ to □ iron phase-transformation. Subsequently thermal expansion is dominating the superimposed sintering effect resulting again in a positive slope of the dilatometer curve until the final temperature 1140°C was reached for isothermal soak. In the graphite doped sample, contraction induced by the phase transformation is much smoother and starts already at lower temperatures compared to the pure iron sample, as can be expected according to the iron carbon phase diagram.

The results of resistivity measurements show a strong decrease in resistivity of about one order of magnitude for the first 300 °C followed by saturation. This was explained as a result of welding of the interparticle contacts by O.Lame and co-authors [6] who performed similar experiments by conventional heating. Thus, this significant change within the material was detected by this unique process setup which is impossible to detect by any other methods.

Above a temperature of about 600 °C to 700 °C, were the start of sintering can be recognized from the dilatometer curve, a second decrease in resistivity comparable in magnitude can be observed. This is followed up by a second saturation phase till the end of the soak time. Thereafter, during the cooling process, the negative slope is measured which can be explained by the intrinsic temperature behavior of the resistivity. This is confirmed by the curve measured during the second heating cycle which is more or less identical as can be easily seen in the right graphs of Figures 2 and 3. Qualitatively, the evolution of the resistivity obtained for the graphite doped sample is similar to the pure iron sample. The main difference was the initial resistivity value which was significantly higher in the graphite doped sample. This demonstrates that carbon is an inhibitor for formation of metallic pressing contacts as prescribed by A. Simchi and co-authors [7]. Additionally the intrinsic temperature behavior as can be seen from the second heating cycle is smaller for the graphite doped sample.

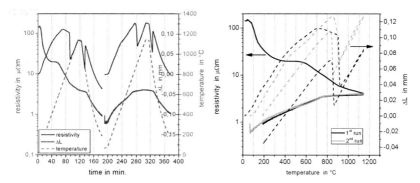

Fig. 2: Results of the dilatometer and resistance measurement with a pure iron sample as a function of time (left) and as a function of temperature (right).

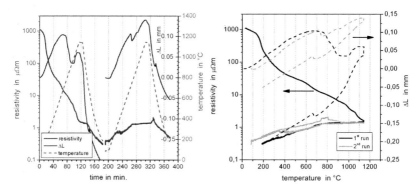

Fig. 3: Results of the dilatometer and resistance measurement with a graphite doped iron sample as a function of time (left) and as a function of temperature (right).

CONCLUSION

The sintering progress of the metal powder compact in the millimetre-wave process has been investigated by application of in-situ resistance measurement. Therefore, this dilatometer set-up with a four-wire test and a high resolution ohm-meter can be used to investigate the sintering behaviour of the metal powder compacts. This test method gives important information about microstructural changes with respect to the inter-particle electrical contacts during the sintering process. A direct comparison of the measured data from two identical subsequent thermal process runs allows differentiating extrinsic effect due to sintering from intrinsic temperature dependent effects. This data indicates that tremendous changes already appear at low temperatures below 300 °C where a strengthening of the material cannot be expected [5].

ACKNOWLEDGMENT
The support of the DFG under the contract GZ TH656/3-1 AOBJ 575894 and of the Hitachi Powdered metals Co., Ltd is kindly acknowledged

REFERENCES
1) R. Roy, D. Agrawal, J. Cheng, and S. Gedevanishvili; *Nature*, **399** (1999) 668-670.
2) J. Cheng, R. Roy, and D. Agrawal; *Journal of Materials Science Letters*, **20** (2001) 1561-1563.
3) Takayama, S.; Link, G.; Miksch, S.; Sato, M.; Ichikawa, J.; Thumm, M.; *Powder Metallurgy*, **49** (2006) 274-280.
4) M. Willert-Porada, H.-S. Park: *Ceramics Transaction*, **111** (2001) 459-470.
5) G. Link, L. Feher, M. Thumm, H.-J. Ritzhaupt-Kleissel, R. Böhme, A. Weisenburger; *IEEE Trans. on Plasma Science*, **27| 2|** (1999) 547-554.
6) Lame O., Bordère S., Denux D., Bouvard D.; *Materials Science and Engineering*, **A363** (2003) 126-133.
7) Simchi A., Danninger H., Gierl C.; *Powder Metallurgy*, **44 |2|** (2001) 148-156.

ROLES OF ELECTROMAGNETIC FIELDS ON MATERIALS PROCESSING AND PERFOR-
MANCE – A THERMODYNAMIC AND KINETIC PERSPECTIVE

Boon Wong
L-3 Communications, Electron Technologies, Inc.
Torrance, California, USA

ABSTRACT
    A classical thermodynamic and kinetic perspective is proposed to provide rationale for non-thermal microwave/RF effects on some "thermally unnatural" processes such as "salt water" dissociation and astounding reaction kinetics.

    Thermodynamics accounting for electromagnetic work predicts that the free energy increase of a system under proper electromagnetic-wave (EMW) irradiation could be as high as the total work done on the system by the electric and magnetic fields of the waves. Nonetheless, the electromagnetic work, sum of the electric and magnetic work derived from Maxwellian electromagnetics, is shown to be a function of the (relative) permittivity and (relative) permeability of the irradiated dielectric. This work input always maximizes at the resonance frequency of a given polarization (magnetization) mechanism. Consequently, at temperature and pressure (or volume), free energy of a system under resonant polarized microwave/RF irradiation may increase so much that it not only enhances the kinetics of a process by activation free energy reduction, but also promotes the feasibility of a "thermally unnatural" reaction.

    Four different types of non-thermal microwave/RF effects on materials processing are discussed and illustrated. Further validation of the proposed rationale to quantitatively support the experimental findings in the field of non-thermal microwave/RF effects is currently in progress.

INTRODUCTION
    Classical thermodynamics was originally formulated and applied to understand and characterize energy changes in fluid systems in the 19th century. Since then, its applications have been mainly focused on systems that only do expansion (PV) work. Unfortunately, this traditional version of applications of thermodynamics has been found to be unable to explain some results and observations in materials produced in the field of microwave/RF processing.[1a] Therefore, the traditional applications of thermodynamics must be appropriately augmented to account for electromagnetic work before it can be used to understand the roles of microwave/RF irradiation on materials processing and performance.

    Some thermodynamics texts, written for chemists and physicists,[2],[3],[4],[5] have already delivered a general approach to augment thermodynamic applications on systems that may involve all kinds of work: expansion and non-expansion work. Unifying the perspectives of both "augmented" thermodynamics and Arrhenius-Eyring reaction kinetics, this paper presents a physical basis for non-thermal microwave/RF effects on some "thermally unnatural" processes reported in literature.

"AUGMENTED" THERMODYNAMICS ACCOUNTING FOR ELECTROMAGNETIC WORK
    Thermodynamics is a fundamental science which underlies first principles of all physical structures and processes. It characterizes and monitors transformations of energy in a system. It also determines trend of any change of the system when it interacts with its surroundings.

1st Law of Thermodynamics
    In classical thermodynamics, the 1st law states that $d\varepsilon_{tot} = dU = 0$ in an isolated system such as the universe, where $d\varepsilon_{tot}$ and dU are the differential changes of the total and internal energies of the

system, respectively. Therefore, the total energy, ε_{tot}, or the internal energy, U, of the universe is conserved.

The total or the internal energy change in a non-isolated, stationary system, on the other hand, according to the 1st law is:

$$d\varepsilon_{tot} = dU = \delta Q + \delta W \tag{1}$$

where δQ and δW are the differential changes of heat and total work, respectively. The symbol δ is used to denote an inexact differential of a path-dependent function such as heat or work. Positive changes of both heat and work are considered to be the heat- and work-energy transfers into the system, whereas negative changes are considered to be energy transfers out of the system into its surroundings.

In addition, the total work in equation (1), which accounts for all kinds of work done by external forces (fields), for example, on a stationary homogeneous system, may include:[2],[3],[4],[5]

$$\delta W = -PdV + V\mathbf{E}\cdot d\mathbf{D} + V\mathbf{H}\cdot d\mathbf{B} + \sum_i \mu_i dn_i + \gamma d\tilde{A} + V\sigma_x de_x + \ldots \tag{2}$$

where
PdV is the expansion work (P: external pressure applied, dV: corresponding differential change in volume of the system),
$V\mathbf{E}\cdot d\mathbf{D}$ is the electric work (V: total volume of the system, $\mathbf{E}$: applied electric field, $d\mathbf{D}$: corresponding differential change of electric displacement or total polarization – electric moment per unit volume),
$V\mathbf{H}\cdot d\mathbf{B}$ is the magnetic work (V: same definition as above, $\mathbf{H}$: applied magnetic field, $d\mathbf{B}$: corresponding differential change of magnetic induction or total magnetization – magnetic moment per unit volume),
$\sum_i \mu_i dn_i$ is the chemical work (μ_i: chemical potential of component i, dn_i: corresponding differential change of its number of moles),
$\gamma d\tilde{A}$ is the surface work (γ: surface tension, $d\tilde{A}$: differential change of surface area),
$V\sigma_x de_x$ is the mechanical work (V: same definition as above, σ_x: applied stress acting in x direction, de_x: corresponding differential change of strain of a solid in x direction).

Now if we consider a stationary, closed, homogeneous system involving only expansion and electromagnetic work, the 1st law, or expression (1), then simply becomes:

$$dU = \delta Q + \delta W = \delta Q - PdV + V\mathbf{E}\cdot d\mathbf{D} + V\mathbf{H}\cdot d\mathbf{B} \tag{3}$$

or

$$dU = \delta Q - PdV + \delta W_{EM} \tag{3a}$$

where $\delta W_{EM} = V\mathbf{E}\cdot d\mathbf{D} + V\mathbf{H}\cdot d\mathbf{B}$, is the differential electromagnetic work done on the system by the electromagnetic field of an electromagnetic wave (EMW) such as a microwave or radiofrequency (RF) wave.

2nd Law of Thermodynamics
Clausius's statement of the 2nd law in its concise mathematical form is:

$$dS \geq (\delta Q/T) \tag{4}$$

where dS is the differential change of the entropy of the system, δQ is the differential change of the heat absorbed by the system, T is the temperature of the surroundings. The equation expression in this mathematical form applies to processes that are reversible (at equilibrium), whereas the inequality expression is for irreversible (spontaneous or actual) changes.

Free Energy – Its Relationship with Electromagnetic Work

Assuming a system involves only expansion (PV) work and electromagnetic (W_{EM}) work, unifying the 1st and the 2nd laws of thermodynamics; i.e., combining expressions (3a) and (4), we obtain:

$$dU + PdV - TdS \leq \delta W_{EM} \tag{5}$$

Expression (5) is the general criteria for spontaneous and equilibrium changes involving expansion and electromagnetic work. Since most reactions occur at constant pressure, $P = P_{external}$ and constant temperature, $T = T_{surroundings}$, expression (5), the criteria for process spontaneity and equilibrium assessment, can further be simplified as shown below:

At constant T and P, expression (5) can be first rewritten as:

$$d(U + PV - TS)_{T,P} \leq \delta W_{EM} \tag{6}$$

Define $G = (U + PV - TS)$, and use it to rewrite expression (6) as

$$dG_{T,P} \leq \delta W_{EM} \tag{6a}$$

where G is called the Gibbs free energy. Expression (6a) relates the change in Gibbs free energy (of a system) under constant temperature and pressure conditions to the change in electromagnetic work input/output. Free energy is the "orderly" energy stored in a system that is available to perform work onto its surroundings.

Similarly, criteria shown in expression (5) can also be simplified for processes that take place at constant temperature and constant volume. Under these conditions, another type of free energy, as first defined by Helmholtz as $A = U - TS$, is used:

$$d(U - TS)_{T,V} = dU - TdS \leq \delta W_{EM} \tag{7}$$

Therefore,

$$dA_{T,V} \leq \delta W_{EM} \tag{7a}$$

According to expressions (6a) and (7a), when a system interacts with an electromagnetic field under constant temperature and pressure (or volume), its maximum increase in free energy is equal to the electromagnetic work input, for the limit of a reversible process. An actual (irreversible) process, however, yields a lower gain in the free energy of the system.

Integrating both sides of expression (6a) with appropriate boundary conditions; i.e., $G_{T,P}$ from G^{Th} to G^{Th+EM}, and W_{EM} from 0 to W_{EM}, we may then obtain the maximum gain in Gibbs free energy of the system under microwave/RF irradiation at constant temperature and pressure:

$$(G^{Th+EM} - G^{Th}) = W_{EM} \tag{6b}$$

where G^{Th+EM} and G^{Th} are the Gibbs free energies of the system under both thermal and electromagnetic-field and under purely thermal conditions, respectively, and W_{EM} is the electromagnetic work done on the system.

Similarly, integrating both sides of expression (7a) with appropriate boundary conditions, we may also obtain the maximum gain in Helmholtz free energy of the system under microwave/RF irradiation at constant temperature and volume:

$$(A^{Th+EM} - A^{Th}) = W_{EM} \tag{7b}$$

where A^{Th+EM} and A^{Th} are the Helmholtz free energies of the system under both thermal and electromagnetic-field and under purely thermal conditions, respectively, and W_{EM} is again the electromagnetic work done on the system.

ELECTROMAGNETIC WORK RESULTING FROM MICROWAVE/RF-DIELECTRIC INTERACTIONS

Microwaves and RF waves, like other EMWs, contain electromagnetic energy that is stored in the electric and magnetic fields of the waves. When a beam of microwaves or RF waves interacts with a dielectric medium, energy transfers may take place between them. On the one hand, some of the electromagnetic energy of the waves may be dissipated, converting to thermal energy, ε_{Th}, due to dielectric and/or magnetic losses during the electric and magnetic polarization cycles. As a result, temperature of the dielectric rises – a phenomenon called thermal microwave/RF effect. On the other hand, another portion of the electromagnetic energy of the waves may do "non-thermal" work, W_{EM}, on the material resulting from polarization and/or magnetization, which may then directly increase the free energy of the dielectric, however, without causing a change in temperature – a phenomenon called non-thermal microwave/RF effect – a key subject of discussions in this paper.

When one-dimensional plane-EMWs interact with an isotropic dielectric, the total electromagnetic energy transfer, ε_{EM}, may be quantified using Maxwellian electromagnetics as shown below:[4],[6],[7],[8]

$$\varepsilon_{EM} = \varepsilon_E + \varepsilon_M = \int_V \int (\mathbf{E} \cdot d\mathbf{D}) dV + \int_V \int (\mathbf{H} \cdot d\mathbf{B}) dV \tag{8}$$

where ε_{EM}, ε_E, and ε_M are the (total) electromagnetic, electric, and magnetic energies, respectively, $\mathbf{E}$ and $\mathbf{H}$ are the electric and magnetic fields of the incident waves, respectively, $d\mathbf{D}$ and $d\mathbf{B}$ are the corresponding differential changes in electric displacement and magnetic induction, respectively, and V is the volume of the dielectric system.

Since mathematically both the permittivity, $\varepsilon*$, and permeability, $\mu*$, of a dielectric are complex functions, according to von Hippel,[6] $\mathbf{D}$ and $\mathbf{B}$ in equation (8) may be expressed as: $\mathbf{D} = \varepsilon*\mathbf{E} = \varepsilon_0(\varepsilon_r - j\varepsilon'')\mathbf{E}$ and $\mathbf{B} = \mu*\mathbf{H} = \mu_0(\mu_r - j\mu'')\mathbf{H}$, respectively. The time-average non-thermal electromagnetic work input (energy stored) into a linear, isotropic dielectric under sinusoidal EMW irradiation may then be approximated by expression (9):

$$\underline{W}_{EM} \approx 1/2V(\varepsilon_0\varepsilon_r E^2 + \mu_0\mu_r H^2) \tag{9}$$

On the other hand, the time-average electromagnetic energy dissipated causing dielectric heating – a temperature rise in the dielectric, during the wave-matter interaction may also be estimated:

$$\varepsilon_{Th} \approx 1/2V(\varepsilon_o\varepsilon''E^2 + \mu_o\mu''H^2) \qquad (10)$$

where ε_r, μ_r, ε'', and μ'' in expressions (9) and (10) are the relative permittivity, relative permeability, (relative) dielectric loss factor, and (relative) magnetic loss factor of the dielectric, respectively. Also, $\varepsilon_o \approx 8.85 \times 10^{-12}$ F/m, $\mu_o \approx 1.26 \times 10^{-6}$ H/m are the permittivity and permeability of vacuum, respectively.

As suggested by expression (9), the non-thermal electromagnetic work done on a dielectric object under microwave/RF irradiation is proportional to the relative permittivity, ε_r, and the relative permeability, μ_r, of the dielectric. Under resonance conditions, e.g., at the resonance frequency of a given polarization or magnetization mechanism, the relative permittivity, or the relative permeability of the dielectric may even reach a maximum value.[6],[9] As a result, the resonance effect during wave-dielectric interaction may greatly enhance the work input, W_{EM}. Subsequently, the free energy of the irradiated dielectric, G^{Th+EM} or A^{Th+EM}, could also significantly increase, as suggested by expressions (6b) and (7b).

CONCEPTS OF PROCESS/REACTION KINETICS

Thermodynamics, i.e., change in free energy of a system, dictates the feasibility of a process; however, it is silent about the rate of the process. On the other hand, reaction kinetics, a branch of physical chemistry, quantifies the rates of reactions.

In classical reaction kinetics,[10],[11] the law of mass action states that the rate of a reaction (process) is proportional to the "active masses" of the reactants.

For a reaction $B + C \rightarrow P$, the rate is:

$$Rate = k[B][C] \qquad (11)$$

where [B] and [C] are the active masses or concentrations (moles/liter) of the reactants B and C, and k is traditionally called the rate constant. With known reactant concentrations, k is a property which may provide full knowledge of the rate.

On the other hand, k is a strong temperature-dependent function. It is governed by the Arrhenius-Eyring relationship derived from the transition-state theory (Figure 1):

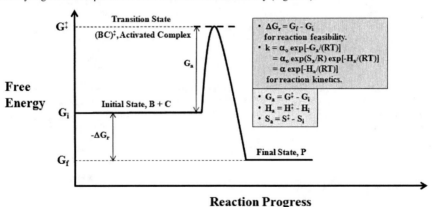

Figure 1. Eyring's transition-state model – reaction-coordinate diagram.

$$k = \alpha_o \exp[-G_a/(RT)] \qquad (12)$$

where α_o is called the frequency factor, R is the gas constant, T is the absolute temperature, and G_a is the activation free energy for the rate-controlling mechanism. In solid-state processes, e.g., sintering, the rate-controlling mechanism is most likely to be a mass transport mechanism.

Activation free energy is a free-energy threshold which must be overcome for every reaction in order to proceed in time. This energy barrier includes the energy required for initial bond breaking among molecules (atoms) of reactants. In other words, reactants can only change into products when they acquire sufficient energy (from an external source) to overcome this barrier. The magnitude of this free-energy barrier, therefore, determines the rate of every thermodynamically feasible process as suggested by equation (12).

When considering Gibbs free energy, $G_a = H_a - TS_a$, the Arrhenius-Eyring relationship, equation (12), can then be rewritten as:

$$k = \alpha_o \exp[S_a/R]\exp[-H_a/(RT)] = \alpha\exp[-H_a/(RT)] \qquad (12a)$$

where $\alpha = \alpha_o \exp[S_a/R]$, which includes both the frequency factor, α_o, and the activation entropy, S_a. H_a is the activation enthalpy. In literature, H_a is commonly approximated to be the empirically determined activation energy, E_a.

Per equation (12), reducing the activation free energy, G_a, will notably increase the rate constant, k, thereby significantly speeding up the reaction rate. Comparing equations (12) and (12a), reducing activation free energy, G_a, also suggests a reduction in the activation enthalpy, H_a (or activation energy, E_a), and/or an enhancement on the activation entropy, S_a, for the reaction mechanism.

Further, according to Figure 1, $G_a = G^{\ddagger} - G_i$. Also, by definition, $G_a = H_a - TS_a$, $G^{\ddagger} = H^{\ddagger} - TS^{\ddagger}$, and $G_i = H_i - TS_i$. Therefore, $H_a = H^{\ddagger} - H_i$, and $S_a = S^{\ddagger} - S_i$, where $G^{\ddagger}$, $H^{\ddagger}$, and $S^{\ddagger}$ are the Gibbs free energy, enthalpy, and entropy of the transition state, respectively, and G_i, H_i, and S_i are the Gibbs free energy, enthalpy, and entropy of the initial state of the system, respectively.

Referring to equations (12) and (12a) again, one may now realize that the rate constant, k and subsequently the rate of reaction can be enhanced by reducing G_a using either one or both of the following two approaches:

(1) Increasing G_i by increasing H_i and/or reducing S_i (disorder) via a proper work input, e.g., an electromagnetic work (W_{EM}), into the initial state of the system (reactants).
(2) Reducing $G^{\ddagger}$ by decreasing $H^{\ddagger}$ and/or increasing $S^{\ddagger}$ (disorder) of the activated complex, thereby changing the path (mechanism) of the reaction, via a proper catalyst addition.

RATIONALE FOR PROCESS FEASIBILITY AND KINETICS UNDER MICROWAVE/RF IRRADIATION

From discussions of previous sections, free energy of a homogeneous dielectric system under (polarized) EMW irradiation may be significantly higher than the free energy of the same system under purely thermal conditions. This concept of gain in free energy due to electromagnetic work input may now be generalized and applied to an irradiated, multi-component, heterogeneous dielectric system (see appendix). The maximum Gibbs free energy gain in an irradiated, heterogeneous dielectric under known polarized irradiation at constant temperature and pressure is:

$$(G^{Th+EM} - G^{Th}) = G^{Th+EM} - [\Sigma_\beta\Sigma_i{}^\beta(\mu_i n_i) + \Sigma_j\gamma_j\tilde{A}_j]_{T,P} = W_{EM}[\Sigma_\beta{}^\beta(V\varepsilon_r), \Sigma_\beta{}^\beta(V\mu_r)] \rightarrow F(\omega, \omega_o) \qquad (13)$$

where G^{Th+EM} and G^{Th} are the Gibbs free energies of the dielectric under both irradiation and thermal conditions and under purely thermal conditions, respectively. $\Sigma_\beta\Sigma_i{}^\beta(\mu_i n_i)$ and $(\Sigma_j\gamma_j\tilde{A}_j)$ are the total chemical and surface energies of the multi-component, heterogeneous material, respectively, and W_{EM} is the total electromagnetic work done on the entire heterogeneous system. Similar expression holds for the maximum gain in Helmholtz free energy of the material under (polarized) irradiation at constant temperature and volume.

Expression (13) emphasizes that the free energy gain in an irradiated dielectric system is a strong function of the product summations of both relative permittivity and volume, and relative permeability and volume of each individual phase, β, in the material. Permittivity and permeability of each phase in turn vary with the (angular) frequency, ω, of the incident waves, and more importantly, reach their maximum values at the resonance frequencies, ω_o.

Let us now apply the proposed thermodynamic and kinetic perspective to provide rationale for non-thermal microwave/RF effects on materials processes and performance. Some previously reported "thermally unnatural" processes or phenomena occurring under resonant polarized EMW (microwave/RF) irradiation are also used as examples for illustrations and support.

Increasing Solubility

At identical temperature (~21 °C), pressure (~1 atm), volume (~500 ml), and dissolution time (~20 h), Brooks (Mortenson)[1b] reported that the solubility of NaCl in resonant-EMW-irradiated water was significantly higher than the solubility in water which had been just kept under purely thermal conditions. Mortenson had then developed a theory for this solubility increase based on resonance science.[1b],[12]

Now, the phenomenon of solubility increase may also be reasoned using free energy analysis based on the "augmented" thermodynamics presented in this paper. As shown in Figure 2, in their

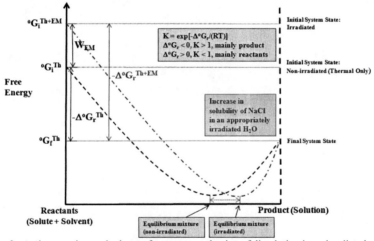

Figure 2. An increase in standard-state free energy reduction of dissolution in an irradiated solute-solvent system. A higher solubility in a properly irradiated system relative to that in a purely thermal system is thermodynamically predicted.

standard states at P = 1 atm and T = 298 K, the initial equilibrium Gibbs free energy of the irradiated system, $^oG_i^{Th+EM}$, was higher than the initial free energy, $^oG_i^{Th}$, in the non-irradiated system by the amount of the electromagnetic work input, W_{EM}. Therefore, the reduction of free energy of the dissolution process in the irradiated system, $\Delta^\circ G_r^{Th+EM}$, became larger, i.e., more negative, than the energy reduction of the process, $\Delta^\circ G_r^{Th}$, in the non-irradiated system. As a result, the equilibrium constant of dissolution, $K^{Th+EM} = \exp[-\Delta^\circ G_r^{Th+EM}/(RT)]$, of the irradiated system was notably larger than the equilibrium constant, $K^{Th} = \exp[-\Delta^\circ G_r^{Th}/(RT)]$, of the non-irradiated system. A right-shift of the equilibrium mixture towards the product then occurred in the irradiated system (Figure 2), and an increase in solubility of the solute (NaCl) in the irradiated solvent (water) was subsequently observed.

The above non-thermal electromagnetic-wave effect on the solubility of the dissolution process conducted under constant temperature and volume conditions may also be interpreted using Helmholtz free energy analysis.

Promoting Process Feasibility

In thermodynamics, a process in a system would favorably proceed forward forming products from reactants if and only if the free energy of the system at its initial reactant state is higher than at its final product state, i.e., change in free energy of the process is negative. A scenario shown in Figure 3 below illustrates that the forward-process of a system under purely thermal conditions is not feasible,

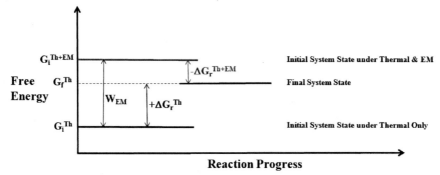

Reaction Progress

Figure 3. Promotion of process feasibility under microwave/RF irradiation. A "thermally unnatural" process could be promoted to become feasible by resonant polarized microwave irradiation. Polarization work done on the final product(s) due to EMW irradiation is assumed to be negligible.

i.e., $\Delta G_r^{Th} > 0$. However, when irradiated with proper EMWs such as resonant polarized microwaves/RF-waves, this system could gain enough free energy via the electromagnetic work input, W_{EM}, from the waves, thereby promoting its initial-state free energy from G_i^{Th} to a much higher level, G_i^{Th+EM}. On the other hand, if the final-state product(s) of the process has a negligible polarizability to the incident waves, its free energy, G_f^{Th}, will remain unchanged during the process. Under this circumstance, as long as the initial-state free energy level of the irradiated system, G_i^{Th+EM}, increases to a level higher than the level of the final (product) state, G_f^{Th}, free energy change will then be favorable to the forward-process, $\Delta G_r^{Th+EM} < 0$. Consequently, the "thermally non-feasible" process will become feasible and favorably proceed forward forming product(s).

One of the impacts of the above non-thermal microwave/RF effect is the production of thermally unstable high-temperature products at relatively low temperatures in the materials industry. A good example: the formation of high-temperature hexagonal $BaTiO_3$ phase under microwave irradiation at as low as 300 °C with no soak time, reported by Agrawal.[13]

Enhancing Process Kinetics

Free-energy increase in a system under resonant polarized EMW irradiation could also enhance the kinetics of a thermodynamically feasible process without necessarily changing the state of equilibrium of the process. The enhancement of process kinetics may be visualized using a reaction-coordinate diagram (Figure 4 shown below).

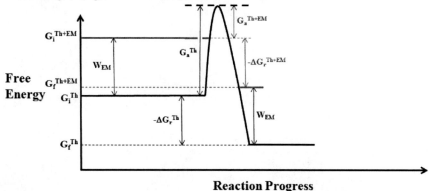

Reaction Progress

Figure 4. Reaction-coordinate diagram showing process kinetics enhancement under resonant polarized microwave/RF irradiation. Note: equilibrium state remains unchanged; $\Delta G_r^{Th} = \Delta G_r^{Th+EM} < 0$.

The process itself shown in the diagram is thermodynamically feasible even under purely thermal conditions ($\Delta G_r^{Th} < 0$). However, the relatively high activation free energy of this purely thermal process, G_a^{Th}, gives a very low rate constant, k, at temperature per equation (12). Consequently, the kinetics of the process under purely thermal conditions is slow.

Under resonant polarized microwave/RF irradiation, however, the activation free energy of the forward-process, G_a^{Th+EM}, may significantly be reduced due to the contribution of the electromagnetic work input, W_{EM}, to the system (as illustrated schematically in the figure). As a result, the rate constant, k, may greatly increase and the kinetics of the process may become astounding.

The above rationale supports numerous published process kinetics enhancements resulting from the use of microwave irradiation. Examples include sintering kinetics enhancements for WC-Co composite,[13] alumina,[14] and oxide ceramics.[15]

Both Promoting Process Feasibility and Enhancing Process Kinetics

Let me illustrate this non-thermal microwave/RF effect on materials processes using a published example: Kanzius's serendipitous discovery of the dissociation of "salt water" under a polarized RF beam.[1a] This thermally unanticipated process was later convincingly confirmed and demonstrated by some fine experiments conducted at PSU.[1c] Based on the thermodynamic and kinetic perspective presented in this paper, a rationale for this "thermally unnatural" dissociation process is proposed. Three scenarios are discussed as follows:

Under purely thermal conditions, dissociation of deionized water into hydrogen and oxygen, causing the liquid to "burn" has neither been observed nor anticipated. This process is thermodynamically non-feasible, i.e., free energy change of the reaction $\Delta G_r^{Th} > 0$, as illustrated in the LHS-plot of Figure 5.

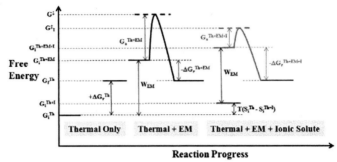

Figure 5. Reaction-coordinate diagram illustrating both feasibility promotion and kinetics enhancement of water dissociation caused by polarized RF irradiation and proper ionic solute addition.

When deionized water was irradiated by a polarized RF beam at ambient temperature,[1c] the free energy of the water may have tremendously increased from G_i^{Th} to G_i^{Th+EM} due to the electromagnetic work input by the incident RF waves as shown in the center-plot of Figure 5. This large free energy gain, which resulted in a $\Delta G_r^{Th+EM} < 0$ for the process, may have then promoted the feasibility of water dissociation – changing it from non-feasible to feasible. However, the activation free energy, G_a^{Th+EM}, for dissociation in this scenario was still too large. Therefore, the rate constant, k, was still too low, and the reaction rate at temperature was still much too slow to be "noticeable". As a result, dissociation of deionized water under RF irradiation is neither "kinetically expected" nor was experimentally observed.[1c]

However, when an adequate amount of NaCl, a highly ionic solute, was added to water forming a solution – "salt water",[1c] the activation free energy for water dissociation under proper irradiation, $G_a^{Th+EM+I}$, was eventually further reduced to a low enough level resulting in a noticeable reaction rate (RHS-plot of Figure 5). This enormous reduction of activation free energy for "salt water" dissociation may be reasoned as follows:

The presence of an adequate amount of strongly charged Na^+ and Cl^- ions caused the (polar) water molecules in the solution to align, significantly decreasing the molecular disorder, i.e., the entropy, S_i^{Th+I}, of the solution relative to that of deionized water (S_i^{Th}). The initial free energy of "salt water" under purely thermal conditions, G_i^{Th+I}, may have thus increased from G_i^{Th} with the amount of $T(S_i^{Th} - S_i^{Th+I})$ as indicated in the figure. Furthermore, "salt-ion-induced polarization" may have also re-oriented the water molecules throughout the entire solution to favor the formation of a lower-energy activated complex – an easier path for O-O and H-H bond formation from water molecules during dissociation. Consequently, a reduction of the transition-state potentials, enthalpy, $H_I^{\ddagger}$, and free energy, $G_I^{\ddagger}$, of the process may have been attained, as illustrated by the lower-activation reaction path shown in the RHS-plot of Figure 5. At this point, however, "salt water" dissociation under purely thermal conditions was still not thermodynamically feasible due to the positive free energy change of the reaction, $(G_r^{Th} - G_i^{Th+I}) > 0$.

When "salt water" was subsequently exposed to resonant polarized RF irradiation, the tremendous W_{EM} input may have then promoted the feasibility of the process with a negative $\Delta G_r^{Th+EM+I}$. At the same time, this work input also could have further reduced the activation free energy to a sufficiently low magnitude, $G_a^{Th+EM+I}$, as shown in the figure. Consequently, "salt water" dissociation under irradiation not only occurred favorably, but also kinetically proceeded, forming hydrogen and oxygen, at a noticeable rate.[1a],[1c] A schematic of the RF-irradiation-assisted dissociation of "salt water" is illustrated below:

SUMMARY

A classical thermodynamic and kinetic perspective is proposed to provide rationale for non-thermal microwave/RF effects. Non-thermal effects initiated during microwave/RF-matter interactions are believed to dictate the feasibility and kinetics of many "thermally unnatural" processes observed, developed, or applied in the materials industry.

Thermodynamics accounting for electromagnetic work predicts that the free energy increase of a system under proper EMW irradiation could be as high as the total work done on the system by the electric and magnetic fields of the waves. Nonetheless, the electromagnetic work performed by the waves on a material, according to electromagnetics, is a function of the (relative) permittivity and (relative) permeability of the dielectric medium. This work always maximizes at the resonance frequency of a given polarization (magnetization) mechanism. At temperature and pressure (or volume), free energy of a system under resonant polarized microwave/RF irradiation may increase so much that it not only enhances the kinetics of a process by activation free energy reduction, but also promotes the feasibility of a "thermally unnatural" reaction.

Four different types of non-thermal microwave/RF effects on materials processes and performance are discussed and illustrated in this paper. They are namely: (1) increasing solubility, (2) promoting process feasibility, (3) enhancing process kinetics, and (4) both promoting process feasibility and enhancing process kinetics.

CONCLUDING REMARK

This paper was mainly written to provide a follow-up rationale for the observations on non-thermal microwave/RF effects reported and discussed at the MS&T'09 Conference in 2009, including those results presented by the late Professor R. Roy and the Materials Research Group at the Pennsylvania State University.[1f] One key requirement for the occurrence of these reported non-thermal microwave/RF effects was the application of resonant polarized electromagnetic waves.

Although non-thermal microwave/RF effects under non-resonance conditions may be negligible, this paper argues that these effects are considerable when the incident polarized irradiation is at resonance with the dielectric material system. For the scenario of a dielectric system irradiated under resonance condition, the relative permittivity, ε_r, of the material for example may abruptly increase, hence, the time-average polarization work input, i.e., the first term of expression (9), may then rise many folds greater than that under non-resonance condition even though the magnitude of the electric field, E, of the incident waves may remain "low".

Further validation of the proposed rationale to quantitatively support the experimental findings in the field of non-thermal microwave/RF effects is necessary and is currently in progress by this author. More results and discussions will be reported on this very important topic of materials processing.

APPENDIX
FREE ENERGY GAIN IN IRRADIATED, MULTI-COMPONENT, HETEROGENEOUS DIELEC-
TRICS
    Unifying the 1st law, the 2nd law, and the definition of Gibbs free energy, the maximum Gibbs free energy change of a multi-component, heterogeneous dielectric system under both thermal and polarized irradiation conditions (at constant temperature and pressure) is obtained:

$$dG^{Th+EM} = \{\sum_\beta\sum_i{}^\beta(\mu_i dn_i) + \sum_j\gamma_j d\tilde{A}_j + \sum_\beta[{}^\beta(V\mathbf{E}\cdot d\mathbf{D}) + {}^\beta(V\mathbf{H}\cdot d\mathbf{B})]\}_{T,P}$$

or

$$dG^{Th+EM} = [\sum_\beta\sum_i{}^\beta(\mu_i dn_i) + \sum_j\gamma_j d\tilde{A}_j + \sum_\beta\delta^\beta W_{EM}]_{T,P} \qquad (A1)$$

where $\sum_\beta\sum_i{}^\beta(\mu_i dn_i)$ and $\sum_\beta\delta^\beta W_{EM} = \sum_\beta[{}^\beta(V\mathbf{E}\cdot d\mathbf{D}) + {}^\beta(V\mathbf{H}\cdot d\mathbf{B})]$ are the sums of differential changes in chemical work and electromagnetic work done on individual phases, β ($\beta = 1, 2, 3, 4 \dots$), in the system, respectively, $\sum_j\gamma_j d\tilde{A}_j$ is the sum of differential changes in surface energies of individual interfaces, j ($j = 1, 2, 3, 4 \dots$), in the system.
    Therefore, the total (equilibrium) Gibbs free energy of the irradiated material at constant temperature and pressure is:

$$G^{Th+EM} = [\sum_\beta\sum_i{}^\beta(\mu_i n_i) + \sum_j\gamma_j\tilde{A}_j + W_{EM}]_{T,P} \qquad (A2)$$

where W_{EM} is the simplified symbol for $\sum_\beta{}^\beta W_{EM}$. It is the total electromagnetic work done on the entire system.
    For linear, isotropic dielectrics under sinusoidal EMW irradiation, the time-average non-thermal electromagnetic work is:

$$\underline{W_{EM}} \approx \sum_\beta[{}^\beta(0.5V\varepsilon_o\varepsilon_r E^2) + {}^\beta(0.5V\mu_o\mu_r H^2)] = 0.5\varepsilon_o E^2\sum_\beta{}^\beta(V\varepsilon_r) + 0.5\mu_o H^2\sum_\beta{}^\beta(V\mu_r) \qquad (A2a)$$

    On the other hand, the maximum Gibbs free energy change of the same heterogeneous dielectric under purely thermal conditions at constant temperature and pressure is:

$$dG^{Th} = [\sum_\beta\sum_i{}^\beta(\mu_i dn_i) + \sum_j\gamma_j d\tilde{A}_j]_{T,P} \qquad (A3)$$

    Therefore, the total (equilibrium) Gibbs free energy of the material under purely thermal conditions at constant temperature and pressure is:

$$G^{Th} = [\sum_\beta\sum_i{}^\beta(\mu_i n_i) + \sum_j\gamma_j\tilde{A}_j]_{T,P} \qquad (A4)$$

    Now, comparing equations (A2) and (A4), subsequently combining with equation (A2a), the maximum gain in Gibbs free energy of the irradiated, linear, isotropic dielectric relative to its non-irradiated (purely thermal) counterpart is:

$$(G^{Th+EM} - G^{Th}) = (W_{EM})_{T,P} \approx [0.5\varepsilon_o E^2\sum_\beta{}^\beta(V\varepsilon_r) + 0.5\mu_o H^2\sum_\beta{}^\beta(V\mu_r)]_{T,P} \qquad (A5)$$

Applying a similar approach, the maximum Helmholtz free energy gain in an irradiated, multi-component, heterogeneous, linear, isotropic dielectric relative to its non-irradiated (purely thermal) counterpart, at constant temperature and volume, is:

$$(A^{Th+EM} - A^{Th}) = (W_{EM})_{T,V} \approx [0.5\varepsilon_o E^2 \sum_\beta{}^\beta (V\varepsilon_r) + 0.5\mu_o H^2 \sum_\beta{}^\beta (V\mu_r)]_{T,V} \qquad (A6)$$

REFERENCES

[1] ACerS, AIST, ASM, and TMS, *Conference Proceedings*, MS&T'09, Pittsburgh, PA (2009).

    [1a] R. Roy and M. L. Rao, The Birth of a New Field of Materials Science: Resonant Polarized Radiation Interactions with Matter, 563-572.

    [1b] J. Brooks (Mortenson), Einstein's Hidden Variable: PartB – The Resonance Factor, 585-596.

    [1c] M. L. Rao, G. P. Flanagan, R. Roy, T. Slawecki, and S. Sedlmayr, Dramatic Structuring of Liquid Water using Polarized Microwave, Radiofrequency Radiation, and Crystal-induced Epitaxy, 655-666.

[2] A. D. Buckingham, *The Laws and Applications of Thermodynamics*, Pergamon Press, Oxford (1964).

[3] G. N. Lewis and M. Randall, revised by K. S. Pitzer and L. Brewer, *Thermodynamics*, 2nd edition, McGraw-Hill Book Company, New York (1961).

[4] E. A. Guggenheim, *Thermodynamics*, 4th edition, North-Holland Publishing Company, Amsterdam (1959).

[5] D. S. Lemons, *Mere Thermodynamics*, Johns Hopkins University Press, Baltimore (2009).

[6] A. von Hippel, *Dielectrics and Waves*, Artech House, Boston (1954).

[7] R. E. Collin, *Foundations for Microwave Engineering*, McGraw-Hill Book Company, New York (1966).

[8] J. F. Nye, *Physical Properties of Crystals*, Clarendon Press, Oxford (1957).

[9] www.doitpoms.ac.uk/tlplib/dielectrics/variation.php, Department of Materials Science and Metallurgy, University of Cambridge, Cambridge, UK.

[10] W. J. Moore, *Physical Chemistry*, 3rd edition, Prentice-Hall, Inc., New Jersey (1962).

[11] G. M. Barrow, *Physical Chemistry*, 2nd edition, McGraw-Hill Book Company, New York (1966).

[12] J. Mortenson, The Fall and Rise of Resonance Science, 2864-2875, *Conference Proceedings*, MS&T'10, ACerS, AIST, ASM, and TMS, Houston, TX (2010).

[13] D. Agrawal, Microwave Sintering of Ceramics, Composites, Metals, and Transparent Materials, *Journal of Materials Education*, vol. 19(4, 5 & 6), 49-57 (1997).

[14] K. H. Brosnan, G. L. Messing, and D. K. Agrawal, Microwave Sintering of Alumina at 2.45 GHz, *J. Am. Ceram. Soc.*, 86, [8], 1307-1312 (2003).

[15] M. A. Janney and H. D. Kimrey, Diffusion-controlled Processes in Microwave-fired Oxide Ceramics, (1990).

Composites

ALUMINA-BASED COMPOSITES REINFORCED WITH TITANIUM NANOPARTICLES

Enrique Rocha-Rangel, José A. Rodríguez-García
Universidad Politécnica de Victoria, Avenida Nuevas Tecnologías 5902
Parque Científico y Tecnológico de Tamaulipas, Tamaulipas, México, 87138

Sergio Mundo-Solís, Juliana G. Gutiérrez-Paredes
Departamento de Posgrado, ESIME-Azc, IPN, Av. de las Granjas # 682
Col. Santa Catarina, México D. F., 02250

Elizabeth Refugio-García
Departamento de Materiales, Universidad Autónoma Metropolitana
Av. San Pablo # 180, Col Reynosa-Tamaulipas, México, D. F., 02200

ABSTRACT

In this study it was explored the effect of titanium nanoparticles additions on the fracture toughness of Al_2O_3-based composites. Mixtures of alumina with different titanium amounts (0.5, 1, 2, 3 vol. %) were high energy grinded in a Symoloyer mill during 8h. After milling, the powder was constituted by very fine particles of 200 nm average sizes, presenting good titanium and alumina particles sharing. Microstructure observations by electron microscopy of uniaxial compacted and pressureless sintered samples, illustrate dense composite materials with fine and homogeneous distribution of titanium in the alumina-matrix. From fracture toughness measurements, that was estimated by the fracture indentation method, it has determined that titanium additions improved fracture toughness until more than 50% with respect to the fracture toughness of monolithic-alumina. This behavior can be due to the formation of metallic bridges by titanium in the alumina-matrix.

INTRODUCTION

Ceramic materials possess favorable mechanical properties such as: high hardness, high compressive strength, good chemical and thermal stability and a high elastic modulus. However, its applications as structural material have been limited by its low fracture toughness and consequently high fragility. Al_2O_3 ceramics can be toughened with the incorporation of fine metallic particles[1] for this reason it has been prepared different Al_2O_3 systems such as; Al_2O_3/Al^2, Al_2O_3/Cr^3, Al_2O_3/Cu^4, Al_2O_3/Ni^5, Al_2O_3/Ag^6 and $Al_2O_3/TiAl^7$. These composites have been successfully fabricated by different techniques such as: direct oxidation of a metal[7], metal infiltration of a ceramic preform[9-10], reactive metal penetration[11-12], hot pressing[13-15] and thermal spray processing[16]. However, most of these processes are costly, present low productivity and they are complex in their procedures and control. From these studies authors have been commented that reinforcement models indicate that is very important the size of metallic inclusion as well as its homogeneous distribution in the ceramic matrix in order to get composites with good toughness properties. In general, for having the best microstructures and mechanical properties, a ceramic-based composite has to be carefully prepared, starting from the powder synthesis in order to obtain good sintered product. With these considerations in mind, simple and cheaper processes are now development for the production of high amounts of ceramic-metal composites. High-energy ball milling combined with pressureless sintering can be a substitute low-cost method for the production of composites, and conventional powder-techniques can be applied for forming and densification. The aim of this study is to synthesize Al_2O_3-based composites reinforced with different amounts of titanium using the powder techniques in order to determine the effect of titanium on the fracture toughness of the alumina-matrix.

EXPERIMENTAL PROCEDURE

Starting materials were: Al_2O_3 powder (99.9 %, 1 μm, Sigma, USA) and Ti powder (99.9 %, 1-2 μm, Aldrich, USA). Final titanium contents in the produced composites were: 0.5, 1, 2 and 3 vol. %. Powder blends of 20 g were prepared in high energy mill (Simoloyer) with ZrO_2 media, the rotation speed of the mill was of 400 rpm, and studied milling time was 8 h. With the milled powder mixture, green cylindrical compacts 2 cm diameter and 0.2 cm thickness were fabricated by uniaxial pressing, using 270 MPa pressure. Then pressureless sinter in an electrical furnace was performed under 10 cm^3/min argon flux, at two different temperatures (1500 and 1600 °C) during 1 h. The microstructure was observed by scanning electron microscopy (SEM) equipped with an analyzer by energy dispersive spectroscopy (EDS). Fracture toughness was estimated by the fracture indentation method[17], (in all cases ten independent measurements per value were carrying out). A flow drawing of the experimental procedure is showed in Figure 1.

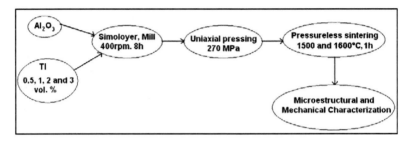

Figure 1. Flow drawing of the experimental procedure.

RESULTS AND DISCUSSION
Powders

SEM powder's picture took in sample with 3 vol. % Ti of milled powders during 8 is presented in Figure 2. In this figure it is observed very fine powders; due to their very small size, agglomerate themselves. With the help of different photographs, it was estimated an average size of the powder minor than 200 nm. The morphology of powders tends to be rounded, situation that must be due to the type of grinding them. In this way it is possible to say that the high energy milling used in this stage has an important effect on the reduction of the powders since the original size of them was among 1 and 2 μm.

Figure 2. Al_2O_3-3 vol. % Ti powder mixture after 8 h milling in a Symoloyer mill.

Figure 3 shows X-ray diffraction pattern of the powders mixture after milling stage for the sample with 3 vol. % Ti. In this figure it can be observed the main starting components of the mixture; Al_2O_3 and Ti. On the other hand, in the pattern it cannot observe a peak of another crystalline constituent, this means that powder mixture has not be contemned during the mechanical milling. For the other compositions similar observations were carry out.

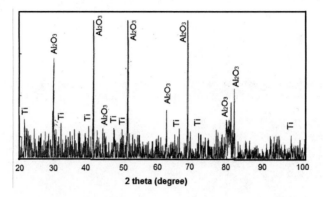

Figure 3. X-ray diffraction pattern of the powder mixture after milling stage. Sample with 3 vol. % Ti.

Sintered samples

Figure 4 shows X-ray diffraction pattern of sample with 3 vol. % Ti sintered at 1600°C during 1 h. In this figure it can be observed the original components Al_2O_3 and Ti. On the other hand, in the pattern it cannot observe a peak of another crystalline constituent, as can be some titanium oxide; this means that argon flux used during sintering helps to protect the oxidation of metallic titanium. Similar X-ray diffraction pattern were obtained for other samples and other sintered temperature (1500°C).

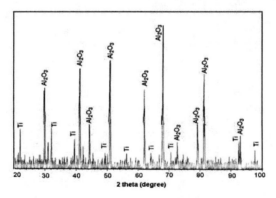

Figure 4. X-ray diffraction pattern of the sintered sample at 1600°C during 1h. Sample with 3 vol. % Ti.

Microstructure

Figure 5. Shows SEM's microstructures of samples sintered at 1600 °C during 1 h as a function of vol. % Ti in the sample. In all pictures it can be seen fine and homogeneous microstructures, with the presence of two different phases, on the basis of (EDS) analysis, it was deduced that the gray phase corresponds to the alumina matrix, whereas the small white and brighter phase corresponds to the titanium reinforcement added to the ceramic matrix. The metallic phase is localized principally at intergranular positions. The main metallic particle size is very small and minor to 1 μm. Judging from the trend disclosed by the Al_2O_3/Ti composites, it can be noticed that the microstructures have no cracks, however the presence of some porosity is observed, although this porosity is small thus suggesting that the addition of titanium to the ceramic helped in the diffusion process in order to obtain well consolidated bodies.

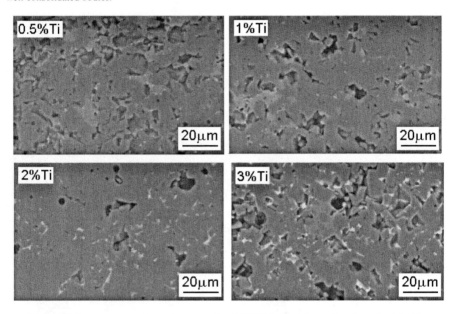

Figure 5. Microstructures of samples sintered at 1600°C during 1 h as a function of vol. % Ti.

Figure 6 presents the microstructure of Al_2O_3-3 vol. % Ti composite after sintered during 1h at 1500°C (top) and 1600°C (down). In both figures they are visibly two different phases as constituents of the microstructure, the main gray phase that corresponds to the matrix and a second phase represented by white particles that match up with the metal aggregated to the matrix. Second phase particles are localized at intergranular positions of the matrix. With the help of an equipment of energy dispersive spectroscopy mounted in the SEM, they were realized punctual chemical analysis in several points of microstructures, resulting that gray phase corresponds to alumina; whereas, the white phase keep up a correspondence with titanium. Spectra resulting from this analysis are also displayed in figure 6. As it can be observed in the figure 6 titanium is homogeneously distributed in the matrix and presents sizes inferior than 5 μm. In some cases titanium tends to join forming and insipient metallic net localized principally at the grains' boundaries of the matrix. Also, in this pictures it is appreciated the typical porosity present in materials processed by powders techniques, from it is noted that the porosity of sintered sample at 1500°C is higher than the porosity of sintered sample at 1600°C, this is because the diffusion phenomena are better at high temperatures in these kind of materials.

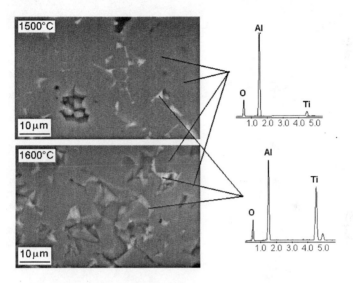

Figure 6. Energy dispersive spectroscopy analysis realized in samples
Al$_2$O$_3$-3 vol. % Ti sintered at 1500 and 1600°C during 1 h.

Fracture toughness

The results of fracture toughness measurements through the fracture by indentation method are presented in Figure 7. One of the most important observations in this figure is the enhancement of fracture toughness with the increments of titanium in the composites. On the other hand, there is a positive effect when sinter temperature rise, because fracture toughness is improved when samples are sintered at higher temperatures (1600°C). These behaviors can be explained by the finest metallic particle size achieved during mechanical milling, situation than let that diffusion phenomena during sinter occurs easily because the distances for diffusion are shorter between particles, at the same time at high sintering temperatures the activation energy necessary for atoms migration can be reached without problems. Several authors have been documented that enhancements in fracture toughness in these kinds of composites may be due to plastic deformation of the metallic phase, which forms crack-bridging ligaments[1]. Situation than is very probable to be present in these composites, because as it has been observed in microstructure of figures 5 and 6, there is a homogeneous distribution of titanium in the ceramic matrix, being these particles responsible for stopping the advance of cracks when the composite are under stress actions.

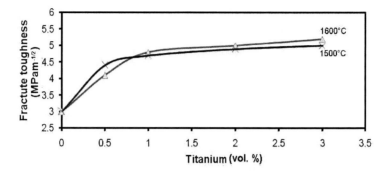

Figure 7. Fracture toughness as a function of the amount of Ti in the composite, for samples sintered at different temperatures.

CONCLUSIONS

o Microstructure of composites fabricated presented a homogeneous distribution of metallic particles in the matrix, these particles are localized in intergranular positions, and in certain zones some of those metallic particles presented a semi-constant metallic network.

o The refined and homogeneous incorporation of titanium in a ceramic matrix (Al_2O_3) improves its fracture toughness. Alternatively, increments in sintering temperature and in the amount of titanium in the matrix are reflected as enhancements of the fracture toughness of the same one.

o From the fracture toughness measurements and microstructure observations, it can be commented that the toughening mechanism in Al_2O_3/Ti composites is due to plastic deformation of the metallic phase, which forms crack-bridging ligaments.

ACKNOWLEDGMENT
Authors would thank CONACYT for financial support for the development of this study through project 132406 as well as Universidad Politécnica de Victoria, ESIME Azc.-IPN and Universidad Autónoma Metropolitana for technical support.

REFERENCES
[1]O.L. Ighodaro and O.I. Okoli, Fracture Toughness Enhancement for Alumina Systems: A Review. *Int. J. Appl. Ceram. Technol.,* 313-323 (2008).
[2]Konopka K. and Szafran M. Fabrication of Al_2O_3-Al composites by infiltration method and their characteristics. *J. Mater. Proc. Technol.,* **175,** 266-270 (2006).
[3]C. Marci and P. Katarzyna, Processing, microstructure and mechanical properties of Al_2O_3-Cr nanocomposite, *J. Eur. Ceram. Soc.,* **27,** 2-3 1273-1277 (2007).
[4]J.G. Miranda Hernández, A.B. Soto Guzmán and E Rocha Rangel, Production and Characterization of Al_2O_3-Cu Composite Materials, *J. Ceram. Proc. Res.,* **7,** 311-314 (2006).
[5]M.I. Lieberthal and K. Kaplan, Processing and properties of Al_2O_3 nanocomposites reinforced with sub-micron Ni and $NiAl_2O_4$ *Mater. Sci. Eng.,* **A302,** 1 83-91 (2001).

[6]J. Lalande, S. Scheppokat, R. Jansen and N. Claussen, Toughening of alumina/zirconia ceramic composites with silver particles. *J. Eur. Ceram. Soc.*, **22**, 13 2165-2168 (2002).

[7]N. Travirskya, I. Gotmanb and N. Claussen, Alumina-Ti aluminide interpenetrating composites: microstructure and mechanical properties, *Mater. Lett.*, **57**, 22-23 3422-3424 (2003).

[8]M. S. Newkirk, A. W. Urquhart, H. R. Zwicker and E. Breval, Formation of Lanxide™ Ceramic Composite Materials, *J. Mater. Res.*, **1**, 81-89 (1986).

[9]S. Wu, A. J. Gesing, N. A. Travitzky and N. Claussen, Fabrication and Properties of Al-Infiltrated RBAO-Based Composites, *J. Eur. Ceram. Soc.*, **7**, 277-281 (1991).

[10]C. Toy and W. D. Scott, Ceramic-Metal Composite Produced by Melt-Infiltration, *J. Am. Ceram. Soc.*, **73**, 97-101 (1990).

[11]S. M. Naga, A. El-Maghraby and A. M. El-Rafei, Properties of Ceramic-Metal Composites Formed by Reactive Metal Penetration, *Am. Cerm. Bull.*, **86**, 9301-9307 (2007).

[12]R. E. Loehman, K. Ewsuk and P. Tomsia, Synthesisi of Al_2O_3/Al Composites by Reactive Metal Penetration, *J. Am. Ceram. Soc.*, **79**, 27-32 (1996).

[13]E. Rocha and C. Vilchis, Fabrication and Consolidation of $NiAl$/Al_2O_3 Composites Using Simultaneously In-Situ Displacement Reactions and Hot Pressing, *in Proceedings of 5th International Conference on High-Temperature Ceramic Matrix Composites – HTCMC5*, American Ceramic Society, 347-352 (2004).

[14]N. Claussen, M. Knechtel, H. Prielipp and J. Rodel, A Strong Variant of Cermets, *Ber. Dtsch. Keram. Ges.* 301-303 (1994).

[15]S.J. Ko, K.H. Min and Y.D. Kim, A study on the fabrication of Al_2O_3/Cu nanocomposite and its mechanical Properties, *Journal of Ceramic Processing Research.* 192-194 (2002).

[16]S. Sampath, H. Herman, N. Shimoda and T. Saito, Thermal Spray Processing of FGM's, *MRS Bull.* 27-31 (1995).

[17]A. G. Evans and E. A. Charles, Fracture Toughness Determination by Indentation, *J. Am. Ceram. Soc.*, **59**, 371-372 (1976).

FABRICATION OF ZrO_2-SiC COMPOSITES FROM NATURAL ZIRCON ORE BY CARBOTHERMAL REDUCTION

Xu Youguo, Huang Zhaohui[*], Fang Minghao, Liu Yan-gai, Ouyang Xin, Yin Li

School of Materials Science and Technology
China University of Geosciences (Beijing)
Beijing 100083, PR China

ABSTRACT

ZrO_2-SiC composite powders were synthesized from natural zircon ore by carbothermal reduction in immerged carbon coke and the effects of synthesis temperature, carbon content, different reducing agents and oxide additives on the phase composition of the products were investigated. The phase composition of the products was characterized by X-ray diffraction. The results show that ZrO_2-SiC composite powders could be prepared from natural zircon ore by carbothermal reduction between 1500 °C and 1600 °C for 4 hours and the ideal synthetic requirement is 1600 °C using carbon black as reducing agent. The phase composition of zirconia in the products with 8 wt% CaO, Y_2O_3 or TiO_2 addition exists in the form of tetragonal zirconia, cubic zirconia and monoclinic zirconia respectively.

INTRODUCTION

Zircon is one of the important zirconium silicate mineral in nature[1]. Zircon is one of fundamental raw materials for manufacturing zirconia[2,3]. Zirconia is one of the most promising ceramic materials for high-temperature structural applications, piezoelectric original, high-temperature optical components, oxygen sensing element, fuel cells owing to its attractive mechanical, thermal, electrical and optical properties[4-7]. In order to further improve its strength and toughness, many methods have been proposed, including second-phase strengthening and whisker strengthening[8,9], which can lead to an improvement in the mechanical response of the material at both room and elevated temperatures. For example, Al_2O_3-SiC composites possess better room and high temperature strengths and toughness as compared to monolithic alumina[10]. In addition, composites display higher resistance to thermal shock and to high temperature creep.

Ceramic composite materials are generally prepared by physical mixing of the matrix powders, e.g. ZrO_2, Al_2O_3 or Si_3N_4, etc., with whiskers, e.g. SiC whisker. They are not homogeneous due to their shape and size mismatch. The whiskers also cause self agglomeration and interlocking problems. The whiskers are expensive due to their preparation from costly organic precursors. On the other hand, in-situ preparation of SiC whiskers by carbothermal reduction of silicate minerals is anticipated to be a simple and cheaper process besides better homogeneity and dispersion of SiC whiskers. Thus, in-situ preparation of ZrO_2-SiC whisker composite powders by carbothermal reduction of zircon is anticipated to be a process for obtaining the homogeneous mixtures by a single step cheaper process as compared to the physical mixing individual components.

We present in this paper some preliminary results on the carbothermal reduction of zircon in immerged carbon coke in order to obtain ZrO_2-SiC composites. We have presented the effect of

[*]Corresponding author: Tel.: +86-10-82322186; Fax: +86-10-82322186;
E-mail: huang118@cugb.edu.cn (Huang Zhaohui)

various reaction parameters, e.g. temperature (1450, 1500, 1550 and 1600 °C, respectively), mol ratios of carbon to zircon (1:3 and 1:6), nature and source of carbon (graphite, carbon coke and carbon black) and oxide additives (8 wt% CaO, Y$_2$O$_3$ or TiO$_2$, respectively) on the synthesis products phase from zircon carbothermal reduction.

EXPERIMENTAL PROCEDURE

Zircon powder (720 nm<d$_{50}$<1 μm, Shandong Province, China) with the chemical composition shown in Table 1 was directly used as the precursor for the reaction. Graphite (d<1 μm, w$_C$>99.0 wt%, Shandong Province, China), carbon coke (d<1 μm, w$_C$>80.0 wt%, Shanxi Province, China) and carbon black (d<1 μm, w$_C$>99.0 wt%, Shandong Province, China) were used as the reducing agent. CaO(AR, Guangdong Shantou Longxi chemical plant Co., Ltd., China), Y$_2$O$_3$(AR, inner Mongolia Jin Yu Jia Industry Co., Ltd., China) and TiO$_2$(AR, Beijing Yili Fine Chemicals Co., Ltd., China) were selected as oxide additives. The overall reaction equation for carbothermal reduction reaction from zircon could be expressed as the follow Eq. (1) and (2). According to Eq. (1) and (2)[11], the raw material ratio relationship was designed.

Table 1. Chemical Composition of Zircon by XRF

Oxide name	ZrO$_2$	SiO$_2$	HfO$_2$	Al$_2$O$_3$	Y$_2$O$_3$	Fe$_2$O$_3$	CaO	TiO$_2$	Cr$_2$O$_3$
Percentage (wt%)	62.63	33.17	2.72	0.93	0.21	0.12	0.09	0.08	0.05

$$ZrSiO_4(s) + 3C(s) = ZrO_2(s) + SiC(s) + 2CO(g) \qquad (1)$$

$$ZrSiO_4(s) + 6C(s) = ZrC(s) + SiC(s) + 4CO(g) \qquad (2)$$

The starting materials were weighed in terms of theoretical proportion relationship, and then mixed together in a ball milling at 200 rpm for 6 hours, using zirconium oxide balls with a diameter of 5 mm as mixing medium and alcohol as milling solution. The mixed powders were subsequently fully dried at 110 °C for 24 hours, and then were shaped into cylinders with 20 mm in diameter and 5 mm in thickness under vertical pressure of 60 MPa. The green samples were dried at 70 °C for 24 hours and were buried in carbon coke in a high-temperature closed corundum crucible. The samples buried in carbon coke were heated at 1450 °C, 1500 °C, 1550 °C and 1600 °C in the high-temperature furnace for 4 hours, respectively. After the desired synthesis temperature and time were reached, the system was cooled to room temperature. The phase composition of the products were monitored via D/max-rA X-ray diffraction (XRD, Rigaku Corporation, Japan) with Cu Kα radiation source (λ=1.5406 Å) at 40 kV and 40 mA using a 2θ step of 0.02°.

RESULTS AND DISCUSSION

The effects of synthesis temperature and carbon content on the phase composition of the products

Fig.1 shows the XRD patterns of natural zircon by carbothermal reduction for 4 hours at various temperatures (from 1450 °C to 1600 °C every 50 °C). It indicates that zircon can not be reduced by carbon black at 1450 °C even the sample with 6 mol of carbon black per mol zircon. The diffraction peaks of m-ZrO$_2$ appeared at 1500 °C while the peaks of β-SiC were present in the products at 1550 °C.

The X-ray diffraction intensities of m-ZrO$_2$ and β-SiC strengthen and that of zircon phase weaken gradually with the synthesis temperature increasing from 1500 °C to 1550 °C. When the samples with 6 mol of carbon black per mol zircon were heated at 1550 °C for 4h (Fig.1b), zircon phase nearly vanishes and the products are mainly m-ZrO$_2$ and β-SiC. Zircon could be completely transformed to m-ZrO$_2$, c-ZrO$_2$, ZrC and β-SiC at 1600 °C. The diffraction of β-SiC at 1600 °C is much more obvious than that at 1550 °C and part of m-ZrO$_2$ transform into c-ZrO$_2$. In addition, most of ZrO$_2$ obtained from zircon was further reduced to ZrC by carbon black at 1600 °C even the sample with 3 mol of carbon black per mol zircon.

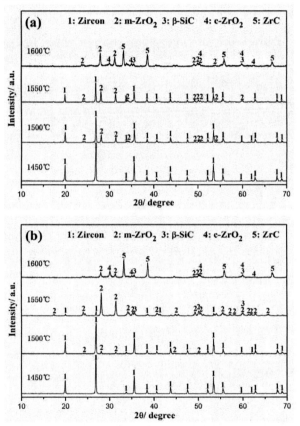

Figure 1. XRD patterns of the products synthesized at different temperatures. a: 3 mol of carbon black per mol zircon according to Eq. 1; b: 6 mol of carbon black per mol zircon according to Eq. 2.

The overall chemical reaction of zircon with carbon can be stated as the Eq. 1 in which silica part of zircon would react with carbon to form SiC. The reaction of silica with carbon is known to proceed in the following steps[12]:

$$SiO_2(s) + C(s) = SiO(g) + CO(g) \tag{3}$$

$$SiO(g) + 2C(s) = SiC(s) + CO(g) \tag{4}$$

$$SiO_2(s) + 3C(s) = SiC(s) + 2CO(g) \tag{5}$$

The reaction (4) which is a solid-vapour reaction is known to be responsible for the formation of SiC mostly in whisker form. It also indicated that SiC could not be formed directly in the C-Si-O system. Due to the gaseous nature of SiO, the experiments carried out in merged carbon coke would reduce product rate for SiC inside of the samples. The initial forming temperature for β-SiC is 1550 °C. Raising the reaction temperature can enhance vibrational energy, chemical response capability and diffuse capability in the solid particle, thus it can speed up the reaction rate and extent between carbon black and zircon which is favorable to synthesis the ZrO2-SiC composites. In summary, the optimized synthesis temperature for β-SiC forming was 1600 °C, but ZrO2 obtained was easily carbonized to ZrC by carbon black.

The effects of reducing agents on the phase composition of the products

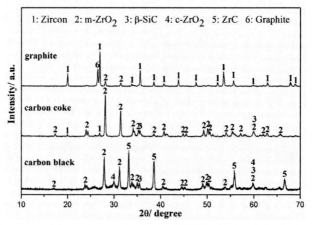

Figure 2. XRD patterns of the products synthesized at 1600 °C with 3 mol of reducing agent per mol natural zircon ore

Reducing agents would have vital impacts on the carbothermal reduction of natural zircon. Fig. 2 shows the XRD patterns of the products synthesized at different sort of reducing agents (carbon black, carbon coke and graphite) with 3 mol of reducing agent per mol natural zircon. In the products of the reaction between carbon black and zircon, the significant diffraction spectrums were m-ZrO2, ZrC, SiC and c-ZrO_2. There is no obvious diffraction spectrum of zircon. In addition to this, most of ZrO_2 obtained from zircon was further reduced to ZrC. When the reducing agent was carbon coke, the significant diffraction spectrums in the products were m-ZrO_2 and SiC, but there also were feeble zircon diffraction spectrums. And zirconia was mainly in the form of m-ZrO_2. In the products of the reaction between graphite and zircon, the conversion rate of zircon was lowest and there is much more significant diffraction spectrum of zircon and graphite. Among the reaction of zircon carbothermal reduction, the best reducing agent was carbon black and the worst was graphite.

The effects of oxide additives on the phase composition of the products

Pure ZrO_2 exists in three different crystal structures, i.e., monoclinic (m-ZrO_2), tetragonal (t-ZrO_2) and cubic (c-ZrO_2). Recently, scientists around the world have done a lot of fruitful research on the phase transformation mechanism and stability of zirconia[13-15]. Doping with such an **ionic** forms zirconia solid solution and apparently leads to form fully stabilized ZrO_2 (FSZ) or partially stabilized ZrO_2 (PSZ) which were widely used in ceramic materials, catalytic materials, oxygen sensors and high-temperature fuel cells and so on. It is harsh to obtain PSZ or FSZ under the formal condition. In particular, the technology for the introduction of low-cost action, such as Y^{3+}, Ca^{2+} and Mg^{2+}, was vital and essential. In this paper, the influences of 8 wt% CaO, Y_2O_3 and TiO_2 on the phase composition for the zircon carbothermal reduction with 3 mol of carbon black reducing agent per mol zircon at 1600 °C and the XRD patterns of the products with above oxide additions was shown in Fig. 3.

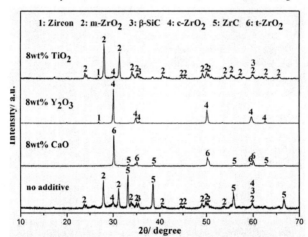

Figure 3. XRD patterns of the products of natural zircon ore carbothermal reduction with different oxide additions

Fig. 3 indicated that doping with oxide additions affect the phase form of zirconia which obtained in the reaction of zircon carbothermal reduction and ZrO_2 existed mostly simplex in the form of t-ZrO_2, c-ZrO_2 and m-ZrO_2 with 8 wt % CaO, Y_2O_3 and TiO_2 respectively. There are two possible reasons for the difference of zirconia phase transformation. Firstly, amorphous SiO_2 got by decomposition of zircon adsorbed on the surface of m-ZrO_2 and made zirconia diffusion difficult, avoiding the nucleation and growth of the ZrO_2 crystalline particles. Under the condition of carbothermal reaction at 1600 °C, amorphous SiO_2 affect the bands on the surface of ZrO_2 and make part m-ZrO_2 transform to t-ZrO_2 or c-ZrO_2. TiO_2 should have the similar effect mechanism, but the influence is weak. Secondly, the m-ZrO_2 is a much more open structure and could tolerate more defects in it, thus CaO and Y_2O_3 addition could dissolve in the lattice of m-ZrO_2. Generally, CaO and Y_2O_3 addition cause defects in the lattice[15,16], such as:

$$CaO \xrightarrow{ZrO_2} Ca_{Zr}^{'} + O_O + V_O^{..} \tag{6}$$

$$Y_2O_3 \xrightarrow{ZrO_2} 2Y_{Zr}^{'} + 3O_O + V_O^{..} \tag{7}$$

So in these CaO and Y_2O_3 doped zirconia in the reaction of zircon carbothermal, the t-ZrO_2 and c-ZrO_2 phase stability is due to the appearance of oxygen vacancies. In addition, there are not obvious ZrC diffraction peaks in the XRD patterns of samples with oxide additive. It indicated that doping with oxide additives such as CaO, Y_2O_3 and TiO_2 could restrain ZrO_2 carbonization. In summary, m-ZrO_2, t-ZrO_2 or c-ZrO_2 could be obtained directly in the process of the fabrication of ZrO_2-SiC composites by zircon carbothermal reduction with appropriate oxide additions.

Conclusions

ZrO_2-SiC composite powder could be successfully synthesized from zircon by carbothermal reduction in immerged carbon coke. The optimized synthesis temperature for β-SiC forming was 1600 °C, and ideal reducing agent was carbon black. Doping with oxide additives such as CaO, Y_2O_3 and TiO_2 in the zircon carbothermal reduction could affect the phase composition of ZrO_2 in the ZrO_2-SiC composite powder. t-ZrO_2, c-ZrO_2 and m-ZrO_2 in the ZrO_2-SiC composite powder could be obtained respectively when zircon carbothermal reduction was carried out with 8 wt% CaO, Y_2O_3 and TiO_2 addition.

ACKNOWLEDGMENTS

The authors greatly appreciate National Natural Science Foundation of China for financial support (Grant Nos. 51032007 and 50972134). We also thank the Fundamental Research Funds for the Central Universities (Grant Nos. 2010ZD12 and 2011PY0172) and the New Star Technology Plan of Beijing under grant number 2007A080.

REFERENCES
[1]A. Nemchin, N. Timms, R. Pidgeon, T. Geisler, S. Reddy, and C. Meyer, Timing of Crystallization of the Lunar Magma Ocean Constrained by the Oldest Zircon, *Nat. Geosci.*, **2[2]**, 133-36 (2009).
[2]N. A. Mohammed and A. M. Daher, Preparation of High-purity Zirconia from Egyptian Zircon: an Anion-exchange Purification Process, *Hydrometallurgy*, **65[2-3]**, 103-07 (2002).
[3]S. Souza and B. S. Terry, Production of Stabilized and Non-stabilized ZrO_2 by Carbothermic Reduction of ZrSiO4, *J. Mater. Sci.*, **29**, 3329-36 (1994).
[4]M. J. Andrews, M. K. Ferber, and E. Lara-Curzio, Mechanical Properties of Zirconia-based Ceramics as Functions of Temperature, *J. Eur. Ceram. Soc.*, **22[14-15]**, 2633-39 (2002).

[5]R. C. Garvie, R. H. Hannink, and R. T. Pascoe, Ceramic Steel?, *Nature*, **258**, 703-04 (1975).

[6]R. H. J. Hannink, P. M. Kelly, and B. C. Muddle, Transformation Toughening in Zirconia-containing Ceramics, *J. Am. Ceram. Soc.*, **83[3]**, 461-87 (2000).

[7]Y. J. He, L. Winnubst, A. J. Burggraaf, H. Verweij, P. G. T. vanderVarst, and B. With, Influence of Porosity on Friction and Wear of Tetragonal Zirconia Polycrystal, *J. Am. Ceram. Soc.*, **80[2]**, 377-80 (1997).

[8]H. Awaji, S. M. Choi, and E. Yagi, Mechanisms of Toughening and Strengthening in Ceramic-based Nanocomposites, *Mech. Mater.*, **34[7]**, 411-22 (2002).

[9]K. N. Xia and T. G. Langdon, The Toughening and Strengthening of Ceramic Materials through Discontinuous Reinforcement, *J. Mater. Sci.*, **29[20]**, 5219-31 (1994).

[10]B. C. Bechtold and I. B. Cutler, Reaction of Clay and Carbon to Form and Separate Al_2O_3 and SiC, *J. Am. Ceram. Soc.*, **63[5-6]**, 271-75 (1980).

[11]P. K. Panda, L. Mariappan, V. A. Jaleel, and T. S. Kannan, Preparation of Zirconia and Silicon Carbide Whisker Biphasic Powder Mixtures by Carbothermal Reduction of Zircon Powders, *J. Mater. Sci.*, **31**, 4277-88 (1996).

[12]N. S. Jacobson and E. J. Opila, Thermodynamics of Si-C-O System, *Metall. Trans. A*, **24[5]**, 1212-14 (1993).

[13]C. J. Ho and W. H. T, Phase Stability and Microstructure Evolution of Yttria-stabilized Zirconia during Firing in a Reducing Atmosphere, *Ceram. Int.*, **37[4]**, 1401-07 (2011).

[14]S. Tekeli and A. Gural, Sintering, Phase Stability and Room Temperature Mechanical Properties of c-ZrO_2 Ceramics with TiO_2 Addition, *Materials & Design*, **28[5]**, 1707-10 (2007).

[15]G. R. Duan, X. J. Yang, A. Q. Lu, G. H. Huang, L. D. Lu, and X. Wang, Comparison Study on the High-temperature Phase Stability of CaO-doped Zirconia Made Using Different Precipitants, *Mater. Charact.*, **58[1]**, 78-81 (2007).

[16]Y. Du, Z. P. Jin, and P. Y. Huang, Thermodynamic Assessment of the ZrO_2-$YO_{1.5}$ System, *J. Am. Ceram. Soc.*,**74[7]**, 1569-77(1991).

MANUFACTURE AND APPLICATIONS OF C/C-SIC AND C/SIC COMPOSITES

Bernhard Heidenreich
DLR – German Aerospace Center
Institute of Structures and Design
Pfaffenwaldring 38-40
D-70569 Stuttgart

ABSTRACT

Carbon fibre reinforced SiC, the most widely used CMC materials, can be divided up into two main types, C/SiC and C/C-SiC, mainly differing in their microstructural design. C/SiC materials are characterized by C fibres, embedded almost separately in SiC matrix. Therefore, fibre preforms are used and the SiC matrix is built up by Chemical Vapour Infiltration (CVI) or Polymer Infiltration and Pyrolysis (PIP). In contrast, C/C-SiC materials are based on porous Carbon/Carbon (C/C) preforms. Thereby, not the single fibres, but dense C/C bundles, consisting of several hundreds of C fibres, embedded in a C matrix, are again embedded in a SiC matrix, built up by Liquid Silicon Infiltration (LSI). In this presentation, the influences of the manufacturing methods on the morphology and properties of the resulting CMC materials, as well as on the feasibility of components are shown, and an overview of current applications is given.

INTRODUCTION

Ceramics offer high thermal and chemical stability, hardness and abrasive wear resistance. However, as the strength of bulk ceramics is determined by the size and distribution of imperfections, like pores, impurities or microcracks, the acceptable strength values are inversely proportional to the volume of the component, generally resulting in thick walled, heavy structures. The most promising method to improve the fracture behaviour of ceramics, and to increase the acceptable strength level of structural parts, is the embedment of continuous fibres. First studies in the late 1970s showed, that the integration of C or SiC fibres into ceramic matrices lead to a new class of very interesting ceramic materials with unique properties. These CMC materials (Ceramic Matrix Composites) combine the outstanding properties of bulk or monolithic ceramics with novel qualities, like quasi-ductile fracture behaviour and high fracture toughness, very low coefficients of thermal expansion (CTE) and extreme thermal shock stability, completely unusual for ceramics.

Thereby it was found that the interphase between the brittle fibres and the brittle matrix is the key issue for the development of non brittle CMC materials [1]. Weak interfaces enable the fibres to locally separate from the matrix, and the CMC material can capitalize on the typically high tensile strength and fracture elongations of the fibres ($\sigma_t > 2000$ MPa; $\varepsilon > 1.5$ %), which are significantly higher than that of bulk ceramics (SiC: $\sigma_t = 400$ MPa; $\varepsilon < 0.05$ %). This leads to a high crack resistance and a pseudo plastic material behaviour, due to energy dissipating effects obtained by crack bridging, crack deflection and fibre pull-out (Fig. 1). In contrast to monolithic ceramics, where cracks irresistibly run through the entire part, in CMC materials, arising cracks are stopped at fibre matrix interfaces or at micro-cracks within the matrix. Therefore the damage area is limited and large structures with high tensile loads can be realized.

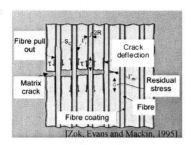 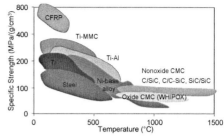

Figure 1:Left: Schematic illustration of energy absorbing phenomena in CMC materials.
Right: Weight specific strength of various structural materials as a function of
temperature

Amongst all available design materials, fibre reinforced ceramics offer the highest specific strength and stiffness at temperatures beyond 900 °C and therefore CMC's are the most promising materials for lightweight structures with high thermal and mechanical loads (Fig. 1). State-of-the-art are carbon fibre reinforced carbon materials (C/C, Carbon / Carbon) which are widely used, e.g. for heating elements and charging rigs in high temperature furnaces, as well as for aircraft brake disks and nozzles in rocket propulsion. However, the main drawback of C/C is its low oxidation resistance, limiting the use of C/C to maximum temperatures of 450 °C for long term applications in air.

To improve the oxidation resistance, various CMC materials based on non-oxide ceramic matrices have been proposed since the 1980s [2, 3]. Up until now, long fibre reinforced SiC based on C and SiC fibres have been developed successfully. Due to their high thermal stability and relatively low cost, compared to SiC fibres, C fibres are mainly used for industrial applications.

Various manufacturing methods for C fibre based SiC composite materials have been developed, leading to characteristic C/SiC and C/C-SiC materials and specific material properties. Moreover, the manufacturing methods are mainly determining the feasibility and cost of CMC components and structures and finally, due to economic aspects, their application areas.

MANUFACTURING METHODS

In principal, all the different manufacturing methods for C/SiC and C/C-SiC materials are based on three main process steps:

- Build up of a C fibre or CFRP preform
- Build up of weak fibre / matrix interphase
- Build up of ceramic matrix

The fibre preform can be used as dry preform or is built up by the manufacture of a CFRP (Carbon Fibre Reinforced Plastic) preform. Therefore, generally all types of carbon fibre yarns can be used. Due to cost aspects, high tenacity (HT) fibres, e.g. Tenax HTA, are most common. Improved material properties can be obtained by the use of intermediate modulus (IM, e.g. Torayca T 800) and high modulus (HM, e.g. Torayca M40) fibres.

Common fibre preforms are usually built up by stacking of 2D woven fabrics. Unidirectional or 1D fibre layers can be used as cross ply laminates or in combination with fabric layers. Additionally, fibre rovings are used for filament wound or braided and stitched 3D preforms. For highly sophisticated components even so called "4D" fibre preforms based on multiaxial oriented fibres have been developed. Randomly oriented fibre reinforcements can be achieved by so-called short fibres which in fact are cut fibres with typical lengths in the range of 2 mm to about 40 mm and above. As a principle for CMC manufacturing, the fibre preforms should be made in near net shape geometry, especially for thin walled lightweight structures, in order to avoid waste and reduce cost for final machining.

In almost all manufacturing processes, a weak fibre / matrix interphase is obtained by fibre coating. Typically, a thin layer of pyrolithic C (pyC), with a thickness of 0.1 to 1 μm, is applied on the fibres via chemical vapour deposition or infiltration (CVD/CVI), prior to the build up of the SiC matrix or CFRP preform. Alternatively, the C fibres are embedded in a dense C matrix via polymer infiltration and pyrolysis (PIP).

Since the late 1970's, various methods for the build up of SiC-based matrices have been investigated. For industrial production of C/SiC or C/C-SiC materials and components, currently, 3 characteristic methods are used, leading to three main process routes:

1. Deposition of gaseous SiC-precursors   ⇨   CVI (Chemical Vapour Infiltration)
2. Pyrolysis of Si-polymers   ⇨   PIP (Polymer Infiltration and Pyrolysis)
3. Reaction of liquid Si with C   ⇨   LSI (Liquid Silicon Infiltration)

In the CVI-method, the SiC-matrix is built up in one process step, using a gaseous precursor. Thereby the fibre preform is infiltrated with the gaseous precursor and SiC-crystals are deposited on the C-fibre filaments, simultaneously. Using polymer precursors, preform infiltration and forming of the SiC-matrix is achieved by several, separated processes. In the PIP-process, also called LPI (Liquid Polymer Infiltration), the ceramic matrix is directly derived by the pyrolysis of the preceramic polymer, which is infiltrated as a liquid precursor in the fibre preform in a previous process step. In the LSI process, a C-precursor is infiltrated in the fibre preform and the resulting CFRP preform is pyrolysed, similar to the PIP-process. However, during pyrolysis the polymer matrix is converted to a C matrix in a first step and subsequently the final SiC matrix is built up in an additional process step by an infiltration of molten Si and a chemical reaction of Si with the C-matrix and C-fibres.

In the following, the manufacturing methods are described in detail and their main advantages and disadvantages are compared.

Chemical Vapour Infiltration (CVI)

The distinctive mark of the CVI processes is the deposition of the ceramic matrix from a gaseous precursor, infiltrated in a porous fibre preform. The CVI method was derived from chemical vapour deposition (CVD), a well established technique for applying thin ceramic coatings [4]. Originally developed for the manufacture of C/C materials, the CVI process could be adapted to deposit SiC matrices, using a defined mixture of methyltrichlorsilane (MTS) and H_2 as process gas.

$$CH_3SiCL_{3,\ gaseous} + H_{2,\ gaseous} \rightarrow SiC_{solid} + 3HCl_{gaseous} + H_{2,\ gaseous}$$

The CVI process is divided into three main process steps:

1. Manufacture of a carbon fibre preform
2. Deposition of fibre coating (interphase)
3. Deposition of the SiC matrix

In the first step, a porous preform made of carbon fibres is made in near net shape geometry. High fibre contents between 40 and 60 % can be achieved by densifying dry fabrics in a heat resistant ceramic mould, during the first step of the CVI densification process. Costly ceramic moulds can be avoided by using free-standing, geometrically stable, dry preforms and so called stabilized preforms.

Dry preforms are built up by special weaving and braiding methods (2.5 D fabrics, 3D and "4D" preforms) or by stitching of 2 D fabric laminates as well as by dry filament winding or braiding. For stabilized preforms, 2D fabrics or cut fibres are coated with a C- or SiC polymer precursor, laminated or filled in a mould, where they are densified to a certain fibre volume content in the range of 15 -25 % (short fibres) up to 30 – 50 % (fabrics), as well as formed to near net shape geometry. After curing the precursor, the preform is demoulded and pyrolysed, leading to a highly porous but stable preform.

The structure of the preforms porosity is an all important parameter for the CVI-process. A translaminar open pore system of evenly distributed, interconnected pores or microchannels with minimum diameters of 10 to 500 μm is necessary for a homogeneous densification. Suitable pore structures can be obtained by multi-dimensional fibre structures. Unidirectional or cross ply laminates based on 1D fibre preforms lead to small pores which cannot be filled easily and therefore are not common for CVI.

Fibre coating is essential to obtain a weak interphase between the fibres and the ceramic matrix. Typically, a thin layer of carbon with a thickness between 0.1 and 1 μm is applied by a CVD-process, prior to the forming of the fibre preform or, more economically, by a first CVI process step, just before the deposition of the SiC-matrix.

In the third step of the CVI-process, the porous fibre preform is set inside an infiltration chamber, and heated to a temperature typically between 800 and 1100 °C. The gaseous precursor passes through the infiltration chamber and infiltrates the porous fibre preform. Parallel to the infiltration, β-SiC is formed out of the precusor and deposited on the hot C-fibres.

One main advantage of the CVI-process is the feasibility of a highly pure and fine grained SiC matrix, leading to high ceramic content and excellent mechanical properties in the resulting C/SiC material. However, to achieve this, the molar ratio $\alpha = [H_2] / [CH_3SiCl_3]$ has to be controlled thoroughly and a homogeneous build up of the matrix has to be ensured. Homogeneity and deposition rate are determined by the process parameters, i. e. temperature T and gas pressure p. High T and p leads to high deposition rates but also to high inhomogeneities, especially in thick walled parts. Thereby, the precursor is mainly reacting at the outer surface of the part, leaving impoverished, less reactive precursor for the centre area. This again leads to a higher deposition rate on the surface, and finally to a sealing of the surface before the matrix build up in the centre is completed.

Two main process variations are used, isothermal / isobaric CVI (ICVI) [5] and forced CVI (FCVI). For the latter the temperature / pressure gradient CVI (GCVI) [6] is very common. Using ICVI, all preforms are heated uniformly ($T_{max.} = 800 – 900$ °C), and the process gas infiltrates the preforms solely by diffusion at a constant pressure of about 50 – 100 hPa. Thereby, even very large parts and complex shaped structures can be manufactured. For the production of high quantities of parts, large furnaces can be used and differently shaped preforms can be processed in one batch. The main drawback of the ICVI-process is the long process duration. Additionally, the maximum wall thickness, which can be manufactured in one infiltration step is limited to about 3 mm. Thicker parts of up to a maximum of about 30 mm can be obtained by using multiple (2 – 4) CVI densification steps. Thereby, the preform has to be machined after each intermediate densification step, to reopen the pores on the surface. This leads to overall process durations of several days for thin walled components up to several weeks or months for thick walled parts.

In contrast to ICVI, where the deposition of the SiC matrix is controlled only by diffusion, the mass transfer of the GCVI process is forced by a pressure gradient. Additionally a temperature gradient is imposed by heating one surface of the preform to a temperature between 1100 and 1200 °C and cooling the opposite surface to lower temperatures of 800 – 900 °C. Thereby, matrix deposition rates are significantly higher compared to ICVI, leading to short process durations of about 2 – 3 days for thin plates of up to 5 mm wall thickness and about 7 days for thick parts of up to 30 mm wall thickness [7]. However, to keep the process efficient, the densification is stopped at a residual open porosity of about 10 to12 %, similar to ICVI. To ensure the temperature and pressure gradients, each preform has to be mounted in a ceramic tool with integrated cooling system (Fig. 2) and the process generally is limited to simple parts like tubular structures or flat plates with constant wall thicknesses.

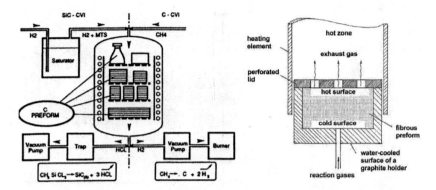

Figure 2: Schematic overview of a ICVI-facility (left) [5] and experimental set-up of the temperature / pressure gradient CVI (GCVI) process [8].

Polymer Infiltration and Pyrolysis (PIP)

The characteristic of the PIP process is the infiltration of the fibres or fibre preforms by a liquid polymer and the subsequent conversion of the polymer into a ceramic matrix via pyrolysis. The PIP technique is one of the most advanced methods for the manufacture of large and complex shaped CMC structures in aerospace industry [9]. Like CVI, PIP is a further well known process for the manufacture of C/C materials. For the production of C/SiC materials, the carbon precursors were replaced by preceramic polymers, and the process parameters were adapted to the chemistry of these new precursor systems.

The PIP process can be subdivided into 3 main process steps (Fig. 3):

1 Coating of the carbon fibres
2 Manufacture of a CFRP preform
3 Pyrolysis of the CFRP preform

In the first process step, the carbon fibres or fibre preforms, are coated via continuous or discontinuous CVD or CVI processes. Usually a thin (0.1 – 0.3 μm) layer of pyrocarbon is deposited onto the fibres, to avoid a high fibre/matrix bonding in the final C/SiC material.

For the manufacture of the CFRP preforms, well known methods, like wet filament winding and fibre placement, vacuum assisted polymer infiltration (VAP, developed by EADS) or resin transfer moulding (RTM) as well as hand lay-up can be used in principal. However, these techniques have to be adapted to the use of preceramic precursors, typically polycarbosilane or polysilane polymers. To increase the ceramic yield, submicron SiC particles or reactive fillers usually are added to the liquid precursors. However, in order to avoid fibre damage by the particles, the maximum fibre volume content is limited.

Wet filament winding can be used to manufacture CFRP preforms directly as well as for the build up of prepreg layers with unidirectional fibre orientation. Subsequently, these 1D prepregs are cut to size and stacked in predetermined orientations, forming laminates for flat plates or even complex shaped structures. Subsequently, the filament wound or laminated structures are consolidated and cured in an autoclave at maximum temperatures between 90 °C and 200 °C and at pressures in the range of 1 to 2 MPa.

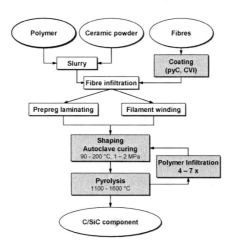

Figure 3: Schematic overview of the manufacture of C/SiC materials via PIP.

In the third process step, the CFRP preform is pyrolysed in inert gas atmosphere or vacuum. During pyrolysis, the polymer bonds break, the organic side chains separate, and the polymer matrix is converted into the final ceramic matrix. The morphology of the ceramic matrix is influenced by the process temperature. Thereby, heating rates are limited to avoid delaminations in the CFRP preform and to obtain a low porous SiC matrix. Low maximum temperatures of about 1100 °C lead to a highly porous, amorphous matrices, whereas low porous, crystalline SiC matrices are obtained at maximum process temperatures of up to 1600 °C.

Due to the high volume contraction of the polymer matrix during pyrolysis, the resulting C/SiC materials are highly porous after the first pyrolysis step. To achieve sufficient mechanical properties, the porosity has to be reduced by multiple densifying cycles, each consisting of a polymer infiltration and a pyrolysis step. To obtain C/SiC materials with open porosities below 10 %, typically five to seven impregnation steps are necessary.

The main advantages of the PIP process are the low process temperatures used for the build up of the ceramic matrix and the good control of the resulting SiC matrix. Additionally, unidirectional (1D) fibre reinforcement is possible, resulting in C/SiC materials with very high strength values parallel to the fibre orientation. Similar to the CVI process, the fibres are not influenced or damaged during the SiC matrix build up, and, in combination with a single fibre embedment, leads to high mechanical properties. Using filament wound preforms, very large and precise CMC structures like nozzles (Ø 1.3 m, l = 3.6 m, t = 1.4 - 3.6 mm) for space propulsion systems have already been realised [10].The disadvantages of the PIP process are high costs for the preceramic precursors and the very long process times of several weeks up to months, due to the multiple infiltration and pyrolysis steps needed.

Liquid Silicon Infiltration (LSI)

The LSI process is based on the experiences from the manufacture of reaction bonded SiSiC materials, as well as on the manufacture of C/C materials via PIP and has been developed since the late 1970s [11], in order to provide a more economic CMC manufacturing method compared to the highly sophisticated CVI and PIP processes. The characteristic of the LSI process is the infiltration of molten silicon in a porous C/C preform and the exothermal reaction of the molten Si with C, forming a dense SiSiC matrix in a single process step.

$$Si_{liquid} + C_{solid} \rightarrow SiC_{solid} \ [\Delta H = -68 \ kJ/mol]$$

The LSI process can be subdivided in four main process steps (Fig. 4):

1. Coating of carbon fibres or fibre preforms (most processes)
2. Manufacture of a CFRP preform
3. Pyrolysis of the CFRP preform to a C/C preform
4. Siliconization of the C/C preform

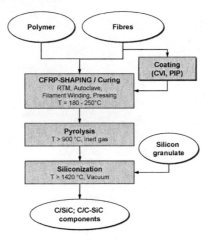

Figure 4: Schematic overview of the manufacture of C/C-SiC materials via LSI.

In the first step, a CFRP preform is made, using common technologies from the plastics industry, like RTM, autoclave technique, wet filament winding or warm pressing, which have been adapted to the use of high carbon yield precursors, like phenolic resins. Almost all C-fibre types, from low cost high tenacity to intermediate modulus and high modulus fibres, can be used. For the build up of the fibre preform, 2D fabrics are common for highly loaded CMC structures. 1D or unidirectional fibre layers can be used as cross plies or in combination with fabric layers. Similar to CVI, fibre preforms, solely based on 1D fibre orientation, are not suitable for LSI, due to their highly dense and inhomogeneous microstructure after pyrolysis, which prevents a reliable Si melt infiltration. Filament winding and braiding is used for the build up of dry fibre preforms. Low cost C/SiC and C/C-SiC materials can be obtained by cut fibres, randomly oriented in warm pressed CFRP preforms.

In contrast to CVI and PIP, where the SiC matrix build up does not influence the carbon fibres, the molten silicon is highly reactive to the carbon matrix as well as to the fibres. Therefore, a direct contact of Si melt and C fibres has to be avoided carefully. Additionally, as mentioned before, a weak embedding of the brittle fibres in the brittle matrix is mandatory to obtain characteristic CMC properties, like high strength, fracture toughness and thermal shock resistance. To ensure both, fibre protection and weak fibre matrix interface, three different methods are used for industrial production, up to now, i.e. fibre coating via CVI, fibre embedding in carbon matrix via PIP, and in situ fibre embedding in carbon matrix (Fig. 5).

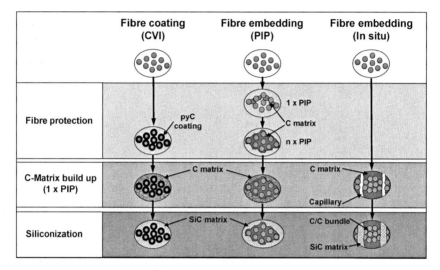

Figure 5: Schematic overview of different methods for ensuring fibre protection and weak fibre matrix bonding in the LSI manufacturing method. Left: Fibre coating via CVI. Middle: Embedding the C-fibres by multiple PIP. Right: In situ fibre embedding due to strong fibre matrix bonding.

Using CVI for fibre coating, a thin layer, (~ 0.1 μm) of pyrolithic carbon is deposited on each fibre filament, resulting in a C/SiC material with the filaments mainly embedded individually in the SiC matrix. Fibre protection via PIP is widely used for the manufacture of short fibre based C/SiC brake disks [12]. Thereby, endless fibre bundles are impregnated with phenolic resin, which is cured and pyrolysed, embedding the fibre filaments in a dense carbon matrix, resulting in a C/C like raw material, and finally, after siliconization, to C/C bundles embedded in SiSiC matrix. Time consuming and costly fibre coatings are not necessary at all, if particularly suitable precursors are used for the manufacture of the CFRP preform. Thereby a strong fibre matrix bonding, leads to a segmentation of each fibre bundle into dense C/C bundles during pyrolysis, and again, C/C bundles embedded in SiSiC matrix are obtained after siliconization. Thereby, only a small amount of fibres at the outer surface of the C/C-bundles come in contact with the molten Si and are converted to SiC.

In the second step, the CFRP preform is pyrolysed in inert gas atmosphere, resulting in a highly porous C/C preform. The pyrolysis is the most critical process step. During the conversion of the polymer to the carbon matrix, high volumes of gaseous products are developed. These gases has to be released safely out of the parts, in order to avoid the build up of internal pressure, which can lead to delaminations and complete destruction of the CFRP or C/C preform, finally. Due to the volume shrinkage of the matrix of 50 – 60 Vol.-% on the one hand and the geometrical stability of the fibre preform on the other hand, internal stress is built up, leading to microcracks or micro-channels and a porous C/C preform, required for the subsequent Si infiltration. However, these stresses can also lead to delaminations and distortions, especially in curved or complex shaped parts with anisotropic fibre architectures. Thereby, the use of e.g. 2D fibre fabric laminates, lead to an anisotropic contraction of about 10 % in the wall thickness, perpendicular to the laminate, and only 0.5 % in the in plane direction, parallel to the fibres, resulting in high interlaminar stresses. To avoid or reduce delaminations and distortions, the maximum heating rates during pyrolysis are limited, and high temperature stable supporting elements or structures made of graphite or C/C are used.

In the third and last process step, molten silicon is infiltrated in the porous C/C preform by capillary forces. In parallel to the infiltration the silicon reacts with the carbon, i. e. carbon fibre and carbon matrix, in the contact areas, forming the SiC matrix. Depending on the microstructure of the C/C preform, the infiltrated Si is not converted completely, but free Si remains, especially in large pores or wide micro-channels, resulting in a dense SiSiC matrix.

Due to a low and predictable contraction rate during pyrolysis, almost no geometrical change during siliconization and the fast and homogeneous Si infiltration process, structural parts can be realized in near net shape technique with almost no restrictions to size and geometry. Up to now, C/C-SiC parts with wall thicknesses of less than 1 mm up to 60 mm and diameters of up to a about 740 mm could be realized successfully. Due to the low porosity and the resulting high interlaminar shear strength of the LSI-materials, even sharp edged, tiny and filigree geometries, like screw threads can be obtained via grinding, milling and turning. Additionally, complex shaped, thin walled structures can be realized via in situ joining, where different subcomponents are assembled in the C/C stage and joined permanently to an integral structure during the subsequent siliconization step [13]. Very large structures can be built up without using a furnace, thereby only limiting the parts size, by the joining of C/C-SiC components in an active soldering process, using Si as solder.

Process duration is determined by the complexity of the part and the pyrolysis process, mainly. CFRP preform manufacture takes some hours to one day for simple plates and up to one week for large and complex shaped structures like nose caps for spacecraft thermal protection systems. For the siliconization, typically one to 2 days are needed, depending on the complexity of the parts. Process durations for pyrolysis are determined by the geometry (wall thickness, size, shape) of the part as well as by its fibre architecture. For parts based on 2 D fibre fabrics the cycle time varies from about 5 days for thin walled plates, up to 10 days for thick walled plates or large and complex shaped structures. Due to the random orientation of short fibre based preforms, the risk of delamination is reduced, and the pyrolysis can be shortened to 1 to 3 days, using higher heating rates.

For fabric based materials, process duration for CFRP manufacture and pyrolysis can be reduced dramatically by integrating the curing of the polymer in a fast pyrolysis process. Continuous processes for pyrolysis and siliconization, which are more economic compared to state of the art batch processes, are already used in serial production or will be introduced in the near future. First investigations of new heating methods based on microwaves for CFRP manufacture and pyrolysis as well as direct sintering for siliconization showed a high potential for significant reductions of overall process duration.

The main advantages of LSI are the short manufacturing times and the variability of the process. Thereby the material properties can be adjusted in a wide range to specific requirements in different application areas. Taking into account that every process step is done only once and that there are almost no limits for the parts geometry, process duration is significantly lower, compared to all the other CMC manufacturing processes, especially for the manufacture of thick walled structures and complex shaped components. Additionally, low cost raw materials like phenolic polymer precursors and Si-alloys can be used and costly fibre coating can be avoided. The main disadvantages are the high process temperatures for siliconization and the anisotropic contraction during pyrolysis, resulting in high inner stresses and risk of delaminations, especially for curved or tubular structures.

PROPERTIES

Material composition and microstructure

C/SiC and C/C-SiC materials are multiphase materials consisting of C fibres and β-SiC matrix and, in the case of LSI derived materials, additional C as well as Si matrix. Material composition and microstructure are mainly determined by the used fibre preform and the manufacturing method, especially the way the SiC matrix is built up and the weak fibre matrix bond is obtained.

Weak C interphases based on CVD/CVI fibre coatings lead to so called C/SiC materials, characterized by single C fibre filaments embedded in SiC matrix, whereas in C/C-SiC materials the carbon fibres are embedded in C matrix, leading to dense C/C bundles, which again are embedded in SiC or SiSiC matrix (Fig. 6).

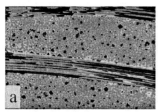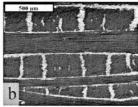

Figure 6: SEM-views showing typical microstructures of carbon fibre reinforced SiC. a: C/SiC material manufactured via CVI (MT Aerospace) with single C-fibre filaments (dark grey) embedded in the SiC matrix (light grey). Porosity is shown in black. b, c: C/C-SiC materials based on LSI and 2D fabrics (b, DLR C/C-SiC XB) and short fibres (c; SGL Sigrasic 6010 GNJ), showing dense C/C bundles (black) embedded in SiC matrix (dark grey) with residual Si (light grey).

Generally, C/SiC materials obtained by CVI and PIP show low C content, i.e. C fibre and pyC fibre coating, and high SiC contents ($\phi_C \approx 40$ -55 Vol.-%, $\phi_{SiC} \approx 35 - 50$ Vol.- %), depending on the fibre content. In contrast, comparable C/C-SiC materials based on high fibre contents are characterized by a high C content (C fibre and matrix) of up to 76 Vol.-% and a significantly lower SiC content ($\phi_{SiC} \approx 16$ Vol.- %) as well as some residual Si ($\phi_{Si} < 4$ Vol.- %). However, the LSI process offers the possibility to vary the material composition in a wide range. For short fibre reinforced materials, highly porous C/C preforms can be used, leading to low C-contents of about 30 Vol.- % and maximum contents of SiC (≈ 60 Vol.-%) and Si (≈ 10 Vol.-%) [14].

Due to economic reasons, the open porosity of CVI and PIP materials is in the range of 8 – 15 %, whereas the very effective melt infiltration processes, lead to relatively dense C/SiC and C/C-SiC materials with open porosities well below 4 %.

C/SiC and C/C-SiC materials are lightweight materials, with a typical density of 1.9 to 2.4 g/cm³, slightly higher than that of CFRP ($\rho = 1.5$ g/cm³) and magnesium ($\rho = 1.74$ g/cm³) but lower than aluminium ($\rho = 2.7$ g/cm³). Due to the usually high fibre content of about 60 Vol.-%, as well as due to the residual C-matrix, 2D reinforced LSI materials offer the lowest density, despite their generally low open porosity. In comparison, C/SiC materials manufactured via CVI and PIP offer a sole SiC matrix and are limited to fibre volume contents of about 50 Vol.-% at most, leading to higher SiC contents and therefore slightly higher densities in the range of 2.1 to 2.3 g/cm^3. Due to their high contents of SiC and Si, short fibre reinforced LSI materials show highest densities of up to 2.4 g/cm^3.

Mechanical properties
The mechanical and thermal properties of C fibre reinforced SiC materials are strongly dependent on the fibre type as well as on fibre content and manufacturing method. Highest mechanical properties with tensile strength values of up to σ_t = 340 MPa and 450 MPa, as well as fracture elongations of up to ε = 0.7 % and 0.5 % can be obtained via CVI or PIP using fabric (2D) and unidirectional (1D) fibre preforms, respectively (Table II). This can be explained by the very homogeneous interphase of almost each single fibre filament and by the low process temperatures, which do not influence the fibres at all, in contrast to the LSI process, where a certain amount of fibres is converted to SiC, leading to a reduced amount of load carrying fibres in the final CMC material as well as to low strength and strain values of σ_t = 80 - 190 MPa and ε = 0.15 – 0,35 % for 2D fabric reinforced materials, depending on the used C-fibres. However, due to the very low open porosity of LSI based materials, the interlaminar shear strength of C/C-SiC is at a similar level to CVI materials but is generally higher compared to PIP materials.
The mechanical strength of short fibre reinforced C/SiC or C/C-SiC materials strongly depends on the fibre length and fibre orientation in the part. Generally, fibre length of about 6 mm, lead to low tensile strength of σ_t = 20-30 but high strain of ε = 0.3 %. Using longer fibres, the mechanical properties increase significantly from a bending strength of σ_b = 60 MPa (l_F =10 mm) to σ_b = 120 MPa (l_f = 40 mm) [15]. However a high scattering of the results is observed, caused by varying fibre orientations in the test specimen.
One main advantage of C fibre reinforced SiC materials is the general stability of the mechanical properties at high temperatures, in contrast to metals, where strength values are significantly decreasing with increasing temperatures [16].

Thermal properties
Due to the use of carbon fibres, which generally offer a low thermal expansion in parallel to the fibre axis, the coefficient of thermal expansion (CTE) for carbon fibre reinforced SiC materials usually is low. With 2D reinforced materials, very low CTE between -1 x $10^{-6}K^{-1}$ and 3 x $10^{-6}K^{-1}$ parallel to the fibres are obtained at ambient temperature. Perpendicular to the fibres, the CTE is dominated by the matrix and therefore is about twice as high (CTE $_\perp$ = 2.5 x $10^{-6}K^{-1}$ to 5 x $10^{-6}K^{-1}$) as parallel to the fibres. Thermal conductivity is generally low compared to metals and bulk ceramics, and, similar to the CTE, anisotropic, offering about twice as high values in the in plane direction compared to the perpendicular direction ($\lambda_{||}$ = 11 – 28 W/mK; $\lambda_\perp$ = 5 – 12 W/mK; at ambient temperature). Highest thermal conductivities of up to 40 W/mK are obtained by short fibre reinforced materials, characterized by very high contents of SiC and Si (λ_{SiC} = 100 - 160 W/mK; λ_{Si} = 148 W/mK; at ambient temperature) [14].

In Table I, the mechanical and thermal properties of typical LSI derived C/C-SiC and C/SiC materials are compared to C/SiC materials based on CVI and PIP. The values can be used as a rough orientation, but cannot be compared directly, due to different evaluation methods used.

Table I: Typical properties of C/SiC and C/C-SiC materials based on 2D fabrics and various manufacturing methods.

		CVI		LPI		LSI	
		C/SiC	C/SiC	C/SiC	C/C-SiC	C/C-SiC	C/SiC
Manufacturer		SPS (SNECMA)	MT Aerospace	EADS	DLR	SKT	SGL (9)
Density	g/cm³	2.1	2.1 - 2.2	1.8	1.9 - 2.0	> 1.8	2 / 2.4
Porosity	%	10	10 - 15	10	2 - 5	-	2 / <1
Tensile strength	MPa	350	300 - 320	250	80 - 190	-	110 / 20-30
Strain to failure	%	0.9	0.6 - 0.9	0.5	0.15 - 0.35	0.23-0.3	0.3
Young's modulus	GPa	90 - 100	90 - 100	65	50 - 70	-	65 /20-30
Compression strength	MPa	580 - 700	450 - 550	590	210 - 320	-	470 / 250
Flexural strength	MPa	500 - 700	450 - 500	500	160 - 300	130 - 240	190 / 50
ILSS	MPa	35	45 - 48	10	28 - 33	14 - 20	-
Fibre content	Vol.%	45	42 - 47	46	55 - 65	-	-
CTE Coefficient of thermal expansion ∥	10^{-6} K^{-1}	3(1)	3	1.16(4)	-1 - 2.5(2)	0.8-1.5(4)	-0.3 / 1.8 (5)
CTE Coefficient of thermal expansion ⊥		5(1)	5	4.06(4)	2.5 - 7(2)	5.5-6.5(4)	-0.03–1.36 (6) / 3 (7)
Thermal conductivity ∥	W/mK	14.3-20.6(1)	14	11.3-12.6(2)	17.0-22.6(3)	28 - 35	23–12 (8) /
Thermal conductivity ⊥		6.5 - 5.9(1)	7	5.3 - 5.5(2)	7.5 - 10.3(3)	12 - 22	40-20 (8)
Specific heat	J/kgK	620 - 1400	-	900-1600(2)	690 - 1550	-	-

∥ and ⊥ = Fibre orientation; (1) RT - 1000 °C ; (2) RT - 1500 °C; (3) 200 - 1650 °C; (4) = RT - 700 °C; (5) 1200 °C; (6) 200 – 1200 °C; (7) 300 – 1200 °C; (8) 20 °C – 1200 °C; (9) values for fabric/short fibre reinforced material

APPLICATIONS

Due to their extraordinary performance but generally high cost, C/SiC and C/C-SiC materials are typically used when there is no alternative material which can meet the requirements. Therefore these materials were initially developed and introduced in aerospace applications. However, due to the development of low cost manufacturing methods, like the LSI process, new application areas beyond aerospace could be opened up for economically viable, high performance products.

Carbon fibre reinforced SiC materials were initially developed for thin walled, lightweight structures in thermal protection systems of reusable spacecraft. Thereby, maximum temperatures of 1800 °C up to 2200 °C, high heating rates of several hundred Kelvin per second and high thermal gradients are obtained locally during the critical re-entry phase, lasting about 20 minutes. One of the first C/SiC structure, manufactured via Liquid phase infiltration was the nose cap for the Buran, a former Soviet Union spacecraft.

For the X-38 demonstrator, which was planned to be the future crew return vehicle for the international space station (ISS), a front end system, consisting of a nose cap and adjacent nose skirts as well as highly loaded, lightweight steering flaps (Fig. 7) were developed by a German consortium in the TETRA program (1998 – 2002) [17, 18]. Currently, nose caps will be used in several European research programmes like EXPERT [19]. Novel TPS systems, based on facetted structures and flat plats, offer higher efficiency at supersonic velocities and significantly reduced manufacturing cost. Thereby extreme thermal loads with maximum temperatures above 2 000 °C are obtained at the leading edges (Fig.7).

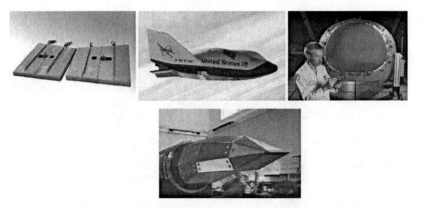

Figure 7: C/SiC (CVI) body flaps (1,5 x 1,5 x 0,15 m³; m = 68 kg, MT Aerospace, left) and C/C-SiC (LSI) nose cap (ca. 740 x 640 x 170 mm³; t = 6 mm; m = 7 kg; DLR; right) for X38 spacecraft (NASA, middle). Facetted TPS structure for the Sharp Edge Flight Experiment (SHEFEX) with C/C-SiC (LSI) nose tip and panels (DLR; below).

As one of the first serial products, C/SiC nozzle flaps for the SNECMA M88 jet engine have been produced since 1996 via CVI [20]. Up to now, several thousand flaps have been produced by Snecma Propulsion Solid (SPS, Fig. 8). Compared to the former used superalloy flaps (Inconel 718), the so called Sepcarbinox® flaps provide weight savings of about 50 %. Further applications were investigated for engine divergent seals for the F 100 engine and for novel systems with moveable vector nozzles [21].

C/SiC and C/C-SiC jet vanes for solid propellant rocket motors are manufactured by e.g. SPS and Nammo AS. Jet vanes are used for thrust vector control systems and are implemented in the exhaust jet stream, where they are exposed to extreme high mechanical and thermal loads (Fig. 8). Thereby, maximum temperatures of up to 3100 K and temperature gradients of several thousand Kelvin per second and about 200 K/mm are obtained. Additionally, the exhaust jet is very abrasive, due to e.g. Al powder, used in the propellant, finally hitting the leading edges in form of Al_2O_3 particles or droplets with velocities of up to 2000 m/s. Compared to metallic jet vanes, which have to be made by refractory metals like tungsten, the use of C/SiC or C/C-SiC composites offers weight savings of up to 90 %.

Figure 8: C/SiC (CVI) outer flap for M 88-2 aeroengine (Snecma Propulsion Solide, left). C/C-SiC (LSI) jet vane for thrust vector control (TVC) of rockets (right).

Due to its excellent temperature and thermal shock stability, high wear and corrosion resistance as well as high specific strength at high temperatures, C/C-SiC materials are well suited for high performance friction applications. Automotive brake disks based on short fibre materials and LSI-techniques were introduced in the automobile market by Porsche and former DaimlerChrysler right at the beginning of this millennium. New factories for large scale production were built up in Europe by SGL Carbon and Brembo / DaimlerChrysler. Since 2009, ceramic brake discs are produced by the joint venture Brembo SGL Carbon Ceramic Brakes. Compared to cast iron brake disks, weight savings of up to 50 % and service life of up to 300 000 km can be obtained. In another automotive application, clutch discs based on fabric reinforced LSI-materials have been used in high end sports and race cars of Porsche and Audi, enabling a safe transfer of high torques above 1000 Nm in a compact and lightweight system (Fig. 9).

C/C-SiC brake pads for emergency brakes in high performance elevators have been manufactured by FCT Ingenieurkeramik GmbH since 2004, offering extremely low wear and constant friction coefficient at maximum service temperatures of about 1200 °C. In contrast to brake applications, where usually high coefficients of friction (COF) are welcome, low COF is required for the emergency runners for the Transrapid in Shanghai. In the case of malfunction of the electromagnetic hovering system, the train has to glide on the concrete driveway at a maximum starting velocity of up to 500 km/h. Thereby, high abrasive resistant C/C-SiC sliding elements, manufactured by Schunk Kohlenstofftechnik, provide low wear to ensure a safe transport to the service station (Fig. 9).

Figure 9:   C/SiC (LSI) automotive brake disc (left), clutch rotor disk (middle; both Porsche) and sliding elements (right) in a runner system for the Transrapid in Shanghai (SKT).

Due to the carbon fibre reinforcement, C/C-SiC materials are used for components requiring extremely low thermal expansion and high geometrical stability. As a typical example, highly precise telescope structures are needed in laser communication systems for satellites. In an experiment, using two satellites (NFIRE, TerraSAR-X), launched in 2007, high data transfer rates of up to 5.5 Gbit/s could be demonstrated. Thereby, C/C-SiC telescope tubes (Fig. 10) with a low CTE in axial direction $(0 \pm 0,1 \cdot 10^{-6}\ K^{-1})$ ensured a constant position of the mirrors at varying temperatures in the range of -50 °C and +70 °C [21]. Compared to other low CTE materials, C/C-SiC is not hygroscopic and does not swell, like CFRP. It offers higher fracture toughness than Zerodur or SiC and significantly lower density compared to Invar, as well as high specific stiffness, enabling thin walled lightweight structures.

Figure 10: In situ joined C/C-SiC telescope tube (∅ 140 mm, l = 160 mm, t = 3 mm; left, Zeiss Optronik / DLR) for the laser communication terminals in the satellites TerraSAR-X and NFIRE (right; TESAT SPACECOM).

SUMMARY

Carbon fibre reinforced SiC materials are combining the advantages of ceramics, like high temperature, chemical and abrasive stability, with the lightweight performance of CFRP materials and a quasiductile fracture behaviour and damage tolerance similar to grey cast iron.

However, high manufacturing costs, caused by expensive raw materials, low matrix deposition rates and therefore long process durations, as well as by expensive investment, needed for the high temperature facilities, were limiting their use to highly specialized applications in space and military aeronautics. With the development of new manufacturing processes, based on a fast build up of the SiC matrix via silicon melt infiltration in low cost C/C preforms, new application areas beyond aerospace could be opened. Compared to highly sophisticated CVI and PIP materials, these so called LSI materials offer lower, but, for many applications, sufficient strength, as well as low porosity, high shear strength and thermal conductivity. Additionally, material properties can be varied in a wide range and there are almost no process limits, regarding geometry and size of the parts. The start of serial production of brake disks for automobiles was an important milestone and breakthrough of this material class.

Ongoing developments concentrate on a further decrease of process durations and costs, as well as on cost efficient machining methods for large scale production. However, to enter demanding markets with high safety standards, e.g. in civil aeronautics, reliable methods for design, quality assurance and material as well as component qualification are a must. Thereby, life cycle prediction is in the main focus, especially under consideration of inevitably material inhomogeneities or defects, either caused by the manufacturing process or by service use.

REFERENCES

[1] D.C. Phillips: *Fibre reinforced Ceramics*, Handbook of Composites, Vol. 4, 373 et sqq., edited by A. Kelly and S.T. Mileiko, Elsevier Science Publishers, 1983.

[2] R. Kochendörfer: *Heiße Tragende Strukturen aus Faserverbund-Leichtbauwerkstoffen;* DGLR-Annual meeting, Berlin, 1987.

[3] R. Naslain, F. Langlais: *Tailoring Multiphase and Composite Ceramics*, ed. By R.E. Tressler et al, Material Science Research, Vol. 20, 145 – 164, 1986

[4] D.P. Stinton, T.M. Besman, R.A. Lowden: *Advanced ceramics by chemical vapour deposition techniques,* American Ceramic Society Bulletin, Vol. 67, 350 – 355, 1988.

[5] R. Naslain, F. Langlais: *CVD-Processing of Ceramic-Ceramic Composite Materials*, Material Science Research, Vol. 20, 145 – 164, 1986.

[6] A. Mühlratzer, M. Leuchs: *Applications of Non-Oxide CMC*, High Temperature Ceramic Matrix Composites, Eds: W. Krenkel, R. Naslain, H. Schneider, Wiley-VCH, Weinheim, Germany, 288 – 289, 2001.

[7] M. Leuchs, A. Mühlratzer: *CVI-Verfahren zur Herstellung faserverstärkter Keramik- Herstellung, Eigenschaften, Anwendungen*, Keramische Verbundwerkstoffe, Ed. W. Krenkel, Wiley-VCH, Weinheim, Germany, 163 – 173, 2003.

[8] D.P. Stinton, A.J. Caputo, R.A. Lowden: *Synthesis of Fibre-Reinforced SiC Composites by Chemical Vapour Infiltration, Journal of* American Ceramic Society Bulletin, Vol. 65/2, 347 - 350, 1986.

[9] R.P. Boisvert, R.J. Diefendorf: *Polymeric Precursor SiC matrix composites*, Ceram. Eng. Sci. Proc., 9, 873 – 880, 1988.

[10] S. Schmidt et al: *Ceramic Matrix Composites: A Challenge in Space Propulsion Technology Applications*, JACT, Vol. 2 [2], 85 – 96, 2005.

[11] C.C. Evans, A.C. Parmee, R.W. Rainbow: *Silicon Treatment of Carbon Fiber-Carbon Composites*, Proceedings of 4th London Conference on Carbon and Graphite, 231 - 235, 1974.

[12] Winnacker, Küchler: *Chemische Technik: Prozesse und Produkte, Eds. R. Dittmeyer et al, Volume 8*, Wiley-VCH, Weinheim, Germany, 11166 -1173, 2005

[13] R. Kochendörfer, W. Krenkel: *CMC Intake Ramp for Hypersonic Propulsion Systems*, in: Ceramic Matrix Composites I: Design Durability and Performance, Eds. A.G. Evans, R. Naslain, Ceramic Transactions, Vol. 57, 13 – 22, 1995.

[14] Data sheet from SGL Carbon Group, SIGRASIC 6010 GNJ – Faserverstärkte Keramik für Bremsscheiben, 2005

[15] B. Heidenreich, R. Renz, W. Krenkel: *Short Fibre reinforced CMC Materials for High Performance Brakes*, High Temperature Ceramic Matrix Composites, Eds: W. Krenkel, R. Naslain, H. Schneider, Wiley-VCH, Weinheim, Germany, 809 - 815, 2001.

[16] D. Desnoyer, A. Lacombe, J.M. Rouges: *Large Thin Composite Structural Parts*, Proceedings of International Conference Spacecraft Structures and Mechanical Testing, Nordwijk, The Netherlands, 1991.

[17] A. Mühlratzer, M. Leuchs: *Applications of Non-Oxide CMC*, High Temperature Ceramic Matrix Composites, Eds: W. Krenkel, R. Naslain, H. Schneider, Wiley-VCH, Weinheim, Germany, 288 – 289, 2001.

[18] H. Hald et al: *Developmentof a Nose Cap System for X 38*, in. Proceedings of International Symposium Atmospheric Reentry Vehicles and Systems, Arcachon, France, 1999

[19] C. Zuber et al: *Manufacturing of the CMC nose cap for the EXPERT spacecraft*, ICACC 2010, 24.- 29. Jan. 2010, Daytona Beach, USA, 2010.

[20] E. Bouillon et al: *Engine Test Experience and Characterization of Self Sealing Ceramic Matrix Composites for Nozzle Applications in Gas Turbine Engines*, Proceedings of ASME Turbo Expo 2003, Power for Land Sea and Air, Atlanta, USA, 2003

[21] L. Zawada et al: *Ceramic matrix composites for Aerospace Turbine Engine Exhaust Nozzles*, High Temperature Ceramic Matrix Composites 5, Eds: M. Singh et al, The American Ceramic Society, Westerville, Ohio, USA, 499 - 506, 2004.

[22] B. Heidenreich,; M. Scheiffele, M. Tausendfreund.; H.-U. Wieland: *C/C-SiC Telescope Structure for the Laser Communication Terminal in TerraSAR-X*, in: High Temperature Ceramic Matrix Composites, (Hrsg.: Krenkel, W.: Lamon, J.), Aviso Verlagsges., Berlin, 2010.

LASER DENSIFICATION OF POROUS ZrB_2 – SiC COMPOSITES

Q. Lonné, N. Glandut[*], and P. Lefort

SPCTS, UMR 6638, CNRS
University of Limoges
12 Rue d'Atlantis, 87068 Limoges, France
[*]nicolas.glandut@unilim.fr

ABSTRACT

Porous ZrB_2 – 30 vol.% SiC ceramic composites were irradiated under a mobile laser beam, and under pure argon. Dense layers were developed on top of the porous samples. The obtained morphologies and microstructures depend on the laser parameters: power density, scanning rate, and number of cycles. The treatment is possible even with traces of oxygen in the reactor. The volatility diagram of the system Si/C/O shows that, at the temperature reached under the laser beam (ca. 3000 K), oxygen forms the gaseous species SiO and does not oxidize the solid substrate. If working in ambient air, the behaviour is very different, with the formation of a surface layer rich in SiO_2. The obtained pellets could find applications in the fields of aerospace and ceramic fuel cells.

INTRODUCTION

Among ultra-high temperature ceramics (UHTCs), ZrB_2 – SiC composites have been widely studied. Indeed, they offer a very interesting set of properties, such as a good resistance to oxidation [1,2] and to thermal shocks [3], making them good candidates for very high temperature applications, particularly in the aerospace field [4]. These composites also have a good electronic conductivity [5] and, as we showed very recently, their oxidation provides a proton conductive surface glassy oxide, able to be used in low-temperature protonic ceramic fuel cells [6].

Nevertheless, their densification remains difficult because of their high melting point [7], whereas a high density is required for many of their applications. For this reason, a laser treatment, which is a well-known promising process for densifying the surface of porous UHTCs, could bring significant progress in this field [8-11].

However, to the best of our knowledge, no significant study has been devoted to the laser densification of porous ZrB_2 – SiC composites, but only to the laser densification of pure ZrB_2 [8] or SiC [11], or to the laser oxidation of dense ZrB_2 – SiC composites [8]. Consequently, it seemed interesting to test the effect of laser treatments on the surface of such ceramics, keeping in mind the potential applications in ceramic fuel cells and aerospace field.

EXPERIMENTAL

ZrB_2 – 30 vol.% SiC composites were obtained from ZrB_2 powder (Grade B, +97 % pure) and α-SiC powder (Grade UF-25, +98 % pure) purchased from H.C. Starck, USA.

Pre-sintering was achieved in a furnace equipped with a graphite resistor (Nabertherm, Germany), under argon (Alphagaz 1, Air Liquide, France). It required a 2.5-hour dwell at 1700 °C, with heating and cooling rates of 10 K min^{-1}, to obtain pre-sintered pellets with a relative density of 60 %. The obtained pellets (9.5 mm in diameter, 3 mm thick) were first polished up to a R_a of ca. 30 µm, then ultrasonically cleaned in acetone, and finally dried in air at 80 °C for 12 hours.

The final laser surface treatments were performed with an ytterbium-doped fibre laser LCF 100 (IPG, Oxford, MA, USA), which main characteristics have already been described in previous papers (continuous wave, TEM$_{00}$, wavelength of 1072 ± 10 nm, maximum power of 100 W) [10, 12]. A high-resolution motorized X-Y table, combined with an axes controller (Newport, CA, USA; models M-ILS100CC and ESP 300), allowed to treat a surface area of 6×6 mm^2.

Laser densification was achieved in a homemade cell, under flowing argon (same grade as above), with a flow rate of ca. 2.8×10^{-4} m^3 s^{-1}. The top of the cell was equipped with a window (CVI, France) made of zinc selenide (ZnSe) allowing the transmission of more than 99.75 % of the laser radiation [12]. The laser pattern used for densifying the surface of the composites under argon is described in Fig. 1. First, path 1→2 was realized and then, path 2→3 immediately overlapping path 1→2. The complete path 1→3 constituted a single cycle.

Microscopic observations were carried out using a Philips XL30 SEM in secondary electron (SE) or back-scattered electron (BSE) mode, with in-situ EDS (Oxford Instruments-INCA, UK). XRD was performed with a Siemens D5000 diffractometer, with a Cu anticathode and a back monochromator. The X-ray patterns were indexed with a DIFFRAC+ EVA software (Bruker AXS) containing the PDF database.

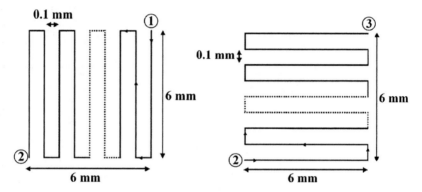

Figure 1 Laser pattern for the surface treatments.

RESULTS AND DISCUSSION

On the basis of two previous studies devoted to the surface densification of porous ZrC by a laser process [10], and to the purification of hot-pressed ZrCO into ZrC [12], surface densifications have been performed on porous ZrB$_2$ – SiC with a power of 90 W and beam diameter of 1 mm. It corresponds to an incident power density of 11.5 kW cm^{-2}. Indeed, power densities lower than 11.5 kW cm^{-2} never allowed to obtain a complete surface densification.

SEM observations, and EDS and XRD analyses of a laser-densified surface are shown on Fig. 2. The cross-section of Fig. 2(a) exhibits a regular, dense crust of ca. 20 μm thick on top of the porous base material. Top-views of the dense layer show two kinds of regions, called "granular" and "eutectic", that were predicted in the pseudo-binary eutectic phase diagram of the ZrB$_2$ – SiC system (Fig. 3) [7]. In Fig. 2(b) is shown a granular region, exhibiting sintered grains of ZrB$_2$ covered with a thin film containing small platelets. The BSE image of Fig. 2(c) represents another surface zone with the features of a lamellar eutectic. The brighter phase is attributed to ZrB$_2$ and the darker to SiC. This characteristic morphology evidences the fact that the sample surface melted under the laser beam.

EDS analyses of these two regions only detected the elements Zr, B, Si and C, and never oxygen, in spite of the presence of traces of this element in the argon used (Figs. 2(d) and (e)). The Si/Zr ratio is low in the granular region, which is consistent with the morphology of the large ZrB$_2$ grains covered with a very thin layer made of Si and/or C (Fig. 2(f)). In the eutectic region, the Si/Zr ratio is higher. This is consistent with the lamellar morphology of Fig. 2(c), where one can see ZrB$_2$ dendrites embedded in SiC. To finish, XRD of Fig. 2(g) identifies only the diffraction peaks of ZrB$_2$, SiC and free carbon on the surface of the samples. It is worth noticing that no carbon was initially detected in the starting powders or the pre-sintered pellets, which proves that its presence is a consequence of the laser treatment.

The power density being fixed at 11.5 kW cm^{-2}, we have studied the influence of two other parameters linked to the total energy received by the sample, i.e., the beam scanning rate and the number of cycles 1→3 (see above).

For the study on the influence of the laser scanning rate, the number of 1→3 cycles was set to one, and the tests were conducted at 0.5, 2 and 3 mm s^{-1}. The surface appears very rugged at the rate of 0.5 mm s^{-1}, with the presence of macropores and severe cracks; lower rates were not tested. Fig. 4(a) shows that the dense layer thickness decreases when the scanning rate increases.

Concerning the study on the influence of the number of cycles, the scanning rate was set to 10 mm s^{-1} because cycling with lower rates gives very rugged surfaces, leading sometimes to the delamination of the densified layer, or even to the pellet fracture. So, laser treatments were performed with 2, 3 or 5 successive cycles without pauses between the cycles. We show on Fig. 4(b) that the dense layer thickness decreases when the number of cycles increases.

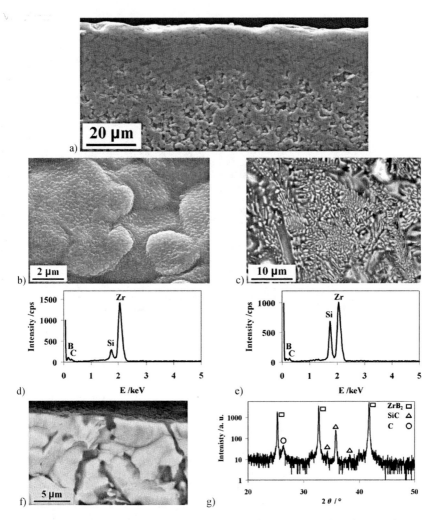

Figure 2 a) SEM observation of a polished cross-section of a laser treated sample under argon. b) Top-view SEM observation of a granular zone. c) Top-view SEM observation in BSE mode of a eutectic zone. d) and e) EDS analyses of a granular and a eutectic zone, respectively. f) Fractured cross-section in BSE mode. g) XRD analysis of the surface of a laser treated sample under argon. The logarithmic scale highlights the carbon peak.
Power density: 11.5 kW cm^{-2}.

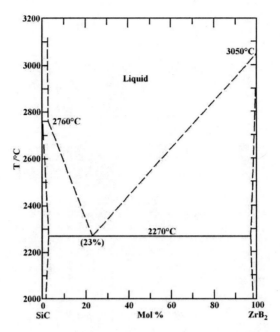

Figure 3 Pseudo-binary eutectic phase diagram ZrB₂ – SiC [7].

Table 1 summarizes the laser parameters used for this study: the number of cycles, the scanning rates, the exposure times, the total incident energies received by the samples, and the average, measured thicknesses of their dense layers. It highlights the fact that, for a single cycle, when the incident energy received by the samples increases (i.e., when the laser scanning rate decreases), the outer crust thickness logically also increases, due to a more important sintering. At the opposite, it is surprising to observe that for a fixed scanning rate, the thickness decreases when the number of cycles increases (i.e. when the total energy received by the sample increases). In fact, it is probable that the thermal conduction of the composite surface strongly changes after the formation of the outer crust, after the first laser passages. A dense/porous composite interface and ZrB₂/SiC interfaces without porosity are created, that might introduce additional thermal contact resistances. Under such conditions, when the laser beam passes again on a zone previously densified, the energy absorbed by the surface poorly diffuses inwards. The heat remains enclosed in the outer part of the crust, the temperature of which necessarily increases, up to an intense volatilisation of SiC, very likely, and less probably ZrB₂.

Another way of stating this is to say that the way the incident energy is provided to the composite surface has an influence on the final thickness of the top dense layer. For example, the samples treated with a single cycle at 2 mm s^{-1} and those treated with 5 cycles at 10 mm s^{-1} receive the same incident energies. However, there is a difference of more than 15 μm between the two thicknesses.

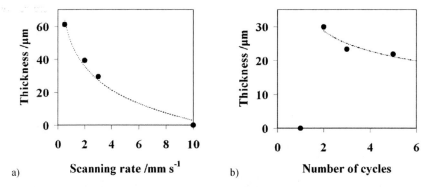

a)

b)

Figure 4 Variation of the thickness of the dense surface layer with: a) the scanning rate (one cycle) and b) the number of cycles (scanning rate of 10 mm s⁻¹). Power density: 11.5 kW cm⁻².

Table 1: Laser parameters used for the study of the surface densifications, with the corresponding exposure times, the total incident energies received by the samples and the average measured thicknesses of their dense layers.

Number of cycles	Scanning rate /mm s⁻¹	t /s	E /kJ	Dense layer thickness /μm ± 2 μm
1	0.5	1488	133.9	61
1	2	372	33.5	39
1	3	248	22.3	30
2	10	148.8	13.4	30
3	10	223.2	20.1	23
5	10	372	33.5	22

Another surprising point to explain is the presence of free carbon at the samples surface after laser treatment. The only possible source is the carbide SiC, which involves its partial conversion. Considering that SiC reacts easily with oxygen [1, 2], one could expect this last element to be found at the surface of the samples after the laser treatments. Indeed, traces of oxygen are present in the argon used and air leaks inside the cell are possible. Nevertheless, oxygen was never detected in the samples. Consequently, in a first approach, a thermodynamic study was carried out in order to understand this behaviour.

The volatility diagram of the SiC/O₂ system has been calculated and drawn in Fig. 5 for a temperature of 3000 K, on the basis of thermodynamic data [13]. This temperature was chosen by considering previous studies under similar conditions [10, 12]. It corresponds also to the present results since it allows the presence of a SiC-based liquid phase, but it remains lower than the melting point of ZrB₂ [7]. The partial pressure of oxygen, P_{O2}, was assumed to be that of the inlet argon, according to the impurities content given by the supplier (around 0.2 Pa; log P_{O2} = -0.7). Oxygen can react with SiC, giving gaseous silicon monoxide (SiO), which lowers P_{O2} to a very low value, calculated as equal to $10^{-7.41}$ Pa, on the basis of the following reaction:

$$SiC(s) + O_2 = SiO(g) + CO(g) \qquad (1)$$

and considering the corresponding equilibrium constant $K_1 = 10^{11.477}$. But, for P$_{O2}$ lower than $10^{-7.41}$ Pa, the oxidation of SiC becomes:

$$SiC(s) + \frac{1}{2}O_2 = Si(g) + CO(g) \tag{2}$$

with an equilibrium constant $K_2 = 10^{5.270}$. And when P$_{O2}$ falls under $10^{-10.076}$ Pa, the carbide decomposes, giving free carbon, according to the reaction:

$$SiC(s) \rightarrow Si(g) + C(s) \tag{3}$$

with an equilibrium constant $K_3 = 10^{-1.134}$. The volatility diagram shows also that the equilibrium partial pressures of gaseous SiO and Si are very high ($> 10^{3.866}$ Pa), which means that the traces of oxygen fall certainly quickly to very low values, allowing the quantitative decomposition of silicon carbide according to (3). This also justifies the presence of free carbon on the samples surface and of a thin film containing silicon, sometimes observed on the pellets (Fig. 2(f)). Indeed, Si (g) vaporises as explained above from the hot area under the laser beam, and probably condenses in the surrounding colder parts.

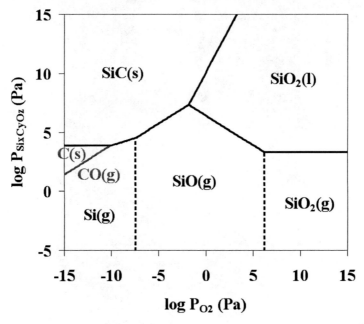

Figure 5 Volatility diagram of the SiC/O$_2$ system at 3000 K. Pressure of reference = 10^5 Pa.

The final structure of a ZrB_2 – SiC composite, after a laser densification in pure argon, with a power density of 11.5 kW cm^{-2}, is:

(i)   a very thin film on top, rich in Si, with traces of free carbon;
(ii)   a ZrB_2-rich, SiC-depleted, ~100 % dense layer containing granular and eutectic microstructures;
(iii) the original non-oxidized porous composite.

It is interesting to remark that a slightly different set of parameters can lead to a rather different morphology. For instance, a treatment under argon with a single 1→3 cycle at a scanning rate of 0.5 mm s^{-1}, with a power of 80 W and a beam diameter of 1 mm (10.2 kW cm^{-2}), leads to the structure of Fig. 6, that is:

(i)   a highly porous ZrB_2-rich, SiC-depleted layer on top;
(ii)   a dense ZrB_2-rich, SiC-depleted layer just underneath;
(iii) the original non-oxidized porous substrate;

characteristic of an insufficient sintering of the outer crust.

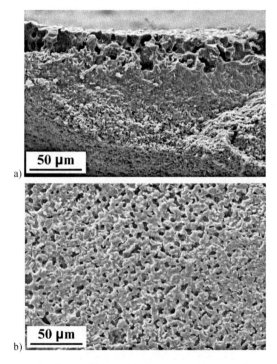

Figure 6 Micrographs of a sample treated by laser in pure argon with a power density of 10.2 kW cm^{-2}: a) fractured cross-section and b) top-view.

GENERAL DISCUSSION, CONCLUSIONS AND OUTLOOK

The laser irradiation of the surface of a porous composite ZrB$_2$ − 30 vol.% SiC can lead to a complete densification, the heat affected zone remaining limited to thicknesses of ca. 40 μm. This is a major difference with the densification in classical furnaces or with hot-pressing treatments (HP or HIP) where the whole piece (surface and bulk) is densified. This could be a drawback for some applications, but for other ones it is a determining advantage, such as for aerospace where it could permit to use pieces much lighter for the same surface behaviour, if the working part of the pieces is the surface.

Another major advantage of the laser treatment is its rapidity, since the pieces can be obtained in only few minutes, when the treatments in furnaces require several hours. Moreover, the energy consumption is by far less important: the treatment described in the present paper requires only ca. 1 kWh cm^{-2} whereas densifications in classical furnaces require much more energy. All this is interesting from a financial point of view, and in a logic of sustainable development. To finish, the cost for a laser device is cheaper than that of most classical sintering equipments.

However, this study shows that the laser densification of a ZrB$_2$ − SiC composite requires very precise conditions: a protective atmosphere, a proper power density, a suitable scanning rate, and only few laser passages.

This work opens also very interesting prospects in the field of reactivity of solids at very high temperatures. For instance, according to the volatility diagram of Fig. 5, liquid silica SiO$_2$ forms at 3000 K at the surface of a ZrB$_2$ − SiC composite, for P_{O2} of about 10^4 Pa. Such ZrB$_2$ − SiC composites, covered by an oxide layer mainly composed of silica has recently been studied [6]. It presented remarkable properties of proton transport at low temperature, making it a good candidate for a new kind of ceramic fuel cell.

On this basis, several tests have been carried out, by irradiating under laser the porous composite ZrB$_2$ − 30 vol.% SiC used in the present study, but by working in air. Fig. 7 presents the first results obtained after 1/2 cycle (path 1→2 of Fig. 1) with a laser power of 60 W and a beam diameter of 0.125 mm (power density of 4.9 kW cm^{-2}). On Fig. 7(a), the surface appears well densified, but with a facies rather different from that observed after densification under argon. There are no longer the "granular" and "eutectic" zones previously obtained with the protective argon atmosphere (compare with Fig. 2(b) and (c)). The cross-section of Fig. 7(b) shows that the surface is covered by a dense and thin layer (about 2 μm), identified as composed of SiO$_2$, while the inner composite seems to be not affected.

These results are very promising, because such features are close to those observed on samples oxidized in a classical furnace, which means that the pieces oxidized under laser beam can potentially be used in a ceramic fuel cell of new generation. A complete study on this topic, complementary to the present one, is in preparation [14].

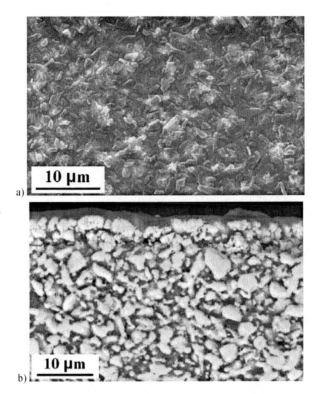

Figure 7 a) Top-view and b) polished cross-section in BSE mode of a sample oxidized by laser in ambient air with a power density of 4.9 kW cm^{-2} showing the thin outer SiO$_2$ layer.

ACKNOWLEDGMENTS

We gratefully acknowledge the *Région Limousin* for Q.L.'s Ph.D. scholarship.

REFERENCES

[1] A. Rezaie, W.G. Fahrenholtz, G.E. Hilmas, Evolution of structure during the oxidation of zirconium diboride-silicon carbide in air up to 1500 °C, J. Eur. Ceram. Soc., Vol 27, 2007, p 2495-24501

[2] W.M. Guo, G.-J. Zhang, Oxidation resistance and strength retention of ZrB$_2$–SiC ceramics, J. Eur. Ceram. Soc., Vol 30, 2010, p 2387–2395

[3] J.W. Zimmermann, G.E. Hilmas, W.G. Fahrenholtz, Thermal shock resistance of ZrB$_2$ and ZrB$_2$ – 30% SiC, Mater. Chem. Phys., Vol 112, 2008, p 140-145

[4] E. Wuchina, E. Opila, M. Opeka, W.G. Fahrenholtz, I. Talmy, UHTCs: Ultra-High Temperature Ceramic Materials for Extreme Environment Applications, Electrochem. Soc., Vol 16, 2007, p 30-36

[5] J.W. Zimmermann, G.E. Hilmas, W.G. Fahrenholtz, R.B. Dinwiddie, W.D. Porter, H. Wang, Thermophysical Properties of ZrB$_2$ and ZrB$_2$ – SiC Ceramics, J. Am. Ceram. Soc., Vol 91, 2008, p 1405-1411

[6] Q. Lonné, N. Glandut, J.-C. Labbe, P. Lefort, Fabrication and Characterization of ZrB$_2$ – SiC Ceramic Electrodes Coated with a Proton Conducting, SiO$_2$-Rich Glass Layer, Electrochim. Acta, Vol 56, 2011, 7212-7219

[7] A.E. McHale et al., Phase Equilibria Diagrams Volume X: Borides, Carbides and Nitrides, The American Ceramic Society, 1994, p 8

[8] D.D. Jayaseelan, H. Jackson, E. Eakins, P. Brown, W.E. Lee, Laser modified microstructures in ZrB$_2$, ZrB$_2$/SiC and ZrC, J. Eur. Ceram. Soc., Vol 30, 2010, p 2279-2288

[9] C.-N. Sun, M.C. Gupta, Laser Sintering of ZrB$_2$, J. Am. Ceram. Soc., Vol 91, 2008, p 1729-1731

[10] A. Bacciochini, N. Glandut, P. Lefort, Surface densification of porous ZrC by a laser process, J. Eur. Ceram. Soc. Vol 29, 2009, 1507-1511

[11] S. Gupta, P. Molian. Design of laser micromachined single crystal 6H-SiC diaphragms for high-temperature micro-electro-mechanical-system pressure sensors, Mater. Des., Vol 32, 2011, p 127-132

[12] F. Goutier, N. Glandut, P. Lefort, Purification of hot-pressed ZrCO into ZrC by a laser treatment, J. Mater. Sci., Vol 46, 2011, p 6794-6800

[13] M.W. Chase Jr., NIST-JANAF Thermochemical Tables, American Institute of Physics and American Chemical Society, 1998

[14] Q. Lonné, N. Glandut, P. Lefort, in preparation.

STRUCTURAL AND COMPOSITIONAL INVESTIGATIONS OF CERAMIC-METAL COMPOSITES PRODUCED BY REACTIVE METAL PENETRATION IN MOLTEN Al AND Al-Fe ALLOY

Anthony Yurcho
Department of Chemical Engineering, Youngstown State University
Youngstown, OH 44555, USA

Klaus-Markus Peters, Brian P. Hetzel
Fireline TCON, Inc.
Youngstown, OH 44505, USA

Raymond Brennan
U.S. Army Research Laboratory
Aberdeen Proving Ground, MD 21005, USA

Matthias Zeller, Timothy R. Wagner
Department of Chemistry, Youngstown State University
Youngstown, OH 44555, USA

Virgil C. Solomon
Department of Mechanical & Industrial Engineering, Youngstown State University
Youngstown, OH 44555, USA

ABSTRACT
Interpenetrating phase composite (IPC) materials consisting of a mixture of Al_2O_3 and various metal phases have been obtained by reactive metal penetration of vitreous silica (SiO_2) preforms with molten Al and an Al-Fe alloy. Morphology, chemical composition and crystalline structure of micro- and nano-sized ceramic and metal phases have been investigated using optical and analytical electron microscopy techniques and X-ray diffraction. It was determined that a three-dimensional network of Al_2O_3 and Al phases formed in the Al-based material, while Al_2O_3, Al, and Al-Fe alloy phases formed in the Al-Fe alloy-ceramic composite. In order to correlate microstructural and compositional findings with macrostructural properties, hardness testing of the composites was performed using the Vickers method.

INTRODUCTION
Interpenetrating phase composites (IPC's) are unique materials containing two or more interlocked ceramic and metallic phases that are both continuous throughout the microstructure.[1] Due to the fine intermixing of two or more phases with individual properties that are distinctly different from one another, it is expected to obtain IPC's with unique properties. In typical metal-ceramic IPCs, such as e.g. the Al_2O_3-Al-system investigated here, the ceramic phase, which encompasses usually around 70% of the composite,[2] provides its inherent high stiffness, its low density and its high strength to the composite, while the continuous metal network gives the ICP its high thermal and electrical conductivity, and a high fracture toughness. Such properties make these materials excellent candidates for replacing traditional materials in a number of applications, such as e.g. high wear/corrosion resistant refractory shapes for handling of molten metals, lightweight vehicle braking components, and high performance military body and vehicle armor.

Several processes have been proposed to manufacture ceramic-metallic IPC materials, including combustion synthesis,[3,4] pressurized melt infiltration,[5] and robotic deposition.[6] Each of these processes suffers from inefficiencies, such as size and shape restrictions, high product porosity, poor reproducibility, or excessive costs. Reactive Metal Penetration (RMP), which is sometimes referred to in literature as a liquid or a liquid-solid displacement reaction, is a viable solution to these processing issues.[1,7-9] RMP processing involves submerging a sacrificial ceramic preform into a molten metal bath for a given length of time and allowing chemical reactions to take place. The resulting product is an IPC that is net-shape or near net-shape according to the perform geometry. The dynamics of the RMP process make it a practical method for producing IPC's on an industrial scale.

The two materials investigated in this study were manufactured at Fireline TCON, Inc. of Youngstown, Ohio, using the TCON[®] RMP process, Figure 1. The first investigated material, used as a reference, was a basic unmodified Al_2O_3-Al-system. Al_2O_3-Al TCON material is produced by the reaction of a clear fused quartz preform with molten (~1200 °C) aluminum, as represented in Equation 1.

$$3SiO_2 + 4Al \rightarrow 2Al_2O_3 + 3Si \qquad (1)$$

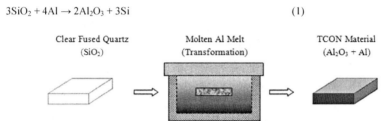

Figure 1. Schematics of the TCON reactive melt penetration process showing the transformation of a clear fused quartz preform into an IPC material.

The second material was prepared by immersion of a clear fused quartz preform into an aluminum-alloy bath containing 7.5 percent iron by weight. Information on Al-Fe-based IPC's prepared by RMP is quite limited, only one scientific report being published so far.[10] The purpose of this research is to provide a detailed microstructural and chemical analysis of the two TCON materials. An understanding of how IPC's micro- and nano-structure development and macroscopic properties are affected by alloying additions, such as Fe, can facilitate the development of composite materials tailored to specific applications.

EXPERIMENTAL PROCEDURE

Clear fused quartz preforms with dimensions 2 in. × 2 in. × 0.25 in. were reacted in two separate molten metal baths. The first was transformed in commercially pure (>99.9wt.%) aluminum for 6 hours, and the other in an aluminum alloy containing 7.5 percent iron by weight for 4.63 hours, Table I. The reaction temperature was set at 1200 °C for both materials. After the transformations were complete the samples were removed from the baths and allowed to cool to room temperature.

Samples for various analytical techniques were cut from the original materials using variable speed MARK V CS600-A and Buehler IsoMet® 1000 saws. For comparative purposes, the Al and Al-7.5wt.%Fe materials were sectioned in similar patterns. Figure 2(b) shows the samples' relative size and the area from which they were acquired. In response to the materials' unique three-dimensional microstructure, an x-y-z coordinate system was established to document results.

Table I. Transformation Conditions for TCON Samples

Name	Starting Material	Metal Bath	Transformation Time	Reaction Temperature
Al	Clear fused quartz (2 in. × 2 in. × 0.25 in)	Pure Al	6 hours	1200 °C
Al-7.5wt.%Fe	Clear fused quartz (2 in. × 2 in. × 0.25 in)	Al - 7.5 wt.% Fe alloy	4.63 hours	1200 °C

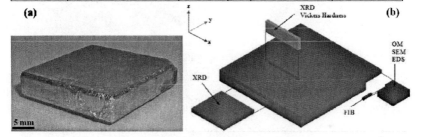

Figure 2. (a) Material synthesized by RMP. (b) Graphical representation of the transformed material showing the areas examined by microscopy, XRD, FIB and Vickers Hardness.

Macroscopic features were investigated using a Nikon SMZ800 stereo optical microscope (OM) and a ZEISS Axiophot compound light microscope for bright field and polarized light microscopy. A JEOL JIB 4500 Multi-beam FIB/SEM System was utilized for scanning electron microscopy analysis and transmission electron microscopy (TEM) sample preparation. The instrument is equipped with both an EDAX™ Apollo SSD EDS detector used to collect X-ray energy dispersive spectroscopy (XEDS) data, and an Omniprobe™ OMP-AUTOPROBE 200.1 nanomanipulator for TEM sample removal and mounting. A JEOL JEM-2100 Scanning/Transmission Electron Microscope operated at 200 kV and equipped with an EDAX™ Apollo XLT SSD energy dispersive spectrometer was used for TEM analysis. Powder X-ray diffraction (XRD) patterns were collected using a Bruker D8 Advance diffractometer with Cu-K_α radiation. The data were collected at room temperature in reflective mode. The data were analyzed and fitted to database patterns using the EVA Application 7.001 software of SOCABIM (1996-2001), distributed by Bruker AXS.

Vickers hardness testing was performed using an indentation hardness tester marketed by Wilson Hardness on the Al and Al-7.5wt.%Fe samples according to ASTM standard C1327 – 08.[11] The machine was calibrated using a standard reference block prior to experimentation, and the samples were cleaned with alcohol to remove surface contaminations. A test load of 1 kgf was used to make 25 indentations in each sample, Fig. 3. Testing was performed at room temperature. Vickers hardness numbers (HV) were computed according to Eq. 2, where P is the load (kgf) and d is the average length (mm) of the diagonals d_1 and d_2.

$$HV = 1.8544(P/d^2) \hspace{4cm} (2)$$

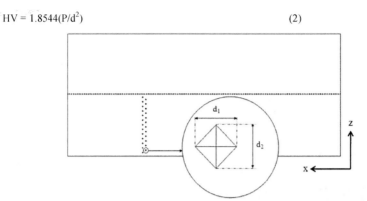

Figure 3. Illustration demonstrating the method used to measure Vickers hardness values for the Al and Al-7.5wt.%Fe samples. Two columns of indentations were made beginning at the edge of the sample and working towards the center. The distance between consecutive indentations in the z-direction is equidistant. The length of each diagonal, d_1 and d_2, was measured in the SEM.

RESULTS AND DISCUSSION

Macroscopic observations

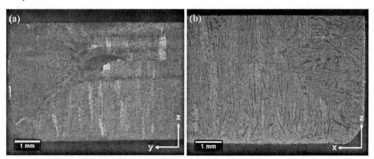

Figure 4. Stereo-OM micrographs showing the polished cross-sections of the Al (a) and Al-7.5wt%Fe (b) samples reveal a Y-shape feature symmetrically positioned relative to the corners and edges of the samples. The symmetry is partially broken due to the fact that the top surface was polished for microscopy investigation. Striations found throughout the bulk of the Al-7.5wt.%Fe sample are not present in the Al sample.

Stereo optical micrographs revealing macroscopic features of the Al and Al-7.5wt.%Fe samples are shown in Fig. 4. Initial observation of the polished cross-sections reveals that both samples contain a distinct Y-shaped boundary feature that is positioned symmetrical relative to the edges and corners of the sample. Macroscopic observation also shows striations throughout the Al-7.5wt.%Fe sample that are absent in the Al sample. In the stereo microscope, the striations appear as dark lines that are generally orientated parallel to each other. The widths of the striations range between 5 and 100 μm and they can reach several hundred microns in length. Other interesting features are bright columns that form perpendicular to the outer edges. The columns can be observed by manipulating the light source of the stereo microscope. By varying the incidence angle of the light rays on the sample surface the contrast of favorably oriented columnar features can be enhanced, and therefore the columns can be readily observed. The columns have different widths, typically on the order of hundreds of microns, and are between 0.5 and 3 mm in length. The directions of both the dark striations and bright columns intersect at the Y-shaped boundary. Since the clear fused quartz preform was homogeneous, it is reasonable to assume that these features are a result of RMP processing.

X-ray diffraction

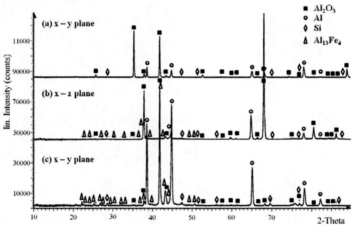

Figure 5. Powder XRD patterns from the Al (A) and Al-7.5wt.%Fe (B) samples. The Al sample was examined in the x-y plane and the Al-7.5wt.%Fe sample was examined in the x-y (bottom) and x-z (top) planes. Both samples show preferential orientation of Al_2O_3 (■ ICDD: 00-010-0173, rhombohedral R$\bar{3}$c, a: 4.758Å, c: 12.991Å)[13] and Al (o ICDD: 00-004-0787, cubic Fm$\bar{3}$m, a: 4.049Å)[14]. The data also show the presence of small amounts of Si (◇ ICDD: 04-001-7247, cubic Fd$\bar{3}$m, a: 5.429Å)[15] in both samples. The pattern suggests the presence of $Al_{13}Fe_4$ (△ICDD: 00-050-0797, orthorhombic Bmmm, a: 7.751Å, b: 23.771Å, c: 4.034Å)[12] in the Al-7.5wt.%Fe sample, but other Al-Fe phases showed similarly well fitting patterns and further analysis is needed for an unambiguous assignment of the Al-Fe binary phase.

Powder XRD patterns from the x-y plane of the Al sample and the x-y and x-z planes of the Al-7.5wt.%Fe sample are shown in Fig. 5. XRD confirmed the presence of Al_2O_3, Al, and Si in both samples. In the Al-7.5wt.%Fe sample, additional diffraction peaks point to the existence of Al-Fe phases, but determination of the exact phase or phases is difficult based on the diffraction pattern alone. An $Al_{13}Fe_4$ (ICDD entry number 00-050-0797)[12] phase has tentatively been selected as the best fit, but other Al-Fe-phases with different compositions showed similar agreement with experimental peaks. XRD did not show any ternary phase peaks. The mismatch between experimental and theoretical peak intensity for the Al_2O_3 and, to a lesser extent, for the Al phase can be explained by preferential orientation of the phases (bright columns in Fig. 4). For example, the c-axis of Al_2O_3 appears to be aligned along the z-direction for both samples. It must be also noted that the degree of disparity in peak intensity of Al_2O_3 is different depending from which plane the diffraction pattern was taken, which further supports the claim of preferential orientation as the cause for the intensity discrepancy. In the case of the Al sample, the Al_2O_3 peaks are well defined but the intensity mismatch is obvious for almost all of the peaks. No preferential orientation of Si was detected in either sample.

Microscopic observations

To better understand the structural properties of the materials, the bright columns observed by optical microscopy were further investigated using SEM. Figure 6 shows several regions that correspond to these columns in the Al sample. The columns' unique reflective properties can be attributed to different orientations of the Al_2O_3 phase, which correlates with the XRD findings. It can be assumed that the RMP process produces Al_2O_3 colonies that are characterized by preferential orientation of their grains. The Al_2O_3 phases stretch mainly in the direction perpendicular to the sample edges and also exhibit some branching. From the two-dimensional micrographs, it is difficult to determine if the early termination of some colonies is due to the intersection with other colonies or if the path continues along a three-dimensional vector. 2D observation of the 3D structure also limits the ability to determine the colony dimensions and the degree by which they vary.

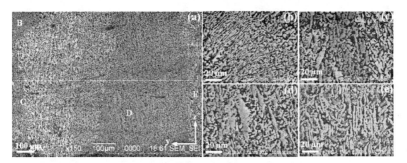

Figure 6. (a) Secondary electron micrographs of the alternating Al_2O_3 columnar-shaped regions in the Al-transformed sample. (b)–(e) Higher magnification micrographs from areas marked in (a).

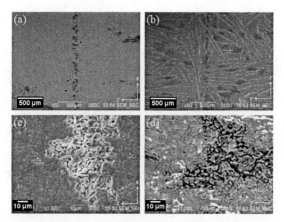

Figure 7. Backscatter electron micrographs showing the series of voids along the centrally positioned boundary in the (a) Al and (b) Al-7.5wt.%Fe samples. (c), (d) The boundary morphology is readily visible in higher magnification micrographs recorded from Al and Al-7.5wt.%Fe samples, respectively.

Scanning electron microscopy was also utilized to investigate the Y-shaped boundary originally observed by stereo-OM. In both samples, the feature is quite uniform in areas away from the sample corners. It is primarily formed by the interface among Al_2O_3 colonies meeting at the intersection of reaction fronts. The feature consists of a network of interconnected voids, as can be seen from the micrographs in Fig. 7.

The formation of the centrally-positioned voids can be explained in relationship with the transformation process during reactive melt penetration and growth of preferentially oriented Al_2O_3 regions. By immersing the preform into molten metal, SiO_2 transforms into Al_2O_3 as described in Equation (1). The transformation proceeds from exterior toward the interior of the sample following the mechanism described by Liu and Koster.[16] Both the Al_2O_3 ceramic and metal phases grow from the exterior toward the interior of the sample. Due to the sample symmetry and the identical transformation velocity (about 2 mm/hr) at any point on the sample surface, the reaction fronts from two opposite surfaces will meet at the center of the sample, where pore accumulation is observed. From these observations, it appears that a direct relationship exists between the interaction of two reaction fronts and pore accumulation,[9] but the exact mechanism of porosity formation needs further clarification.

Micrographs from ultra fine polishing of the Al sample, accomplished by ion milling, are shown in Fig. 8. The FIB polish exposed fine details that could not be observed with mechanical polishing, including precipitated Si particles, micro-scale cracking, and grain boundaries. The secondary electron micrograph shown in Fig. 8(a) clearly indicates the coexistence of ceramic (Al_2O_3) and metal (Al) interpenetrating phases. The presence of elongated Si particles within the Al metal network is demonstrated in Fig. 8(b). XEDS analysis was used to confirm the elemental composition of the above-mentioned phases in the investigated samples. Figure 8(c) is a FIB secondary electron micrograph that has been slightly tilted to induce ion channeling effects.[17] The contrast achieved by tilting the sample reveals grain boundaries in the aluminum phase.

Grain boundaries are not observed in the alumina phase, which signifies that the alumina grains may be continuous or at the very least are larger than the aluminum grains.

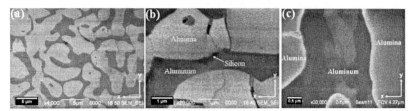

Figure 8. (a) Secondary electron micrograph showing the interpenetrating networks of Al_2O_3 and Al, in the Al-transformed sample. (b) Electron micrograph of an area containing Si particles embedded into the Al matrix. (c) FIB secondary electron micrograph of the FIB polished Al sample tilted to induce ion channeling. Ion channeling exposed submicron-sized grains in the Al phase.

In the Al-7.5wt.%Fe sample, beside the interpenetrating networks of Al_2O_3 and Al, stripe-like features can be observed in both optical and electron micrographs, Figs. 4(b) and 7(b). Figure 9(a) shows the typical morphology of a stripe-like feature, while Fig. 9(b) shows the matrix (an area devoid of 'stripes') morphology of an Al-7.5wt.%Fe-transformed sample. The micrograph in Fig. 9(b) reveals two interpenetrating co-continuous networks of Al metal and Al_2O_3 ceramic.

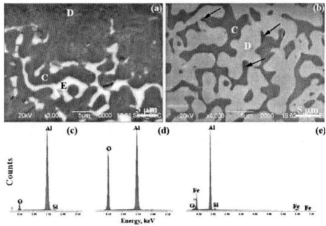

Figure 9. (a), (b) High resolution SEM micrographs of a sample 2 cross-section taken from a stripe-like feature and from the matrix, respectively. (c) – (e) EDX spectra from volumes C, D and E in (a).

Using spot EDS analysis it was determined that the dark gray areas (C) correspond to Al metal, the light gray areas (D) correspond to Al_2O_3 ceramic, while the bright (E) areas correspond to an Fe-Al intermetallic phase. Small particles of precipitated silicon were also observed, as indicated by black arrows. The silicon particles are an expected by-product from the reaction. It is understood that the displaced Si from the SiO_2 preform diffuses away from the reaction front through the aluminum channels.[8] It should be noted that no iron was observed in the investigated volume illustrated in Fig. 9(b).

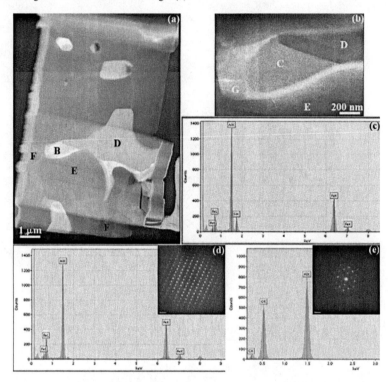

Figure 10. (a) Dark-field scanning-transmission electron micrograph from a striated (Fe-rich) volume of sample 2. (b) High magnification dark-field scanning-transmission electron micrograph of area B in (a). (c) EDS spectra of area C in (b). (d), (e) Electron diffraction patterns and EDS spectra from an Al-Fe phase (area D in (a) and (b)), and Al_2O_3 phase (area E in (a) and (b)), respectively. F denotes pure Al regions.

The XEDS analysis of areas identical to that presented in Fig. 9(a) indicate the presence of Fe-rich phases aside of the expected aluminum, alumina, and precipitated silicon phases. The bright stripes observed in the Al-7.5wt.%Fe material are an Fe-Al phase surrounded by small amounts of Fe-Al-Si and Fe-Al-Si-O phases. The existence of nano-scale sized Fe-Al-Si and Fe-Al-Si-O phases was confirmed by TEM analysis, Fig. 10. Based on the electron microscopy analysis, it is concluded that the plate-like stripes observed in the Al-Fe-transformed sample consist of two phases: Fe-Al and Fe-Al-Si, while the matrix consists of two networks, Al metal and Al_2O_3 ceramic, as reported above.

Examination of the silicon distribution in the XEDS maps shows that silicon appears as concisely defined particles in the ironless regions. Whereas, in the iron-rich regions, the silicon is more dispersed and individual particles are not observed by SEM investigation. This suggests that the iron may have reacted with diffused silicon to form nano-scale phases. Other studies have shown that Fe, Al and Si can form intermetallics such as α-Al_8Fe_2Si, β-Al_5FeSi, or δ-Al_3FeSi_2.[18]

Figure 10 shows the results of an S/TEM investigation of an Fe-rich volume of the Al-Fe-transformed sample. Four different micrometer-sized phases plus several nanoscale features are observed in the dark field STEM micrograph in Fig. 10(a). By combining the results of EDS analysis with the information provided by XRD investigation the micron-sized volumes were identified as an Al-Fe-Si phase (C in Fig. 10(b)), an Al-Fe intermetallic (D in Figs. 10(a), (b)), Al_2O_3 ceramic (E) and Al (F). A nano-scale transition layer (G) is readily observed at the ceramic-metal boundary, Fig. 10(b). Based on EDS analysis the chemical composition of the transition layer was identified as Al-Fe-Si-O. The formation of nano-scale phases may contribute to the superior flexural strength of the Al-alloy-Al_2O_3 composite, as reported by Hemrick et al.[7] Future work involving high resolution TEM combined with electron crystallography will aim to determine the exact composition and structure of these phases.

The results of the Vickers hardness testing are displayed in Table II. The average hardness of the Al and Al-7.5wt.%Fe samples is 468.2 HV and 485.5 HV, respectively. It seems that the Al-7.5wt.%Fe-transformed material have a slightly higher hardness due to the iron rich phases. However, the large standard deviations, 41.2 for Al and 63.2 for Al-7.5wt.%Fe, make the averages statistically indistinguishable. The hardness is an average of the combined phases since the indentation is much larger than the individual grains. There was no attempt to include or exclude the iron-rich regions in the Al-7.5wt.%Fe sample; therefore, the hardness value represents random sampling. Figure 11 plots the hardness values versus their distance from the outer edge. The graph indicates that the materials have a slightly lower hardness less than 1 mm from the edge, while the bulk of the sample contains a more uniform distribution. A common flaw during mechanical polishing is a slight rounding of the sample edges. If this occurred, the resulting uneven surface can explain the decrease in hardness near the edges.

Table II. Vickers Indentation Hardness Results (HV)

	Al	Al-7.5wt.%Fe
Mean	468.2	485.5
Standard Deviation	41.2	63.2
Minimum	397.2	332.5
Maximum	538.1	573.7
Satisfactory Indentations	21	22
Total Indentations	25	25

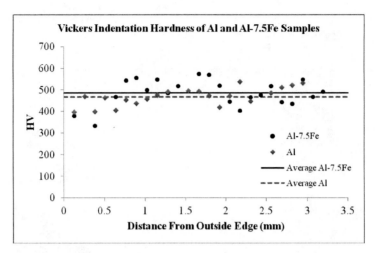

Figure 11. Graphical representation of the HV measurements taken from the Al and Al-7.5wt.%Fe samples.

CONCLUSIONS

Morphology, chemical composition, crystalline structure and hardness of interpenetrating phase composites (IPC's) produced by reactive melt penetration in molten Al and in Al-7.5wt.%Fe alloy were investigated using optical and analytical electron microscopy techniques, X-ray diffraction and the Vickers hardness testing method. It was determined that a three-dimensional interpenetrating network of Al_2O_3 and Al phases formed in the Al-based material, while in Al-Fe alloy-ceramic composite an interpenetrating network of Al_2O_3, Al, and intermetallic Al-Fe and Al-Fe-Si phases were observed. Based on the X-ray diffraction analysis an $Al_{13}Fe_4$ phase has tentatively been selected as the best fit for the observed Al-Fe binary phase, but other Al-Fe-phases with different compositions showed similar agreement with experimental peaks. A nanoscale sized Al-Fe-Si ternary phase, observed in high resolution electron micrographs, showed no peaks in the X-ray diffraction spectra. Further electron diffraction experiments are necessary in order to clarify the crystalline structure of the two intermetallic phases. Dispersed sub-micron sized Si particles were observed throughout the Al-transformed material and in the iron-less regions of Al-Fe-ceramic composite. The Al-Fe-ceramic composite was slightly harder than Al-ceramic, perhaps due to the reinforcement caused by the presence of intermetallic Al-Fe network. This experiment proves that the microstructure and macroscopic properties of IPC's produced by reactive melt penetration can be tailored by altering the molten metal bath composition.

REFERENCES

[1]R. Paul, Microstructural and Chemical Characterization of Interpenetrating Phase Composites as Unique Refractory Materials Produced Via Reactive Metal Penetration, Master's Thesis, Youngstown State University, Youngstown, Ohio, U.S.A. (2007).

[2]G.M. La Vecchia, C. Badini, D. Puppo, and F. D'Errico, Co-continuous Al/Al$_2$O$_3$ composite produced by liquid displacement reaction: Relationship between microstructure and mechanical behavior, *J. Mater. Sci.* **38**, 3567-3577 (2003).

[3]H.J. Feng, J. and Moore, In situ combustion synthesis of dense ceramic and ceramic-metal interpenetrating phase composites. *Metallurgical and Materials Transactions B*, **26(2)**, 265-273 (1995).

[4]Q. Hu, P. Luo, and Y. Yan, Influence of an electric field on combustion synthesis process and microstructures of TiC-Al2O3-Al composites, *Journal of Alloys and Compounds*, **439**, 132-136 (2007).

[5]A. Mattern, B. Huchler, D. Staudenecker, R. Oberacker, A. Nagel, and M.J. Hoffmann, Preparation of interpenetrating ceramic-metallic composites, *Journal of European Ceramic Society*, **24**, 3399-3408 (2004).

[6]C. San Marchi, M. Kouzeli, R. Rao, J.A. Lewis, and D.C. Dunand, Alumina-aluminum interpenetrating-phase composites with three-dimensional periodic architecture, *Scripta Materialia*, **49(9)**, 861-866 (2003).

[7]J.C. Hemrick, M.Z. Hu, K.M. Peters, and B. Hetzel, Nano-Scale Interpenetrating Phase Composites (IPC's) for Industrial and Vehicle Application, (ORNL/TM-2010/80) Oak Ridge, Tennessee, U.S.A. (2010).

[8]M.C. Breslin, J. Ringnalda, L. Xu, M. Fuller, J. Seeger, G.S. Daehn, T. Otani, and H.L. Fraser, Processing, microstructure, and properties of co-continuous alumina-aluminum composites, *Materials Science and Engineering*, **A195**, 113-119 (1995).

[9]V.S.R. Murthy, K. Kawahara, Y. Saito, T. Matsuzaki, and T. Watanabe, Orientation and Grain Boundary Microstructure of Alumina in Al/Al$_2$O$_3$ Composites Produced by Reactive Metal Penetration, *Journal of the American Ceramic Society*, **88(10)**, 2902-2907 (2005).

[10]N. Yoshikawa, A. Hattori, and S. Taniguchi, Growth rates and microstructure of reacted layers between molten Al-Fe alloy and SiO$_2$, *Materials Science and Engineering*, **342(A)**, 51-57 (2003).

[11]ASTM Standard C1327-08 (2008). Standard Test Method for Vickers Indentation Hardness of Advanced Ceramics. (DOI: 10.1520/C1327-08) ASTM International, West Conshohocken, PA.

[12]Aluminum Iron Powder Diffraction File; International Centre for Diffraction Data: Newtown, PA, 2011; PDF# 00-050-0797 (accessed 10 Feb 2011).

[13]Aluminum Oxide Powder Diffraction File; International Centre for Diffraction Data: Newtown, PA, 2011; PDF# 00-010-0173 (accessed 10 Feb 2011).

[14]Aluminum Powder Diffraction File; International Centre for Diffraction Data: Newtown, PA, 2011; PDF# 00-004-0787 (accessed 10 Feb 2011).

[15]Silicon Diffraction File; International Centre for Diffraction Data: Newtown, PA, 2011; PDF# 00-010-0173 (accessed 10 Feb 2011).

[16]W. Liu, and U. Koster, Criteria for formation of interpenetrating oxide/metal-composites by immersing sacrificial oxide performs in molten metals, *Scripta Materialia*, **35(1)**, 35-40 (1996).

[17]L.A. Giannuzzi, B. I. Prenitzer and B.W. Kempshall, Ion-Solid Interaction, in Introduction to Focused Ion Beams, Instrumentation, Theory, Techniques and Practice, Eds. L.A. Giannuzzi, F. A. Stevie, Springer Science+Business Media, Inc., 13-52 (2010).

[18]M. Timpel, N. Wanderka, B.S. Murty, and J. Banhart, Three-dimensional visualization of the microstructure development of Sr-modified Al-15Si casting alloy using FIB-EsB tomography, *Acta Materialia*, **58**, 6600-6608 (2010).

MANUFACTURE AND MECHANICAL CHARACTERIZATION OF POLYMER-COMPOSITES REINFORCED WITH NATURAL FIBERS

Enrique Rocha-Rangel, J. Ernesto Benavides-Hernández, José A. Rodríguez-García
Universidad Politécnica de Victoria, Avenida Nuevas Tecnologías 5902
Parque Científico y Tecnológico de Tamaulipas, Tamaulipas, México, 87138

Alejandro Altamirano-Torres, Y. Gabriela Torres-Hernández, Francisco Sandoval-Pérez
Departamento de Materiales, Universidad Autónoma Metropolitana
Av. San Pablo # 180, Col Reynosa-Tamaulipas, México, D. F., 02200

ABSTRACT

This work shows the experimental results obtained from the mechanical characterization of a biopolymer-composite reinforced with coconut fibers. Firstly, coconut fibers were dried at 80 °C during 2, 3, 4, 5 and 6 h, in order to know the best dried up condition for their best performance under loads action. Results of tensile test realized on fibers, show that sample dried during 3 h, was the sample that reached maximum failure stress of 76.15 MPa. For longer times, this property decrease significantly (4 h = 18.13 MPa, 5 h = 20.31 MPa and 6 h = 19.10 MPa). Specimens of biopolymer-composite reinforced with coconut fibers were fabricated by extrusion techniques using the temperature range between 140 °C and 160 °C. Extrusion temperature demonstrated does not have any important effect on the tensile resistance of the extruded composite. However it was observed rough surface of material when samples are extruded at highest temperature (160 °C). Consequently, the best determined conditions for the fabrication of the composite material were; dried during 3 h, 140 °C extrusion temperature and coconut fiber length of 0.1 – 0.3 cm.

INTRODUCTION

There are several studies about the properties of composites materials epoxy resin-based reinforced with synthetic fibers such as: glass and carbon[1-7], these materials have many uses in the construction, transport and navy industries. However, there are not numerous studies that spook the concerning to composites materials reinforced with natural fibers. On the other hand, persistence of plastics in the environment, the shortage of landfill space, the reduction of petroleum resources, concerns over emissions during incineration, and entrapment by and ingestion of packing plastics by fish, chicken and animals have spurred efforts to develop biodegradable plastics materials. The potential use of natural fibers lie mainly in their excellent mechanical properties, dimensional stability, low weight, high availability as well as economic and especially ecological advantages than they can offer in different applications.

Natural fibers can be able to win the market that currently is dominate by the synthetic fibers, or others materials obtained through nonrenewable resources, these markets include the use of composites for applications as: insulation, packaging, filters, geotextiles and adsorbents[8]. Actually, synthetic polymers are obtained from petroleum derived products (non-renewable resources), consequently they are not degradables. For that reason, it is necessary the study of new formulations of composite materials based on biodegradable polymers[9] and reinforced with natural fibers, in order to obtain composites that could have similar or better mechanical properties than synthetic polymeric composites but with less negative environmental impact[10]. Production methods of biopolymer-composite reinforced with natural fibers are not well established and clearly exposed. For this reason the present study is realized with the intention to propose a processing route of composites resultants from the combination of biopolymer PLA 2002-D and coconut fibers.

EXPERIMENTAL PROCEDURE

Materials used for the manufacture of the desired composites were: pellets of the biopolymer PLA 2002-D (Polylactic acid, Promaplas, S.A de C.V. Mexico) extrusion grade. As reinforced material it was used natural coconut fibers. Coconut fibers were previously dried up in an oven at 80 °C during different times (1, 2, 3, 4, 5 and 6 h) in order to know the best dried conditions of the fibers. Starting materials, were extruded by separated in order to obtain solid pipes, from they were obtained specimens for flexure and tension resistance tests according to ASTM standards[11, 12]. In this step it was used the following extrusion temperatures: 140, 145, 150, 155 and 160 °C. Subsequently, tensile tests were performed at room temperature. All these studies were performed on a universal testing machine (United, model: SSTM-1, USA). They were testing by separated 5 specimens of biopolymer and 5 specimens of the coconut fibers. The tests were carried out with a cross–heat speed of 0.2 in/min, in order to determine their mechanical properties and know the better conditions for the manufacturing of the composite material. Once the better conditions of extrusion of the biopolymer and dried coconut fibers were fixed, it was manufactured the composite materials by mechanical mixing the biopolymer with the followings percentages of coconut fibers: 0, 3 and 6 wt. %. Coconut fibers were previously cut at lengths of 0.1–0.3 cm. Mixed time was 5 minutes, at 30 rpm. The obtained composites were tested in flexure and tensile conditions (5 specimens of each composition were evaluated).

RESULTS AND DISCUSSION

Fibers characterization

Figure 1 (a) shows optical pictures of the superficial texture of coconut fiber dried at 80 °C during 3 h. In this figure it is observed high roughness in the fiber surface than can helps to have good quality adhesion between the biopolymer matrix and the fiber, situation that can improve mechanical properties of the final composite. The figure 2 (b) illustrate a transverse fracture in the coconut fiber, also, in the picture is observed that the fiber is composed by various micro-fibers.

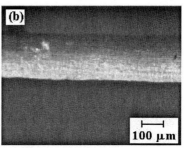

Figure 1. Optical microscopy pictures: a) Superficial texture,
b) The fracture type of coconut fiber after of the tensile test.

Figure 2 illustrates the experimental results obtained from the tensile test made in coconut fibers, as a function of the dried time at 80 °C. In This figure it is observed that there is not a significant variation in the tensile strength in the manner that is increasing the dried time. The specimen without dried treatment present the major tensile strength (84.37 MPa), but has the inconvenient that present high percentage of damp, therefore, the adhesion between the biopolymer matrix and the coconut fiber cannot be good. On the other hand, this damp increase the presence of internal porosity due to that the water contained in the fiber will produce steam during the manufacturing of the composite material.

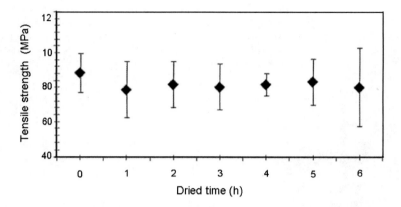

Figure 2. Tensile strength of coconut fibers dried at 80 °C, as a function of dried time.

Biopolymer characterization

In Figure 3 is presented the experimental results obtained from the tensile test made in the biopolymer as a function of extruded temperature. Results of figure 3 show that extrusion temperatures have not an important influence on the tensile strength of the biopolymer. Therefore, it is select for preparing the composite material the extrusion temperature of 140 °C that is the condition in where it is obtained the best tensile strength (52.00 MPa) in the biopolymer.

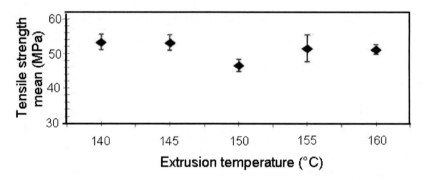

Figure 3. Tensile strength of biopolymer (PLA 2002-D), as a function of extrusion temperature.

Figures 4 and 5 show experimental results obtaining from flexure and tensile tests respectively, made in composite materials as function of coconut fibers wt. %. In both figures it can be observed that composite with 3 wt. % of fibers is the one that shows the best flexure and tensile strength performance. In opposite way the composite with 6 wt. % of coconut fibers present the worst mechanical behavior. Probably this performance is because fibers don not have an arrangement and they are distributed randomly in the material, so high fibers content (6 wt. %) does not get good adhesion with the polymeric matrix and therefore, the stress transmission through fibers is not achieved when the material is under loads action. Another, important observation in the behavior of composite under the action of loads, is that the flexion resistance of the composite in higher than the tension resistance, these behavior can have its explanation in the fact that when composites are worked under the action of flexion stresses, fibers were breaking little by little transmitting between them the stresses very well, in the opposite way, when composites are worked under the action of tension loads, fiber cannot transmitting the stresses and they break when it is reached the maximum stress of each one.

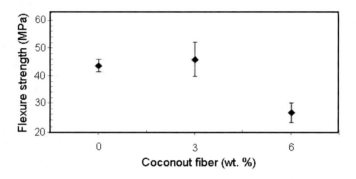

Figure 4. Flexion strength of composite material as a function of wt. % of coconut fibers.

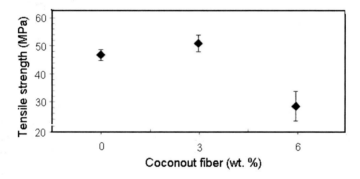

Figure 5. Tensile strength of composite material as a function of % wt of coconut fibers.

Figures 6a and 6b are photographs showing the fracture surfaces of the composites prepared with 3% and 6 wt. % of coconut fiber respectively. Figure 6a shows a good hold between the fibers and the polymer, whereas in figure 6b it is observed the opposite. This is the reason that the material with 3 wt. % of coconut fibers presents better mechanical properties.

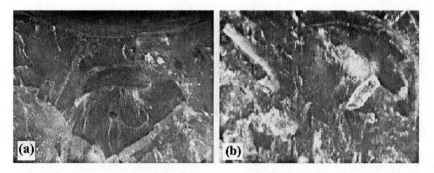

Figure 6. Fracture surfaces of the composites. (a) and (b) prepared with 3 % and 6 wt. % of coconut fiber respectively.

CONCLUSIONS

o   Polymer-composites reinforced with natural fibers were successfully fabricated by the extrusion technique.

o   The results obtained of the mechanical characterization (tensile strength), for the biopolymer and the coconut fibers, shows that the better conditions for the manufacturing of the composite material are: Extrusion temperature of 140 °C and fibers dried time of 3 h.

o   Contents equal to 3 wt. % of coconut fibers in a biopolymer matrix results in better mechanical resistance of the polymeric composite.

o   Composites have better mechanical performance when they work under flexion loads than when they work under tension loads.

ACKNOWLEDGMENT
        Authors would thank PROMEP-SEP for financial support as well as Universidad Politécnica de Victoria and Universidad Autónoma Metropolitana for technical support.

REFERENCES
        1. S.E. Buck, D.W. Lischer y S. Nemat-Nasser, Material Science Engineering, A317, 128-134, (2001).
        2. J.C. Gerdeen, H.W. Lord and R.A.L. Rorrer, Engineering Design with Poymers and Composites, Ed. CRS Press, Taylor & Francis Group, USA (2006).
        3. D. K. Singh, Fundamentals of Manufacturing Engineering, Ed. CRS Press, Taylor & Francis Group, USA (2008).
        4. Sanjay K. Mazumdar, Composites manufacturing; Materials, Product and Process Engineering, CRC Press, USA (2002).

5. E. Rocha-Rangel, J.A. Rodríguez-García, E. Martínez-Peña, E. Refugio-García, A. Leal-Cruz and G. Munive, Estudio de la Resistencia Mecánica de Materiales Compuestos Poliméricos Reforzados con Fibras de Carbono, Journal Avances en Ciencias e Ingeniería, **2**, 4, (2011). In spanish

6. H. Hongwei, L. Kaixi, W. Jianlong, W. Jian, G. Jianyu and L. Renjie, Effects of novolac resin modification on mechanical properties of carbon fiber/epoxy composites. Polymer Composites, **32** (2), 227-235 (2011).

7. D. Kopeliovich, Carbon Fiber Reinforced Polymer Composites, NCHRP Report, http://www.substech.com/dokuwiki/doku.php?id=carbon_fiber_reinforced_polymer_composit es. (2009).

8. T.J. Madera Santana, M. Aguilar Vega and F. Vázquez Moreno, Potencial de las fibras naturales para su uso industrial, Ciencia Ergo Sum, Toluca Estado de México, november (2000). In spanish.

9. "Biodegradable Polymers", Technical Report, Environment and Plastics Industry Council, november 24, (2000).

10. P. Mutjé and M. F. Llop, Desarrollo de materiales plásticos reforzados con fibras lignocelulosicas", Iberoamerican Congress on Pulp and Paper Research (2002). In spanish.

11. ASTM D638; Tensile properties of plastics (2010).

12. ASTM-D790; Standard Test Methods for Flexural Properties of Unreinforced and Reinforced Plastics and Electrical Insulating Materials (2010).

Foreign Object Damage

EFFECTS OF THE MODE OF TARGET SUPPORTS ON FOREIGN OBJECT DAMAGE IN
AN MI SiC/SiC CERAMIC MATRIX COMPOSITE

D. Calvin Faucett, Jennifer Wright, Matthew Ayre, Sung R. Choi[†]
Naval Air Systems Command, Patuxent River, MD 20670

ABSTRACT
    Foreign object damage (FOD) phenomena of a prepreg MI SiC/SiC ceramic matrix
composites (CMC) were assessed using spherical steel ball projectiles in an impact velocity
range of 100 to 340 m/s. The CMC test coupons were ballistically impacted at a normal
incidence angle while supported in three different configurations of full support, partial support,
and cantilever support. Surface and subsurface impact damages, typically in the forms of craters,
fiber breakage, delamination, and cone cracks, were characterized with respect to the mode of
target supports. Quantitative effects of the support mode on impact damage were determined
through residual strength measurements. The cantilever support resulted in the greatest impact
damage, the full support gave rise to the least impact damage, and the partial support yielded the
intermediate degree of damage. The impact morphological observations were all consistent with
the results of post-impact residual strengths with respect to the severity of impact damage
involved.

INTRODUCTION
    The brittle nature of either monolithic ceramics or ceramic matrix composites (CMCs) or
ceramic environmental barrier coatings (EBCs) or ceramic thermal barrier coatings (TBCs) has
raised concerns on structural damage when subjected to impact by foreign objects. This has
prompted the propulsion communities to take into account for foreign object damage (FOD) as
an important design parameter when those materials are intended to be used for aeroengine hot-
section applications. A significant amount of work on impact damage of brittle monolithic
materials has been conducted during the past decades experimentally or analytically [1-14],
including gas-turbine grade toughened silicon nitrides [15-17].
    Ceramic matrix composites have been used and are being considered as enabling
propulsion materials for advanced civilian and/or military aeroengines components. A span of
FOD work has been carried out to determine FOD behavior of some gas-turbine grade CMCs
such as state-of-the-art melt-infiltrated (MI) SiC/SiC [18], N720™/aluminosilicate (N720/AS)
oxide/oxide [19], N720/alumina (N720/A) oxide/oxide [20], and 3D woven SiC/SiC [21].
Hertzian indentation responses were also determined using a SiC/SiC CMC to simulate a quasi-
static impact phenomenon [22]. Unlike their monolithic counterparts, all the CMCs investigated
have not exhibited catastrophic failure for impact velocities up to 400 m/s, resulting in much
increased resistance to FOD. However, the degree of damage that resulted in strength
degradation was still substantial, particularly at higher impact velocities ≥340 m/s when
impacted by hardened steel ball projectiles with a diameter of 1.59 mm.

[†] Corresponding author. Email address: sung.choi1@navy.mil

The mode of target supports has shown significant effects on the degree of impact damage associated. In the case of flexure test configuration, the partial support always resulted in greater impact damage than the full support counterpart, regardless of types of CMCs [18-20]. Furthermore, it was found that the uniaxial configuration of target supports yielded much worse impact damage when compared to the flexural support configurations [23-24].

The current paper is to extend the previous work to determine more specifically the effects of target supports on impact damage using a state-of-the-art prepreg MI SiC/SiC CMC material system. Three different types of target supports in a flexure configuration -full, partial, and cantilever supports- were employed. The impact damage of each support was assessed in conjunction with residual strength and impact morphologies. The three different support configurations employed in this work could be representative of some of hot-section components of aeroengines such as blades, nozzles, and shrouds.

EXPERIMENTAL PROCEDURES

Materials

The CMC material utilized in this work was a commercial, prepreg MI SiC/SiC, manufactured by GE Ceramic Composite Products, CCP (Gen2, Vintage 2011, Newark, DE). The composite was fabricated in a proprietary process with Nicalon™ Type S fibers with a fiber volume fraction of around 0.30. The fiber interface coating was presumably in BN. The CMC was $0°/90°$ cross-plied with a total of 8 plies. Some physical and mechanical properties of the composite are shown in Table 1. Target specimens were machined from a composite panel, measuring 10 mm in width, 47 mm in length, and about 2.0 mm in as-furnished thickness.

Table 1. Basic physical and mechanical properties of target CMC and projectile materials at ambient temperature

Materials		Fiber/matrix interface	Bulk density (g/cm^3)	Fiber volume fraction	Elastic modulus E (GPa)	Flexure strength (MPa)
Target	MI SiC/SiC	BN (?)	3.0	~0.30	300	550±43
Projectile	Hardened chrome steel (SAE52100)	-	7.78	-	200*	>2000*

Note: Flexure strength was determined using five flexure test specimens in a four-point flexure with 20/40 spans.

Foreign Object Damage Testing

A ballistic impact gun, as described elsewhere [15,16,18], was used to carried out FOD testing. Hardened (HRC≥60) chrome steel-ball projectiles with a diameter of 1.59 mm were inserted into a 300mm-long gun barrel. Helium gas and relief valves were utilized to pressurize and regulate a reservoir to a specific level, depending on prescribed impact velocity. Upon reaching a specific level of pressure, a solenoid valve was instantaneously opened accelerating a steel-ball projectile through the gun barrel to impact onto the as-furnished 10mm-wide side of a CMC target. The target specimens were supported in three different configurations such as full, partial, and cantilever supports, as illustrated in Fig. 1. For the partial and the cantilever supports, the span L was kept to be L=20 mm. Three different impact velocities of 150, 200, and 340 m/s were employed for a given target support. Three to four target specimens were utilized at each impact velocity. Three target specimens were additionally used for impact only with which their impact morphologies as well as cross-section features were characterized with respect to different impact velocities and target supports. The use of the three target supports employed was to simulate the structural or operational configurations of gas-turbine airfoil components such as blades, vanes, and shrouds.

Post-Impact Residual Strength Testing

Post-impact residual strength testing was performed to determine residual strengths of impacted target specimens from which the degree of impact damage was assessed more specifically. Strength testing was carried out in four-point flexure with 20mm-inner and 40-mm outer spans using an MTS servohydraulic test frame (Model 312) at a crosshead speed of 0.25 mm/min. The as-received flexure strength of the composite was also determined for a base-line reference with a total of five test specimens using the same test frame and test fixture that were used in post-impact strength testing.

RESULTS AND DISCUSSION

Projectiles

Due to the composite's high density and high elastic modulus (or high hardness), the steel ball projectiles exhibited significant plastic deformation upon impact particularly at the highest impact velocity of 340 m/s, irrespective of the type of target supports. This was consistent with the observations in monolithic silicon nitrides (with Vickers hardness of 15 GPa and elastic modulus of 300 GPa) where the projectiles were severely flattened or fragmented at impact velocities ≥350 m/s [15-17]. The oxide/oxide CMCs (N720/aluminosilicate or N720/alumina), by contrast, exhibited no visible damage or plastic deformation upon impact even at the highest impact velocity of 340 m/s [19, 20], attributed to their open and soft structure as well as lower elastic modulus of 70-80 GPa.

Impact Morphology of CMC Targets

Impact damage on CMC targets included features such as crater formation, fiber/matrix breakage, layer-to-layer breakage, delamination, fiber-tows response/breakage, etc. A typical example of an impact site generated at 340 m/s in cantilever support is shown in Fig. 2, where a

3-D topological image (one quarter) depicts how the damage was formed accompanying crater formation, layer-to-layer breakage, and others. The image was taken by a digital microscope via Extended Focus 3D Synthesis (Model KH-7700, HiRox, Japan). Figure 3 presents the frontal impact damage of target specimens with respect to impact velocity and target support. For a given impact velocity, the dependency of impact damage on the type of support was hardly distinguishable from the impact sites, as can be seen from the figure. The ballistic impact did generate surface and subsurface damage including craters with fiber/matrix breakages, delaminations, and some material removal with their severity being dependent on impact velocity.

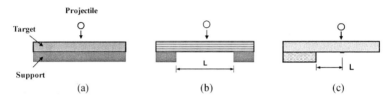

Figure 1. Three different types of target supports used in this work: (a) Full support; (b) Partial support; and (c) Cantilever support. The span L=20mm.

Figure 2. A typical example of a 3-D topological image (one quarter) taken from an impact site generated at 340 m/s in cantilever support in a prepreg MI SiC/SiC CMC impacted by 1.59-mm hardened chrome steel ball projectiles. The impact direction is from top to bottom.

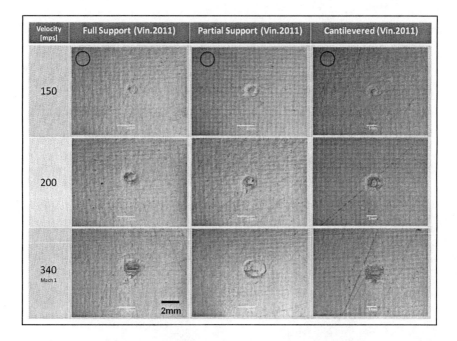

Figure 3. Frontal impact damages with respect to impact velocity for three different types of target supports in a prepreg MI SiC/SiC CMC impacted by 1.59-mm hardened chrome steel ball projectiles. The circles in the figure represent the size of projectiles utilized. All the images are under the same magnification. "Vin": Vintage.

The overall backside damage is depicted in Fig. 4. The backside damage started even at the lowest impact velocity of V=150 m/s as a ring crack configuration[‡] (arrowed in the figure), independent of the type of target supports. However, at V= 200 m/s, the backside damage increased from full support to partial support to cantilever support with increasing backside spalling or scabbing associated with delamination. The situation was amplified as impact velocity increased to V=340 m/s. Compared to the frontal damage (Fig. 3), the backside counterpart was much greater (≥ 5), attributed to the formation of cone cracks as well as to the ballistic dynamic effect, as seen from many monolithic ceramics and CMCs [15-20].

It is very important to note that the degree of the backside damage was greater in cantilever support than in partial support. It was initially speculated that the backside damage would be rather greater in partial support than in cantilever because of the presence of additional tensile stresses in the backside of the targets in partial support. However, that was not the case in reality. Hence, a conclusion presumably to draw is that the backside damage accompanying spalling or scabbing in partial or cantilever support would be mainly due to ballistic, dynamic effect than quasi-static state of stresses related with the support configurations. Unlike the partial support, the cantilever support in terms of support configuration produces no backside tensile stresses because of 'zero' moment of arm at impact sites. It should be noted that no damage was observed from the tensile sides of the cantilever target specimens at L= 0, where a maximum tensile stress was expected due to a cantilever configuration with respect to impact force.

Figure 5 shows the cross-sections of target specimens in association with different impact velocities and support types. The cross-sections provided many important features such as the shape and size of craters, the damage beneath impact sites, the formation of cone-cracks, interlaminar delaminations, the mode of penetration, and the backside damage including scabbing, etc. It can be also observed how the overall damage progressed with respect to impact velocity and support type. A complete penetration by the projectiles would be expected to occur at ≥ 360 m/s. Despite significant damage in both partial and cantilever supports, the target specimens still survived, without any catastrophic failure upon impact, as seen from many 2-D woven CMCs [18-20]. The formation of cone cracks is noted and has been commonly observed in ballistic impact by spherical projectiles in either CMCs, monolithic ceramics, or in glasses [15-21]. A detailed cross-sectional view of a target specimen impacted at 340 m/s in cantilever support is depicted in Figure 6. Multiples cone cracks were formed cutting through matrix as well as fiber tows. The cone angle (α) was approximately $\alpha=100$ deg. Also note how scabbing or spallation took place from the backside via delamination and interaction with cone cracks.

Post-Impact Residual Strength

Most of the impacted target specimens in post-impact strength testing failed from the impact sites. However, an exception to this was the target specimens impacted at the lowest impact speed of 150 m/s, where about one out of four specimens at each target support failed a little away from the impact sites but within the size of the backside damage. Those target specimens that did not fail from the impact sites were included in the data pool since they still

---

[‡] The ring crack configuration is actually the bottom part of a cone crack, which can be seen from Fig. 4.

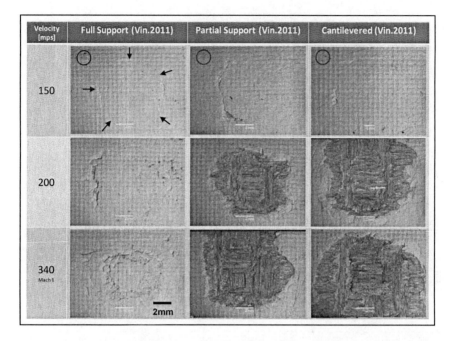

Figure 4. Backside impact damages with respect to impact velocity for three different types of target supports in a prepreg MI SiC/SiC CMC impacted by 1.59-mm hardened chrome steel ball projectiles. The circles in the figure represent the size of projectiles utilized. Backside damage in a form of a ring crack was marked as arrows. All the images are under the same magnification. "Vin": vintage.

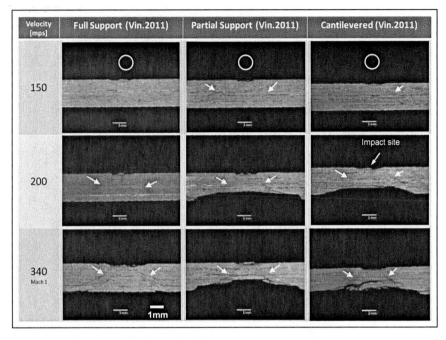

Figure 5. Cross-sectional views of impact damage with respect to impact velocity for three different types of target supports in a prepreg MI SiC/SiC CMC impacted by 1.59-mm hardened chrome steel ball projectiles. The circles in the figure represent the size of projectiles utilized. The arrows indicate cone cracks developed through-the-thickness direction of targets. All the images are under the same magnification. "Vin": vintage.

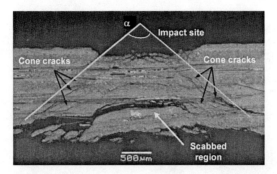

Figure 6. A detailed cross-sectional view of a target impacted at 340 m/s in cantilever support in a prepreg MI SiC/SiC CMC impacted by 1.59-mm hardened chrome steel ball projectiles. Impact sites, cone cracks, and scabbled portion of the material are clearly seen. The angle α is the cone angle, which is $\alpha \approx 100$ deg.

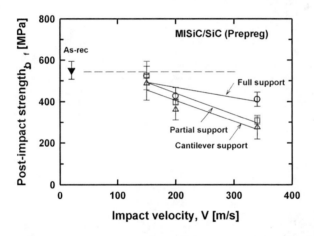

Figure 7. Post-impact strength as a function of impact velocity for three different types of target supports determined in a prepreg MI SiC/SiC CMC impacted by 1.59-mm hardened chrome steel ball projectiles. "As-rec" indicates as-received strength.

failed from the overall impact-damage zone. The results of post-impact residual strength testing for impacted target specimens are summarized in Fig. 7, where residual strength was plotted as a function of impact velocity for three different target supports. The as-received flexure strength (σ_f) of the material is also included, which is $\sigma_f = 550\pm43$ MPa. Despite some inherent data scatter, the overall trend is obvious such that post-impact strength decreased with increasing impact velocity for all types of target supports. At a low impact velocity of V=150 m/s, the difference in residual strength was negligible, as seen from the impact morphologies in Figs. 3-5. At V≥200 m/s, the difference in residual strength was augmented. The greatest strength degradation occurred in cantilever support, the intermediate degradation took place in partial support, and the least strength degradation occurred in full support.

Results of linear regression analysis on post-impact strength vs. impact velocity based on the data in Fig. 7 for the three different target supports are as follows:

$$\sigma_f = -0.47V + 557; \quad r_{coef} = 0.657 \quad \text{For full support}$$
$$\sigma_f = -1.03V + 648; \quad r_{coef} = 0.856 \quad \text{For partial support}$$
$$\sigma_f = -1.04V + 619; \quad r_{coef} = 0.792 \quad \text{For cantilever support}$$

where σ_f is post-impact strength (MPa), V is impact velocity (m/s), and r_{coef} is the coefficient of correlation in regression.

The results of post-impact residual strength were all consistent with the morphological observations with regard to the degree of impact damage as related to strength degradation. The more impact damage yields the more strength degradation and *vice versa*. Therefore, a combined effort to obtain both morphology and residual strength should be pursued if feasible, since it could give more detailed qualitative and quantitative information and insights regarding the nature and degree of impact damage associated.

An important observation acquired from this work is that the overall impact damage or strength degradation in a prepreg MI SiC /SiC CMC was greater in cantilever support than in partial support. Unlike the situation in partial support, there was supposed to be no backside tensile stress statically at the impact point in cantilever support because of zero moment-arm therein. This should have created damage much less than by partial support. However, the outcome was quite opposite and contradictory to our basic understanding of mechanics (i.e., elementary beam theory). This implies that the overall impact behavior in cantilever support would have been governed more by dynamic aspects rather than by static alone. More study should be sought to further validate this finding using different material systems, increased number of test coupons, and using appropriate analytical (dynamics and mechanics) approaches.

Consideration Factors in FOD

Designing aeroengine CMC components to withstand FOD is a highly complex task. Considerations of many affecting factors are needed and can be listed as follows:

- Effect of primary impact variables such as impact velocity and incidence angle
- Effect of the material, geometry, and size of projectiles
- Effect of target material in architectures, material constituents (fiber, matrix, interface coating), and processing routes
- Effect of the type, size, and material of target support
- Effect of operating temperature, environment, and thermal/mechanical loading
- Effect of protective coatings such as impact barrier coatings, and/or thermal/environmental barrier coatings (T/EBCs)

Foreign object damage is complex from an analytical as well as a characterization standpoint. Appropriate FOD modeling should also serve in a feedback loop to characterization (experimental) techniques, one complementing and improving the other.

CONCLUSIONS
    The overall impact damage of the prepreg MI SiC/SiC ceramic matrix composite was dependent not only on impact velocity but also on the type of target supports. The prepreg CMC shows significant impact damage occurring particularly at 340 m/s in either partial or cantilever support. Of the three different types of target supports used, the cantilever support resulted in the greatest impact damage, the partial support yielded the intermediate damage, and the full support gave rise to the least impact damage. The results of post-impact residual strength were all consistent with the impact morphological observations. The outcome that the cantilever support generated more impact damage than the partial support was of importance but is somewhat against our common mechanics concept. This would be of a future study.

Acknowledgements
    The authors acknowledge the support by the Office of Naval Research and Dr. David Shifler.

REFERENCES
1. Wiederhorn, S. M., and Lawn, B.R., 1977, "Strength Degradation of Glass Resulting from Impact with Spheres," J. Am. Ceram. Soc., **60**[9-10], pp. 451-458.
2. Wiederhorn, S. M., and Lawn B. T., 1979, "Strength Degradation of Glass Impact with Sharp Particles: I, Annealed Surfaces," J. Am. Ceram. Soc., **62**[1-2], pp. 66-70.
3. Ritter, J. E., Choi, S. R., Jakus, K, Whalen, P. J., and Rateick, R. G., 1991, "Effect of Microstructure on the Erosion and Impact Damage of Sintered Silicon Nitride," J. Mater. Sci., **26**, pp. 5543-5546.
4. Akimune, Y, Katano, Y, and Matoba, K, 1989, "Spherical-Impact Damage and Strength Degradation in Silicon Nitrides for Automobile Turbocharger Rotors," J. Am. Ceram. Soc., **72**[8], pp. 1422-1428.
5. Knight, C. G., Swain, M. V., and Chaudhri, M. M., 1977, "Impact of Small Steel Spheres on Glass Surfaces," J. Mater. Sci., **12**, pp.1573-1586.

6. Rajendran, A. M., and Kroupa, J. L., 1989, "Impact Design Model for Ceramic Materials," J. Appl. Phys, **66**[8], pp. 3560-3565.
7. Taylor, L. N., Chen, E. P., and Kuszmaul, J. S., 1986 "Microcrack-Induced Damage Accumulation in Brittle Rock under Dynamic Loading," Comp. Meth. Appl. Mech. Eng., **55**, pp. 301-320.
8. Mouginot, R., and Maugis, D., 1985, "Fracture Indentation beneath Flat and Spherical Punches," J. Mater. Sci., **20**, pp. 4354-4376.
9. Evans, A. G., and Wilshaw, T. R., 1977, "Dynamic Solid Particle Damage in Brittle Materials: An Appraisal," J. Mater. Sci., **12**, pp. 97-116.
10. Liaw, B. M., Kobayashi, A. S., and Emery, A. G., 1984, "Theoretical Model of Impact Damage in Structural Ceramics," J. Am. Ceram. Soc., **67**, pp. 544-548.
11. van Roode, M., et al., 2002, "Ceramic Gas Turbine Materials Impact Evaluation," ASME Paper No. GT2002-30505.
12. Richerson, D. W., and Johansen, K. M., 1982, "Ceramic Gas Turbine Engine Demonstration Program," Final Report, DARPA/Navy Contract N00024-76-C-5352, Garrett Report 21-4410.
13. Boyd, G. L., and Kreiner, D. M., 1987, "AGT101/ATTAP Ceramic Technology Development," Proceeding of the Twenty-Fifth Automotive Technology Development Contractors' Coordination Meeting, p.101.
14. van Roode, M., Brentnall, W. D., Smith, K. O., Edwards, B., McClain, J., and Price, J. R., 1997, "Ceramic Stationary Gas Turbine Development – Fourth Annual Summary," ASME Paper No. 97-GT-317.
15. (a) Choi, S. R., Pereira, J. M., Janosik, L. A., and Bhatt, R. T., 2002, "Foreign Object Damage of Two Gas-Turbine Grade Silicon Nitrides at Ambient Temperature," Ceram. Eng. Sci. Proc., **23**[3], pp. 193-202; (b) Choi, S. R., et al., 2004, "Foreign Object Damage in Flexure Bars of Two Gas-Turbine Grade Silicon Nitrides," Mater. Sci. Eng. **A 379**, pp. 411-419.
16. Choi, S. R., Pereira, J. M., Janosik, L. A., and Bhatt, R. T., 2003, "Foreign Object Damage of Two Gas-Turbine Grade Silicon Nitrides in a Thin Disk Configuration," ASME Paper No. GT2003-38544; (b) Choi, S. R., et al., 2004, "Foreign Object Damage in Disks of Gas-Turbine-Grade Silicon Nitrides by Steel Ball Projectiles at Ambient Temperature," J. Mater. Sci., **39**, pp. 6173-6182.
17. Choi, S. R., 2008, "Foreign Object Damage Behavior in a Silicon Nitride Ceramic by Spherical Projectiles of Steels and Brass," Mat. Sci. Eng. **A497**, pp. 160-167.
18. Choi, S. R., 2008, "Foreign Object Damage Phenomenon by Steel Ball Projectiles in a SiC/SiC Ceramic Matrix Composite at Ambient and Elevated Temperatures," J. Am. Ceram. Soc., **91**[9], pp. 2963-2968.
19. (a) Choi, S. R., Alexander, D. J., and Kowalik, R. W., 2009, "Foreign Object Damage in an Oxide/Oxide Composite at Ambient Temperature," J. Eng. Gas Turbines & Power, Transactions of the ASME, Vol. **131**, 021301. (b) Choi, S. R., Alexander, D. J., and Faucett, D. C., 2009, "Comparison in Foreign Object Damage between SiC/SiC and Oxide/Oxide Ceramic Matrix Composites," Ceram. Eng. Sci. Proc., **30**[2], pp. 177-188.
20. Choi, S. R., Faucett, D. C., and Alexander, D. J., 2010, "Foreign Object Damage in An N720/Alumina Oxide/Oxide Ceramic Matrix Composite," Ceram. Eng. Sci. Proc. **31**[2] pp.

221-232.

21. Ogi, K., et al., 2010, "Experimental Characterization of High-Speed Impact Damage Behavior in A Three-Dimensionally Woven SiC/SiC Composite," Composites Part A, **41**[4], pp. 489-498.

22. Herb, V., Couegnat, G., Martin, E., 2010, "Damage Assessment of Thin SiC/SiC Composite Plates Subjected to Quasi-Static Indentation Loading," Composites Part A, **41**[11], pp. 1677-1685.

23. Faucett, D. C., Choi, S. R., 2011, "Foreign Object Damage in An N720/Alumina Oxide/Oxide Ceramic Matrix Composite under Tensile Loading," *Ceramic Transactions* **225**, pp. 99-107, Eds. N.P. Bansal, J.P. Singh, J. Lamon, S.R. Choi.

24. Faucett, D. C., Choi, S. R., 2011, "Effects of Preloading on Foreign Object Damage in An N720/Alumina Ceramic Matrix Composite," Ceramic Eng. Sci. Proc., **32**[2], pp. 89-100.

FOREIGN OBJECT DAMAGE (FOD) IN THERMAL BARRIER COATINGS

D. Calvin Faucett, Jennifer Wright, Matt Ayre, Sung R. Choi*
Naval Air Systems Command, Patuxent River, MD 20670

ABSTRACT
    Foreign object damage (FOD) has been structural or functional issues to thermal barrier coatings (TBCs) in hot-section components of advanced aeroengines. Thermal barrier coatings (TBCs) are ceramics, brittle in nature, and are thus prone to damage/cracking when subjected to impact by foreign objects. A series of FOD tests on ceramic TBCs in aeroengine components were conducted to assess their particle impact behavior. Two different types of electron beam-physical vapor deposition (EB-PVD) and air plasma spray (APS) TBCs were ballistically impacted by 1.6 mm-diameter steel projectiles at three different impact velocities up to 300 m/s. Degree and morphologies of impact damage were characterized and quantified with respect to velocity and component service hour. Natural FOD of airfoil components having occurred during service was also presented and analyzed with Weibull statistics in its damage distribution.

INTRODUCTION
    The brittleness of ceramic materials, either monolithic ceramics or ceramic matrix composites (CMCs) has raised concerns on structural damage when they are used as aeroengine components and subjected to impact by foreign objects. This has prompted the propulsion community to take into account foreign object damage (FOD) as an important design parameter. A large amount of work on impact damage of monolithic ceramic materials has been done in the past decades [1-15]. Also, work on ballistic impact in CMCs has been continued to include a variety of CMC material systems [16-25]. Ceramic environmental barrier coatings (EBCs) in SiC/SiC composites were also assessed in their responses to ballistic impact [26,27].
    Thermal barrier coatings (TBCs) have attracted increasing attention for advanced gas turbine applications because of their ability to provide thermal insulation to engine components. The merits of using ceramic TBCs have been well recognized and include a potential increase in engine operating temperature with reduced cooling requirements resulting in significant improvements in thermal efficiency, performance, and reliability. Zirconia-based ceramics are the most important coating materials because of their low thermal conductivity, relatively high thermal expansivity, strain tolerance, and unique plasma sprayed or EB-PVD microstructure. However, the durability of TBCs under severe thermal and mechanical loading conditions encountered in heat engines remains one of the major issues primarily attributed to lower mechanical properties of the coatings [28]. This, also due to their brittle nature, leads to a challenging issue on their significant susceptibility to FOD by foreign objects ingested into engines or by particles torn from other engine components or by combustion products. A considerable amount of work has been done on the subjects of erosion of TBCs [29-32]. However, despite its significance, work on FOD has been relatively sparse particularly in aeroengine components [33-37].

* Corresponding author; Email address: sung.choi1@navy.mil

This paper presents an experimental work that was conducted to determine FOD behavior of TBCs in actual aeroengine components in service. The components coated with 7-8wt% yittria-stabilized zirconia (7-8wt% Y_2O_3-ZrO_2 or 7-8YSZ) topcoats by EB-PVD or by APS were subjected to ballistic impact by 1.6 mm steel ball projectiles in an impact velocity range of 100 to 300 m/s. Foreign object damage was characterized through impact site and cross-sectional damage morphologies with respect to impact velocity and components service hours. This paper also presents a two-parameter Weibull FOD-size distribution, determined using a large number of actual airfoil components experienced with natural impact in aeroengine service.

EXPERIMENTAL PROCEDURES

Materials/Components
Two different TBCs were used in this work including EB-PVD and APS TBCs. The EB-PVD TBCs were from high-pressure turbine (HPT) airfoil components, while the APS TBCs were from a combustor liner. They were both 7-8wt% YSZ. The coating thicknesses were 120 μm and 1000 μm, respectively, for the HPT airfoils and the combustor liner. The respective bondcoats were Pt-Al and NiCoCrAlY. The HPT airfoils were taken from three different engines having service hours (SH) of 3, 14, and 23[†]. The combustor liner had a service hour of <10. The components are proprietary so that their detailed descriptions were excluded here.

Foreign-Object-Damage (FOD) Testing
A ballistic impact gun, as described elsewhere [13,16], was utilized to conduct FOD testing. A hardened (HRC≥60) chrome steel-ball projectile with a diameter of 1.6 mm was inserted into a 300mm-long gun barrel. Helium gas and relief valves were utilized to pressurize and regulate a reservoir to a specific level, depending on prescribed impact velocity. Upon reaching a specific level of pressure, a solenoid valve was instantaneously opened accelerating a steel-ball projectile through the gun barrel to impact onto TBCs of a target component. The locations of impact employed in the suction side of a HPT EB-PVD airfoil are shown in Fig. 1. A total of three different impact velocities of 150, 200 and 300 m/s were used. For a given impact velocity, four impacts were applied in a random but systematic fashion, away from the both leading and trailing edges. All impact testing was done in a normal incidence angle, regardless of impact locations. The HPT airfoils were chosen from three different aroengines that had service hours of SH= 3, 14, and 23, as aforementioned. A total of four HPT airfoils were used for a given aeroengine.
One segment, measuring 60 mm by 60 mm, was cut from the combustor liner for APS-TBCs impact testing. The number of impacts employed was four for a given impact velocity. The impact velocity and impact angle in the APS combustor were the same as those used in the HPT EB-PVD airfoil testing.

[†] The service hour given is in convention a characteristically, relative numerical quantity but not an absolute value on service.

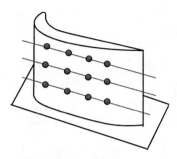

Figure 1. A schematic showing the location of impact in a HPT EB-PVD airfoil applied in this work.

RESULTS AND DISCUSSION

EB-PVD HPT Airfoils

Figure 2 shows the impact sites generated at V=150 and 300 m/s. At a low impact velocity of V=150 m/s, all the impact sites showed only impressions by the projectiles with some ring cracks along the impression peripherals. By contrast, at a high impact velocity of V=300 m/s, the impact sites exhibited complete spallation of the coatings where the coatings delaminated from the bond coats. Figure 3 shows how the coatings impacted at V=300 m/s were removed from the bond coats and the surrounding material at the outer boundary of spallation.

The results of the FOD testing for EB-PVD HPT airfoils are presented in Fig. 4, where impact damage size was plotted as a function of impact velocity for three different service hours. Overall, the damage size increased with increasing impact velocity. At the same time, the scatter of the damage size increased significantly with increasing impact velocity, indicating that FOD response to higher impact velocity (V≥200 m/s) was indeed stochastic (i.e., statistical) rather than deterministic, as observed in many brittle ceramics [13,14]. Because of this stochastic nature, the effect of engine service hour on FOD did not clearly exhibit although some trend seemed to be derivable. This implies that a large number of impacts (or components) for a given engine operation condition would be required to get much improved statistical reproducibility and reliability.

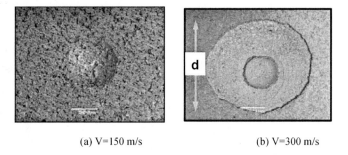

(a) V=150 m/s   (b) V=300 m/s

Figure 2. Typical examples of impact damage generated in EB-PVD TBCs of airfoils by 1.6 mm steel ball projectiles at: (a) V=150 m/s; (b) V=300 m/s.

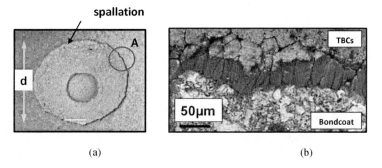

(a)   (b)

Figure 3. Impact damage showing spallation and fracture of EB-PVD TBCs from the bondcoats and surrounding material, impacted by an 1.6 mm steel ball projectile at V=300 m/s: (a) Overall impact site; (b) Details of region 'A'.

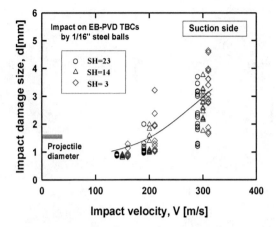

Figure 4. Impact damage size (d) as a function of impact velocity (V) for three different engine service hours of EB-PVD HPT airfoils by 1.6 mm steel ball projectiles. SH: service hour.

Plasma-Sprayed Combustor Liner

Similar to the EB-PVD HPT airfoils observed at a low impact velocity of V=150 m/s, the impact sites of the APS combustor liner showed only impressions in the coatings, regardless of impact velocity. The impressions were in a form of crater in geometry, showing both the diameter and depth of the craters increasing with increasing impact velocity. An example of the impact site at V=300 m/s is presented in Fig. 5, where the overall and the outer boundary of the impact site are shown with a somewhat clear-cut demarcation along the spallation boundary.

Figure 6 is a summary of impact damage size determined with respect to impact velocity. The delamination size was also included in the figure, which will be described later in this section. A monotonic increase in damage size (diameter, d) with increasing impact velocity is noted, which is in contrast with the case of the EB-PVD airfoils shown in Fig. 4. The EV-PVD airfoils were characterized as significant increase in damage size with increasing impact velocity, resulting in delamination at V≥200 m/s.

When the cross-sections of damage sites were examined, the situation became quite different. Figure 7 represents the cross-sectional views of the impact damages induced at impact velocities of V=150 and 300 m/s. The formation of craters is clearly seen with its size increasing with increasing impact velocity. However, significant delamination also occurred *within the coatings* with the delamination crack size several times greater than the frontal impact diameter. Also, note that the delamination took place along the plane above the TBCs and bondcoat interface. At V=300 m/s, some plastic deformation of the metallic substrate was also observed.

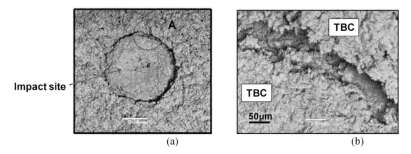

(a) (b)

Figure 5. Typical example of impact damage showing spallation of plasma-sprayed TBCs in the combustor liner, impacted at 300 m/s by 1.6 mm steel ball projectiles: (a) Overall; (b) detailed impact morphology in the region of 'A'.

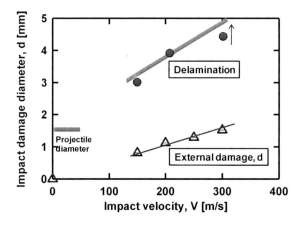

Figure 6. Impact damage size (d) as a function of impact velocity (V) for plasma-sprayed TBCs in the combustor liner, impacted by 1.6 mm steel ball projectiles. The delamination data is also included.

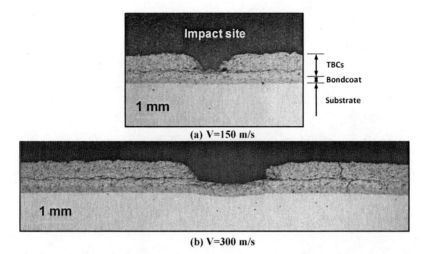

(a) V=150 m/s

(b) V=300 m/s

Figure 7. Cross-sectional views of impact damage showing significant delamination cracks of plasma-sprayed TBCs impacted by 1.6 mm steel ball projectiles at: (a) V=150 m/s; (b) V=300 m/s.

A plausible reason for the delamination of APS TBCs within the material rather than at the interface, as seen from Fig. 7, is illustrated in Fig. 8. When the projectile impacted the coatings, the coatings would be subjected to compression plastic deformation (or densification due to the coatings' inherent porosity) as well as fracture or pulverization. When the deformed projectile pushed away the surrounding coatings to the radial direction, a compression force would be acting to induce shear stress along the coatings. If the shear stress (or mode II energy release rate) is greater than the shear strength (or mode II crack growth resistance) of the coatings, then shear delamination or cracking may occur. Plasma-sprayed coatings are actually in a layered structure because of the nature of processing [28], very analogous to laminate materials such as woven or cross-plied continuous fiber-reinforced CMCs. Due to their layered architectures, woven or cross-plied CMCs exhibited significantly inferior interlaminar shear or interlaminar tensile properties [38], compared to their in-plane counterparts. The same would be true to *layered* PS TBCs so that shear delamination could occur more easily along the layers rather than along the TBCs/bondcoat interface. The uniaxial tensile strength of typical, as-processed APS TBCs is around 15 MPa [28]. Hence, interlamiar shear strength of the APS TBCs would be expected to be much lower than the uniaxial tensile counterpart (=15 MPa). The coating thickness of 1000 μm, somewhat thick, would have some other effects on such a within-the-material delamination.

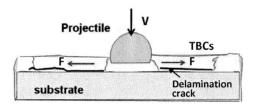

Figure 8. A schematic illustrating how delamination occurs upon impact by a spherical projectile onto plasma-sprayed TBCs. F is the force acting on TBCs by a projectile to the radial direction.

The results of delamination crack size thus determined from cross-sections are also shown in Fig. 6. The delamination crack size was approximately three times the crater size. This result can give an important implication in terms of characterization regarding the nature of impact damage in PS-TBCs: Characterizations of impact damage cannot be completed without appropriate cross-sectional views. This may be true to the EB-PVD airfoils. The cross-sectional views should be made in the EB-PVD airfoils, which will reveal more detailed and clearer impact morphologies concerned, which would be a task of further study. Pertinent NDE (non-destructive evaluation) techniques should be sought as supplementary tools, too.

Natural Foreign object Damage in EB-PVD HPT Airfoils

A total of 30 HPT airfoils coated with EB-PVD TBCs that were foreign-object-damaged in service were examined to determine their damage size and distribution. The results were presented in a two-parameter Weibull scheme as shown in Fig. 9. F is the probability of occurrence, and 'a' is the average damage size (i.e., diameter). A total of 130 damages were identified from the 30 airfoil components. A bi- or tri-modality of damage distribution is evident from the figure. The two-parameter Wibull analysis resulted in a Weibull modulus of m=2.4 and a Weibull median size of a = 2mm. The upper region, which is of great interest, corresponded to the case that was obtained at V=300 m/s in this experiment (see the result of Fig. 4). More FOD data from actual airfoils in service will definitely provide more insightful information and guidelines to design, operation, and service of related TBC systems. However, the statistical data shown in Fig. 9 is still of great value in terms of understating of the reality encountered in the TBC components under actual aeroengine operational conditions.

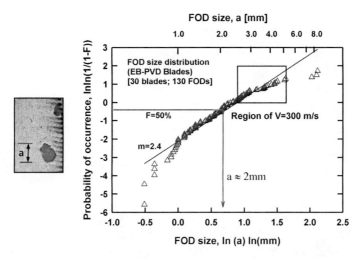

Figure 9. Two-parameter Weibull FOD distribution determined from a total of 30 EB-PVD HPT airfoils in service. Typical damage and its size description are shown in an inset. The boxed, upper region indicates the case that was reproduced in this experiment with at V=300 m/s by 1.6mm steel ball projectiles.

Acknowledgements
This work was conducted through the support of ONR and Dr. D. Shifler.

References
1. J. E. Ritter, S. R. Choi, K. Jakus, P. J. Whalen, R. G. Rateick, "Effect of Microstructure on the Erosion and Impact Damage of Sintered Silicon Nitride," *J. Mater. Sci.*, **26** 5543-5546 (1991).
2. Y. Akimune, Y. Katano, K. Matoba, "Spherical-Impact Damage and Strength Degradation in Silicon Nitrides for Automobile Turbocharger Rotors," *J. Am. Ceram. Soc.*, **72**[8] 1422-1428 (1989).
3. C. G. Knight, M. V. Swain, M. M. Chaudhri, "Impact of Small Steel Spheres on Glass Surfaces," *J. Mater. Sci.*, **12** 1573-1586 (1997).
4. A. M. Rajendran, J. L. Kroupa, "Impact Design Model for Ceramic Materials," *J. Appl. Phys*, **66**[8] 3560-3565 (1989).
5. L. N. Taylor, E. P. Chen, J. S. Kuszmaul, "Microcrack-Induced Damage Accumulation in Brittle Rock under Dynamic Loading," *Comp. Meth. Appl. Mech. Eng.*, **55** 301-320 (1986).
6. R. Mouginot, D. Maugis, "Fracture Indentation beneath Flat and Spherical Punches," *J. Mater. Sci.*, **20** 4354-4376 (1985).

7. A. G. Evans, T. R. Wilshaw, "Dynamic Solid Particle Damage in Brittle Materials: An Appraisal," *J. Mater. Sci.*, **12** 97-116 (1977).
8. B. M. Liaw, A. S. Kobayashi, A. G. Emery, "Theoretical Model of Impact Damage in Structural Ceramics," *J. Am. Ceram. Soc.*, **67** 544-548 (1984).
9. M. van Roode et al.,"Ceramic Gas Turbine Materials Impact Evaluation," ASME Paper No. GT2002-30505 (2002).
10. D. W. Richerson, K. M. Johansen, "Ceramic Gas Turbine Engine Demonstration Program," Final Report, DARPA/Navy Contract N00024-76-C-5352, Garrett Report 21-4410 (1982).
11. G. L. Boyd, D. M. Kreiner, "AGT101/ATTAP Ceramic Technology Development," Proceeding of the Twenty-Fifth Automotive Technology Development Contractors' Coordination Meeting, p.101 (1987).
12. M. van Roode, W. D. Brentnall, K. O. Smith, B. Edwards, J. McClain, J. R. Price, "Ceramic Stationary Gas Turbine Development – Fourth Annual Summary," ASME Paper No. 97-GT-317 (1997).
13. (a) S. R. Choi, J. M. Pereira, L. A. Janosik, R. T. Bhatt, "Foreign Object Damage of Two Gas-Turbine Grade Silicon Nitrides at Ambient Temperature," *Ceram. Eng. Sci. Proc.*, **23**[3] 193-202 (2002); (b) S. R. Choi, J. M. Pereira, L. A. Janosik, R. T. Bhatt, "Foreign Object Damage in Flexure Bars of Two Gas-Turbine Grade Silicon Nitrides," *Mater. Sci. Eng.*, **A 379** 411-419 (2004).
14. S. R. Choi, J. M. Pereira, L. A. Janosik, R. T. Bhatt, "Foreign Object Damage of Two Gas-Turbine Grade Silicon Nitrides in a Thin Disk Configuration," ASME Paper No. GT2003-38544 (2003); (b) S. R. Choi, J. M. Pereira, L. A. Janosik, R. T. Bhatt, "Foreign Object Damage in Disks of Gas-Turbine-Grade Silicon Nitrides by Steel Ball Projectiles at Ambient Temperature," *J. Mater. Sci.*, **39** 6173-6182 (2004).
15. S. R. Choi, "Foreign Object Damage Behavior in a Silicon Nitride Ceramic by Spherical Projectiles of Steels and Brass," *Mat. Sci. Eng.*, **A497** 160-167 (2008).
16. S. R. Choi, "Foreign Object Damage Phenomenon by Steel Ball Projectiles in a SiC/SiC Ceramic Matrix Composite at Ambient and Elevated Temperatures," *J. Am. Ceram. Soc.*, **91**[9] 2963-2968 (2008).
17. R. T. Bhatt, S. R. Choi, L. M. Cosgriff, D. S. Fox, K. N. Lee, "Impact Resistance of Uncoated SiC/SiC Composites," *Mat. Sci. Eng.*, **A476** 20-28 (2008).
18. D. C. Phillips, N. Park, R. J. Lee, "The Impact Behavior of High Performance, Ceramic Matrix Fibre Composites," *Composites Sci.Tech.*, **37** 249-265 (1990).
19. A. R. Boccaccini, S. Atiq, D. N. Boccaccini, I. Dlouhy, C. Kaya, "Fracture Behavior of Mullite Fibre Reinforced-Mullite Matrix Composites under Quasi-Static and Ballistic Impact Loading," *Composites Sci. Tech.*, **65** 325-333 (2005).
20. Y. Leijiang, F. Ziyang, C. Qiyou, "Low Velocity Impact Damage Evaluation of 2D C/SiC Composite Material," *Advanced Materials Research*, **79-82** 1835-1838 (2009).
21. (a) S. R. Choi, D. J. Alexander, R. W. Kowalik, "Foreign Object Damage in an Oxide/Oxide Composite at Ambient Temperature," *J. Eng. Gas Turbines & Power, Transactions of the ASME*, Vol. **131**, 021301 (2009). (b) S. R. Choi, D. J. Alexander, D. C. Faucett, "Comparison in Foreign Object Damage between SiC/SiC and Oxide/Oxide Ceramic Matrix Composites," *Ceram. Eng. Sci. Proc.*, **30**[2] 177-188 (2009).
22. S. R. Choi, D. C. Faucett, and D. J. Alexander, "Foreign Object Damage in An N720/Alumina Oxide/Oxide Ceramic Matrix Composite," *Ceram. Eng. Sci. Proc.*, **31**[2]

221-232 (2010).

23. K. Ogi, T. Okabe, M. Takahashi, S. Yashiro, A. Yoshimura, "Experimental Characterization of High-Speed Impact Damage Behavior in A Three-Dimensionally Woven SiC/SiC Composite," *Composites Part A*, **41**[4] 489-498(2010).

24. V. Herb, G. Couegnat, E. Martin, "Damage Assessment of Thin SiC/SiC Composite Plates Subjected to Quasi-Static Indentation Loading," *Composites Part A*, **41**[11] 1677-1685 (2010).

25. D. C. Faucett, S. R. Choi, "Foreign Object Damage in An N720/Alumina Oxide/Oxide Ceramic Matrix Composite under Tensile Loading," *Ceramic Transactions* **225** 99-107, Eds. N. P. Bansal, J. P. Singh, J. Lamon, S. R. Choi (2011).

26. R. T. Bhatt, S. R. Choi, L. M. Cosgriff, D. S. Fox, K. N. Lee, "Impact Resistance of Evironmental Barrier Coated SiC/SiC Composites," *Mater. Sci. Eng.*, **A476** 8-19 (2008).

27. D. Hass, "Impact and Erosion Resistance of Novel Thermal/Environmental Barrier Coating Systems," presented at the MS&T'11 Materials Science & Technology Conference, October 16-20, 2011, Columbus, OH.

28. S. R. Choi, D. Zhu, R. A. Miller, "Mechanical Properties/Database of Plasma-Sprayed ZrO_2-8wt% Y_2O_3 Thermal Barrier Coating," *Int. J. Appl. Ceram. Technol.*, **1**[4] 330-342 (2004).

29. R. W. Bruce, "Development of 1232°C (2250°F) Erosion and Impact Tests for Thermal Barrier Coatings," *Tribology Trans.*, **41**[4] 399-410 (1998).

30. M. Fathy, W. Tabakoff, "Computation and Plotting of Solid Particle Flow in Rotating Cascades," *Computers & Fluids*, **2**[1-A] 1-15 (1974).

31. X. Chen, M. Y. He, I. Spitsberg, N. A. Fleck, J. W. Hutchinson, A. G. Evans, "Mechanisms Governing the High Temperature Erosion of Thermal Barrier Coatings," *Wear*, **256** 735-746 (2004).

32. J. R. Nicholls, M. J. Deakin, D. S. Rickerby, "A Comparison Between the Erosion Behavior of Thermal Spray and EB-PVD Thermal Barrier Coatings," *Wear*, **233-235** 352-361 (1999).

33. X. Chen, R. Wang, N. Yao, A. G. Evans, J. W. Hutchinson, R. W. Bruce, "Foreign Object Damage in a Thermal Barrier Systems: Mechanism and Simulations," *Mater. Sci. Eng.*, **A352** 221-231 (2003).

34. A. G. Evans, N. A. Fleck, S. Faulhaber, N. Vermaak, M. Maloney, R. Darolia, "Scaling Laws Governing the Erosion and Impact Resistance of Thermal Barrier Coatings," *Wear*, **260** 886-894 (2006).

35. J. R. Nicholls, Y. Jaslier, D. S. Rickerby, "Erosion and Foreign Object Damage of Thermal Barrier Coatings," *Material Science Forum*, **251**[1-2] 935-948 (1997).

36. J. R. Nicholls, R. G. Wellman, "Erosion and Foreign Object Damage of Thermal Barrier Coatings," RTO-MP-AVT-109, 20-1_20-30 (2003); presented at the *RTO AVT Specialist Meetings on "The Control and Wear Military Platforms,"* June 7-9, 2003, Williamsburg, VA.

37. M. W. Crowell, et al., "Experimental and Numerical Simulations of Single Particle FOD-Like Impacts of HPT TBCs," in preparation (2011).

38. S. R. Choi, N. P. Bansal, "Interlaminar Tension/Shear Properties and Stress Rupture in Shear of Various Continuous Fiber-Reinforced Ceramic Matrix Composites," *Advanced in Ceramic Matrix Composites XI, Ceramic Transactions,* **175**, 119-134, Eds. N. P. Bansal, J. P. Singh, W. M. Kriven (2006).

Testing, Evaluation, and Microstructure-Property Relationships

HIGH-TEMPERATURE INTERLAMINAR TENSION TEST METHOD DEVELOPMENT FOR CERAMIC MATRIX COMPOSITES

Todd Z. Engel
Hyper-Therm High-Temperature Composites, Inc.
Huntington Beach, CA, United States of America

ABSTRACT

Ceramic Matrix Composite (CMC) materials are an attractive design option for various high-temperature structural applications. However, 2D fabric-laminated CMCs typically exhibit low interlaminar tensile (ILT) strengths, and interply delamination is a concern for some targeted applications. Currently, standard test methods only address the characterization of interlaminar tensile strengths at ambient temperatures, which is problematic given that nearly all CMCs are slated for service in elevated temperature applications. This work addresses the development of a new test technique for the high-temperature measurement of CMC interlaminar tensile properties.

1.0 INTRODUCTION

Hot structures fabricated from ceramic composite (CMC) materials are an attractive design option for the specialized components of future military aerospace vehicles and propulsion systems because of the offered potential for increased operating temperatures, reduction of component weight, and increased survivability. CMC materials exhibit the high-temperature structural capabilities inherent to ceramic materials, but with significantly greater toughness and damage tolerance, and with lessened susceptibility to catastrophic component failure than traditional monolithic ceramic materials – all afforded by the presence of continuous fiber reinforcement. Some potential applications for CMC materials involve the replacement of high-temperature metals in aeroengine components aft of the combustor; examples include convergent and divergent flaps and seals at the engine nozzle, and blades and vanes in the turbine portion of the engine.

Most CMC laminates are fashioned from an assemblage of two-dimensional (2D) woven fabrics. As a result, the thru-thickness direction typically lacks fiber reinforcement and exhibits significantly lower strengths and toughness than the in-plane directions. For this reason, there is concern that thermostructural components fabricated from 2D fabric may have inadequate interlaminar tensile (ILT) strengths for some applications, thereby increasing their vulnerability to interply delamination when subjected to high thru-thickness thermal gradients, acoustic/high cycle fatigue, impact damage, free edge effects, and/or applied normal loads. For these reasons, it is important to fully characterize the ILT strengths of candidate CMC materials systems to ensure robust design when they are being considered for use in high-temperature structural applications.

Currently, the ILT strength of CMCs is evaluated as per ASTM C1468 [1] ("Standard Test Method for Transthickness Tensile Strength of Continuous Fiber-Reinforced Advanced Ceramics at Ambient Temperature"). This test method, commonly referred to as the flatwise tension (FWT) test, involves the use of square of circular planform test coupons machined from flat plate stock of representative CMC material. The specimens are typically machined from thin-gage plate material; for this reason it is not possible to directly grip the specimen for the application of the thru-thickness normal loading required to initiate ILT failure. Instead, loading blocks are adhesively bonded to the opposing faces of the specimen in order to facilitate tensile loading in the thru-thickness direction. The lack of availability of high-strength, high-temperature structural adhesives currently precludes the applicability of this methodology to elevated temperature testing. A robust test method for performing ILT measurements at elevated temperatures is currently lacking and must be addressed to enable the serious consideration of CMC materials for insertion into high-temperature structural applications.

1.1 Proposed High-Temperature Specimen Configuration

This work targets the aforementioned problem by introducing an alternative ILT specimen design that is conducive to testing at elevated temperatures. The body of the proposed specimen is machined from flat CMC plate stock. V-shaped notches are machined into the thickness, and a state of interlaminar tension is induced between the specimen notches with wedge-type fixtures loaded in compression (Figure 1). Because this configuration only requires the application of compressive loading, it is readily adaptable to testing at high temperatures with the use of monolithic ceramic test fixturing. The conceptual specimen geometry, illustrated in Figure 1, shows the four primary geometric parameters – namely the specimen thickness (t), the notch half-angle (φ), the notch spacing (b), and the specimen width (w).

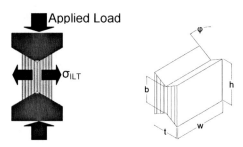

Figure 1. Proposed high-temperature ILT specimen.

The stresses induced in the loaded specimen were determined as a function of specimen geometry using a Strength of Materials analysis approach and assuming: (1) linear-elastic material behavior; (2) small displacement behavior and negligible friction between the specimen and loading fixtures; (3) loading from the fixture is transferred uniformly to the specimen surfaces; (4) effects of the notch tip (i.e. stress concentrations) are ignored. The loaded specimen was decomposed into the individual components of stress in detail in previous work [2]. A formulation for the interlaminar tensile stress of the loaded specimen was derived from this previous analysis, and is provided in Equation 1 as a function of the applied load (P), specimen notch half-angle (φ), specimen width (w), and notch tip spacing (b):

$$\sigma_{ILT} = \frac{P}{wbTAN\varphi} \qquad (1)$$

2.0 EXPERIMENTAL PROCEDURES

2.1 Materials

Three (3) distinct CMC material systems were evaluated under the current work; these were a CG Nicalon fiber-reinforced SiC/SiC laminate with a chemical vapor infiltration (CVI)-derived SiC matrix, a T-300 fiber-reinforced C/SiC laminate with a CVI-derived SiC matrix, and a Nextel 720 fiber reinforced Oxide/Oxide laminate with an alumina matrix derived from sol-gel processing techniques. A summary of the various material systems is provided in Table I.

Table I. CMC Material Systems

Material	Type	Fiber	Matrix	Matrix Processing
1	SiC/SiC	CG Nicalon	SiC	CVI
2	C/SiC	T-300	SiC	CVI
3	Oxide/Oxide	Nextel 720	Alumina	Sol-Gel

2.2 Flatwise Tension Test Setup

Flatwise tension (FWT) testing was used to establish baseline ILT strength properties for all of the CMC laminates tested; these baseline ILT properties were used as a basis to evaluate the correlation of apparent ILT strengths determined using various configurations of the proposed high-temperature test technique. Flatwise tension testing was conducted on round 19mm-diameter specimens extracted normal to the surface of the laminate. Theses specimens were bonded between steel loading dowels, and subsequently subjected to monotonic tensile loading in an Applied Test Systems (ATS) universal test frame under displacement-controlled conditions at a loading rate of 2.5mm per minute. The interlaminar tensile stress (σ_{ILT}) was calculated as a function of the failure load (P) and cross-sectional area of the specimen (A) according to Equation 2:

$$\sigma_{ILT} = \frac{P}{A} \qquad (2)$$

Universal joints were utilized to alleviate undesirable bending stress components during the test. A schematic representation of the test setup is provided in Figure 2.

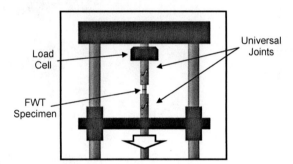

Figure 2. Flatwise tension (FWT) test setup.

2.3 High-Temperature ILT Test Setup

Test fixturing was designed and manufactured to facilitate the wedge loading of the proposed ILT specimen at both ambient and elevated temperatures. A schematic of the fixture assembly, along an image of the fixture components is provided in Figure 3. The compressive load is transferred into the furnace hot zone during high-temperature testing by way of 12.7mm-diameter alumina rods (Figure 4). An aerosol-based Boron Nitride (BN) coating was applied to the specimen v-notch regions prior to testing. The coating served as a compliant interfacial media to promote more uniform loading at the specimen/loading fixture contact interface, as well as to reduce interfacial friction. The coating was implemented during both room temperature and elevated temperature testing. Testing was performed in an Applied Test Systems (ATS) Series 900 universal test frame under a displacement-controlled loading rate of 2.5mm per minute. This typically resulted in tests lasting 3-5 seconds, whereby failure was indicated by a sharp and significant drop in the measured load. The test rate is an important consideration while testing at elevated temperatures; the loading of the specimen should be rapid enough such that specimen fails in fast-fracture, rather than due to time-dependent failure mechanisms (i.e. slow crack growth), such as those demonstrated in interlaminar shear testing by Choi [2].

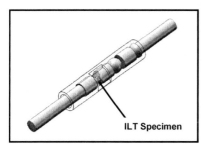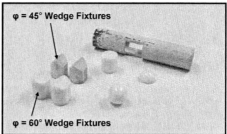

Figure 3. Alumina fixtures for both ambient and elevated temperature ILT testing.

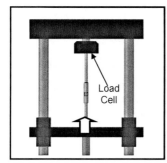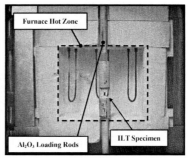

Figure 4. High-temperature ILT test setup.

2.4 Specimen Sizing

Previous development of the proposed test method by the author [3] included a parametric study to evaluate six (6) distinct potential specimen configurations, including two (2) values of the notch half-angle, φ (45° and 60°), and three (3) values of the specimen notch tip spacing to thickness ratio ($b/t = \frac{1}{2}$, 1, and 2). The performance of a particular specimen configuration was evaluated by comparing its resultant apparent ILT strength at room temperature to ILT strengths determined using the ASTM prescribed room temperature flatwise tension test method. This testing was performed upon a 6.4mm-thick SiC/SiC laminate with CG Nicalon fiber reinforcement. The results of this testing demonstrated the best correlation of apparent ILT was achieved with b/t ratios of 1; both notch half-angle configurations (45° and 60°) performed comparably, and no determination could be made as to which was preferred. The results of this preliminary sizing study have been applied to the current work, which focuses on thinner 3.2mm-thick CMC laminates. All specimens tested in the current work had b/t ratios of 1, notch half-angles (φ) of both 45° and 60°, and specimen widths (w) of 12.7mm.

2.5 High-Aspect Ratio Tip Notch

During exploratory testing of the proposed ILT specimen with 3.2mm-thick CMC materials, it was qualitatively observed that failures would occasionally occur away from the prescribed fracture plane between the notch tips (Figure 5). This assumed failure plane is the location of maximum ILT stress in the specimen, and serves as the basis for calculating the resultant ILT strength. This is problematic in a sense that it introduces some ambiguity into the results for a couple of different reasons. First, the fracture may be occurring at regions of lower ILT stress than is used for calculating the ILT strength of the material. The second, and more significant issue is the close proximity of the fracture to the contact interface between the specimen and loading fixtures; it cannot be known with any certainty the influence of any localized contact stresses on the initiation of failure.

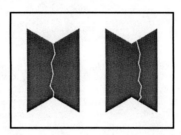

Figure 5. Desirable (left) and undesirable (right) ILT specimen failures.

These concerns were addressed through the introduction of an additional 0.3mm-thick notch feature at the existing notch apex. Two different notch depth-to-width aspect ratios (1:1 and 3:1) for this feature were evaluated. The CVI-based SiC/SiC laminate material with CG Nicalon fiber reinforcement was utilized for this evaluation. Ten (10) specimens with each notch configuration were evaluated under ambient conditions and compared to baseline ILT strengths obtained with FWT testing. These specimens utilized a notch half-angle (φ) of 60°, b/t ratio of 1, and a width of 12.7mm. The results of the testing are shown in Table II. Both the 1:1 and 3:1 tip notch configurations resulted in comparable average ILT strengths at 14.0 MPa and 14.1 MPa, respectively. These numbers correlate well with the average ILT strength of 14.5 MPa obtained with the FWT testing. The 1:1 configuration demonstrated more significant scatter in the data with a standard deviation and coefficient of variation (COV) of 2.3 MPa and 17.0%; this compares to 1.0 MPa and 6.7% for the 3:1 notch configuration and 0.8 MPa and 5.6% for the FWT specimens. The 3:1 tip notch resulted in specimen failures that were confined to the prescribed fracture plane between the specimen notch tips. However, testing of the 1:1 notch configuration resulted in a number of specimen failures that occurred outside of the prescribed fracture area (Figure 6). For this reason, as well as the reduced scatter in the ILT strength results, the 3:1 tip notch configuration was adopted for the remainder of the testing detailed in this paper.

Table II. Tip Notch Aspect Ratio Test Results

Test Configuration	n	ILT Strength		
		Avg. (MPa)	Std. Dev. (MPa)	COV
FWT	10	14.5	0.8	5.6%
1:1	10	14.0	2.3	17.0%
3:1	10	14.1	1.0	6.7%

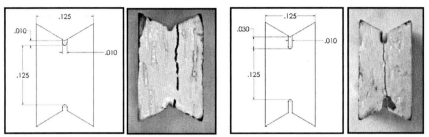

Figure 6. Tip notch aspect ratio configuration of 1:1 (left) and 3:1 (right).

3.0 EXPERIMENTAL RESULTS AND DISCUSSION

3.1 Correlation of Ambient Temperature ILT and FWT Results

The results of flatwise tension testing and ambient testing of the proposed high-temperature ILT specimens were part of the notch tip aspect ratio study reported in §2.5. To reiterate the results, Ten (10) replicates of the SiC/SiC material were tested in flatwise tension, resulting in an average ILT strength of 14.5 MPa, with a standard deviation and COV of 0.8 MPa and 5.6%. Additionally, ten (10) replicates of the proposed ILT test technique were evaluated at ambient temperatures, resulting in an average ILT strength of 14.1 MPa, with a standard deviation and COV of 1.0 MPa and 6.7%.

Six (6) replicates of the C/SiC material were tested in flatwise tension, resulting in an average ILT strength of 11.3 MPa, with a standard deviation and COV of 2.1 MPa and 18.1%. Ten (10) replicates were tested in both the 45° and 60° notch half-angle configurations of the proposed high-temperature ILT specimen at ambient temperatures. The 45° configuration yielded an average ILT strength of 10.6 MPa, with a standard deviation and COV of 2.3 MPa and 22.2%. The 60° configuration yielded an average ILT strength of 11.1 MPa, with a standard deviation and COV of 2.3 MPa and 21.0%.

Thirteen (13) replicates of the Oxide/Oxide material were testing in flatwise tension, resulting in an average ILT strength of 4.9 MPa, with a standard deviation and COV of 0.9 MPa and 18.7%. Ten (10) replicates of the proposed ILT specimen were tested in the 45° configuration, resulting in an average ILT strength of 0.8 MPa, with a standard deviation and COV of 0.2 MPa and 23.3%. Thirty (30) replicates were tested in the 60° configuration, resulting in an average ILT strength of 5.0 MPa, with a standard deviation and COV of 0.8 MPa and 15.8%.

With the exception of the 45° notch half-angle configuration of the Oxide/Oxide composite, all configurations of the proposed ILT specimen demonstrated good correlation of ILT strength results to those obtained using flatwise tension test techniques. In the case of the C/SiC material, both notch half-angle configurations tested (45° and 60°) yielded very consistent results. The 45° notch half-angle configuration of the Oxide/Oxide material did not correlate to either the FWT results, or the results of the alternate 60° configuration. The reasons for this are not immediately clear, and are to be a subject of future investigation. A summary of the measured ILT strength data is provided in Table III. Typical FWT fracture surfaces and failed ILT specimens for the three CMC material systems are provided in Figures 7 and 8, respectively.

Table III. Summary of Room Temperature ILT Results

| Material | Test Type | n | ILT Strength | | |
			Avg. (MPa)	Std. Dev. (MPa)	COV
SiC/SiC	60°	10	14.1	1.0	6.7%
	FWT	10	14.5	0.8	5.6%
C/SiC	45°	10	10.6	2.3	22.2%
	60°	10	11.1	2.3	21.0%
	FWT	6	11.3	2.1	18.1%
Oxide/Oxide	45°	10	0.8	0.2	23.3%
	60°	30	5.0	0.8	15.8%
	FWT	13	4.9	0.9	18.7%

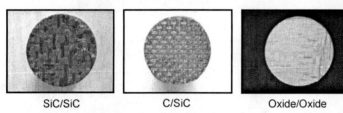

SiC/SiC   C/SiC   Oxide/Oxide

Figure 7. Typical FWT fracture surfaces.

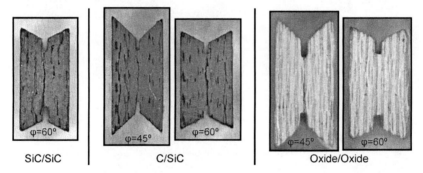

SiC/SiC   C/SiC   Oxide/Oxide

Figure 8. Typical failed high-temperature ILT specimens tested at ambient temperatures.

3.2 High-Temperature ILT Test Results

A limited number of high-temperature ILT tests were performed at 1100°C in air using the C/SiC and Oxide/Oxide materials. Three (3) C/SiC specimens were tested in the 45° configuration, resulting in an average ILT strength of 11.0 MPa, with a standard deviation and COV of 2.1 MPa and 19.1%. Two (2) C/SiC specimens were tested in the 60° configuration, resulting in an average ILT strength of 9.0 MPa, with a standard deviation and COV of 1.8 MPa and 19.9%. Five (5) specimens of the Oxide/Oxide material were tested in the 60° configuration, yielding an average ILT strength of 4.3 MPa, with a standard deviation and COV of 0.9 MPa and 20.7%.

The high-temperature strength results for both materials are within the experimental scatter observed during room temperature testing; therefore, no obvious ILT strength degradation was observed at 1100°C in air for either the C/SiC or Oxide/Oxide materials for the limited data available. More high-temperature data will be generated in future work to confirm this. The high-temperature results are summarized in Table IV, and a comprehensive plot of the ILT test data for the three CMC material systems is provided in Figure 9.

Table IV. Summary of High-Temperature ILT Results

Material	Test Type	n	Temperature (°C)	ILT Strength		
				Avg. (MPa)	Std. Dev. (MPa)	COV
C/SiC	45°	10	25	10.6	2.3	22.2%
	45°	3	1100	11.0	2.1	19.1%
	60°	10	25	11.1	2.3	21.0%
	60°	2	1100	9.0	1.8	19.9%
Oxide/Oxide	60°	30	25	5.0	0.8	15.8%
	60°	5	1100	4.3	0.9	20.7%

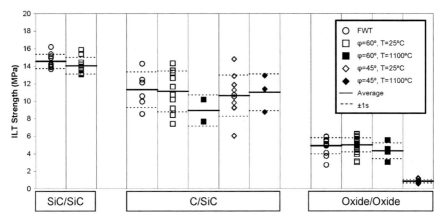

Figure 9. Graphical representation of ILT strength distributions.

4.0 CONCLUSION

A test method was proposed for the interlaminar tension testing of CMC materials at elevated temperatures. The current work borrows from previous development [2] with regard to establishing the appropriate specimen geometry for the test method. A high aspect ratio tip notch was proposed in the current work as a modification to the specimen geometry that would allow for a more clearly defined "gage section" in the specimen that would produce a more repeatable failure mode. This modification was further motivated by a desire to relocate the fracture surface of the specimen away from the contact interface of the specimen and loading fixtures, which could introduce additional undesirable components of localized stress. This proposed specimen configuration was evaluated using a SiC/SiC CMC material tested under ambient conditions; the ILT strength results obtained demonstrated good correlation to baseline ILT strengths obtained via conventional flatwise tension (FWT) test techniques.

Additional testing was performed on C/SiC and Oxide/Oxide materials using the adopted specimen geometry. The majority of this testing was performed at ambient temperatures using two distinct specimen geometries with notch half-angles (φ) of both 45° and 60°. The C/SiC material demonstrated good correlation of observed ILT strengths to those attained during FWT testing for both specimen configurations. The Oxide/Oxide material demonstrated good correlation of observed ILT strengths to FWT test results for the 60° configuration, while the 45° configuration yielded ILT strengths that were significantly lower than those obtained from both the 60° specimen and the FWT testing. The reason for the inconsistent results of the Oxide/Oxide material in the 45° configuration is uncertain at this time.

Elevated temperature testing was performed at 1100°C in air on both the C/SiC and Oxide/Oxide materials. Both materials yielded ILT strengths that were comparable to those observed during room temperature testing; therefore, these materials don't appear to suffer any significant degradation of ILT strength up to 1100°C in air. Additional replicates will be tested in future work to confirm this result, and the ILT properties of these materials will be characterized at various additional temperatures.

The levels of scatter observed in the ILT test data were consistent for both the FWT testing and the proposed high-temperature ILT technique in all three CMC material systems evaluated. It was initially anticipated that the proposed technique would produce some level of increased scatter in the ILT strengths results. Typically, ceramic materials exhibit statistical strength distributions that relate to their size. Provided that the ILT failure mode is a brittle fracture, and that the stressed area of the FWT specimen is seven times that of the high-temperature ILT specimens tested in this work, some discrepancy in the distribution of strengths between the two methods was anticipated. However, this result was not observed. Despite the difference in specimen sizes, the levels of scatter was comparable for both techniques.

ACKNOWLEDGEMENTS
We would like to acknowledge that this work has been funded by NAVAIR under contract number N68335-08-C-0491, and monitored by Dr. Sung Choi.

REFERENCES
[1]ASTM Standard C 1468-06: Standard Test Method for Transthickness Tensile Strength of Continuous Fiber-Reinforced Advanced Ceramics at Ambient Temperature, *Annual Book of ASTM Standards*, Vol. 15.01, ASTM International, West Conshohocken, PA, 2009.
[2]Choi, S.R., "Life Limiting Behavior of Ceramic Matrix Composites Under Interlaminar Shear at Elevated Temperatures," *Processing and Properties of Advanced Ceramics and Composites III*, The American Ceramic Society, Westerville, OH; Ceramic Transactions, **225** (2011).
[3]Engel, T., W. Steffier, and T. Magaldi, "High-Temperature Interlaminar Tension Test Method Development for Ceramic Matrix Composites," *Processing and Properties of Advanced Ceramics and Composites III*, The American Ceramic Society, Westerville, OH; Ceramic Transactions, **225** (2011).

HIGH TEMPERATURE FURNACE DOOR TEST FOR WOLLASTONITE BASED CHEMICALLY BONDED PHOSPHATE CERAMICS WITH DIFFERENT REINFORCEMENTS

H. A. Colorado[a,b], C. Hiel[c, d], H. T. Hahn[a], J. M. Yang[a]

[a]Materials Science and Engineering Department, University of California, Los Angeles, USA
[b]Universidad de Antioquia, Mechanical Engineering. Medellin-Colombia.
[c]Composite Support and Solutions Inc. San Pedro, California, USA.
[d] MEMC-University of Brussels (VUB)

ABSTRACT
        The main goal in this paper is to test the chemically bonded phosphate ceramics (CBPC) composites reinforced with different materials, using a high temperature furnace door test. The CBPCs composites were fabricated by mixing phosphoric acid formulation and Wollastonite powder in a mechanical mixer. CBPCs are ceramic materials consolidated by chemical reactions at room temperature, instead of the firing or sintering processing used in traditional ceramics, which constitutes a good contribution for a clean environment.
Glass, graphite, Basalt and SiC fibers, as well as graphite nanoparticles have been used as reinforcements for the CBPC. The composites samples at 25°C were placed in a furnace at 1000°C. A thermocouple was located in the center of the door. The characterization was conducted by both Optical and X-ray diffraction.

INTRODUCTION
        Chemically bonded ceramics (CBCs) are materials fabricated completely at low temperatures, typically less than 100°C. Thus, for these materials, the bonding takes place by a chemical reaction at low temperatures, as opposed to fusion or sintering at elevated temperatures in traditional ceramics and cements which is a mixture of ionic, covalent, and van der Waals bonding (with the ionic and covalent dominating). CBPCs are a promissory in situ solution for fire resistant applications since it possess excellent fire resistance and thermal insulation; manufacturing is easy, inexpensive and environmentally benign[1].
        CBPC is the result of mixing wollastonite powder ($CaSiO_3$) and phosphoric acid (H_3PO_4) to produce calcium phosphates (brushite $CaHPO_4 \cdot 2H_2O$, monetite ($CaHPO_4$) and calcium dihydrogenphosphate monohydrate ($Ca(H_2PO_4)2 \cdot H_2O$)) and silica for molar ratios between 1 and 1.66[2]. The reaction of CBPC is exothermic and relatively fast, producing air bubbles that are very difficult to remove because of the viscosity of the paste (of around 4000 cp). Therefore, dilute acid will reduce the viscosity and help made void-free samples. An optimized procedure to increase the curing time and pot life, reduce viscosity, and avoid a porous material was recently developed[3]. Little research has been conducted about CBC with reinforcements by fibers[4-5] where thermo and mechanical stability can really be improved.
        The main goal in this paper is to explore the CBPC reinforced as a substitute of materials under thermal shock and fire environments. The most important challenges to solve are the cracking caused by the shrinkage, the phase instabilities and the manufacturing issues to produce samples without pores.

EXPERIMENTAL PROCEDURE

Samples manufacturing

For CBPC samples (without reinforcement), phosphoric acid formulation (from Composites Support and Solutions-CS&S) and Wollastonite powder (M200 from CS&S, see Table 1) were mixed to obtain a 1.2 ratio of liquid/powder.

Table 1 Chemical composition of Wollastonite powder.

Sample	CaO	SiO_2	Fe_2O_3	Al_2O_3	MnO	MgO	TiO_2	K_2O
M100	46.25	52.00	0.25	0.40	0.025	0.50	0.025	0.15

The mixing of Wollastonite with the acidic formulation was conducted in a Thinky Mixing AR-20 apparatus. Both Wollastonite powder and phosphoric acid formulation were cooled to 5°C in a closed container in order to reduce the viscosity and increase the pot life of the resin[3]. The mixing time was 2 min for all samples. A Teflon fluoropolymer mold was used to prevent adhesion of the CBPC and was then covered with plastic foil to preserve humidity and decrease the shrinkage.

CBPC reinforced with fibers were fabricated manually as shown in Figure 1a. About 20wt% of fibers are in all composites reinforced with fiber fabricated in this research. Glass fibers Textrand 225 from Fiber Glass Industries, Graphite fibers Tenax-A 511, Basalt fibers BCF13-1200KV12 Int from Kammemy Vek, and SiC fibers Nicalon were used in this research. For CBPCs reinforced with GNPs, Wollastonite and the acidic solution were mixed for 2min. Then, 1.0wt% GNPs were added to the mixture and mixed for 1min.

All samples were released after 24 hours and then dried at room temperature in open air for 5 days. Samples were dried slowly in order to prevent residual stresses in the electrical furnace to remove the unbonded water content following this sequence: 5 days in open air at room temperature, followed by a thermal process in the furnace of 1 day at 50°C, then 1 day at 100°C and finally 1 day at 200ºC.

Thermal shock tests

For evaluation of the thermal shock stability, thick samples of 1.27 x 20 x 20 cm^3 were fabricated. This shape was chosen because it easily allows the observation damages while the sample is heated up with one side at high temperature and the other in contact with air. Thermal stability was measured by the water quench test. Samples were dried at 110°C and then transferred into an electric furnace at 1000°C for 30min, then quenched in water and left for 3 min, dried following the same process described above, and returned to the furnace at 1000°C for several thermal cycles. This was repeated 5 times and not further damage was observed after the deterioration obtained from the first thermal shock test.

XRD at high temperature

High temperature X-Ray Diffraction (XRD) experiments were conducted on X'Pert PRO equipment, at 45KV and scanning between 10º and 80º. The samples were ground in an alumina mortar and XRD tests were performed at room temperature. Testing was performed fwith a measure at room temperature (for reference), followed by a measure at 1000 °C, and followed an annealing at 1000 º C. The scan step size was 0.0167113º and wavelength $k\alpha_1$ was 1.540598Å.

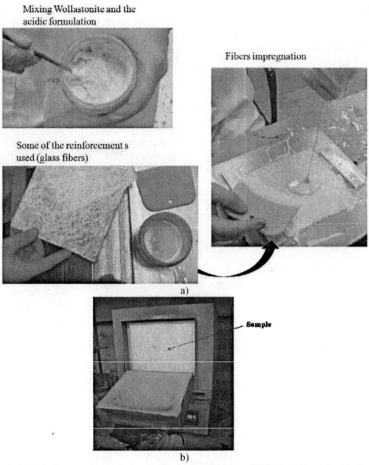

Figure 1 a) Typical ceramic matrix composites manufacturing of CBPC reinforced with fibers, b) typical thick plate sample in the door oven thermal shock test

Other characterization

Three point bending tests (3PBT) were conducted in an Instron 4411 machine over rectangular samples obtained from the thick plates described above. The span length was 80mm, the samples dimensions were 12.5x15x200 mm^3 and the crosshead speed was 2.5 mm/min. All samples were tested after drying at room temperature in open air for 5 days, and others after the following drying process: 50 °C for 24 hours, followed by 105 °C for 24 hours and 200 °C for 24 hours. To see the microstructure, sample sections were progressively ground using silicon carbide papers of 500, 1000 and 2400 in an optical microscope.

ANALYSIS AND RESULTS

High Temperature XRD characterization was conducted over CBPC in order to determine the chemical stability at high temperatures. The thermal process followed was progressively taking data at 25°C and 1000°C with corresponding annealing for 30min. The CBPC at room temperature is a mixture of remaining Wollastonite, brushite and amorphous phases. After the high temperature, new phases appeared as shown in Figure 2.

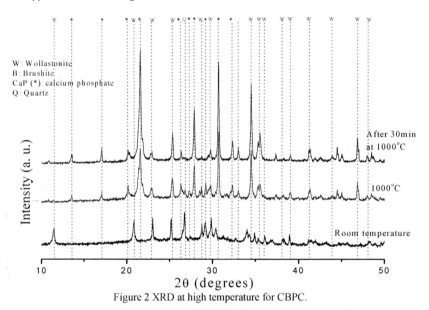

Figure 2 XRD at high temperature for CBPC.

Figure 3 shows cross sectional optical images for different samples fabricated. Figure 3a shows the CBPC before the thermal tests, Figure 3b shows the CBPC reinforced with glass fiber before the tests. Similar images were obtained for all composites reinforced with fibers and GNPs before the thermal shock test. Figures c, d, e and f show images for the CBPC reinforced with glass, basalt, carbon and SiC fibers after the first thermal shock test. The cracks are mainly due to the shrinkage in the phosphate phases during the first heating[6], when they lost the bonded water. This phase transformation corresponds to brushite ($CaHPO_4 \cdot 2H_2O$) transforming to monetite ($CaHPO_4$). Thus, after the first cycle, no much structural changes are expected[2].

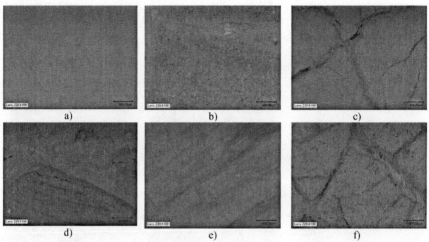

Figure 3 typical images for the CBPC thick plates used in the door oven thermal shock tests here presented, before thermal tests: a) CBPC b) CBPC with glass fibers; after the tests: c) CBPC with glass fibers, d) CBPC with basalt fibers, e) CBPC with carbon fibers and c) CBPC with SiC fibers.

The three point bending results for samples before and after the thermal shock tests are presented below. The decrease in the bending strength has been associated mainly to the deterioration of the ceramic matrix (shrinkage leading micro cracks) and the deterioration of interfaces involved, not only in the bonding fiber-matrix but also in the amorphous matrix and the Wollastonite particles. These results are currently under improvement and will be presented in a future paper.

SAMPLE	Bending Strength [MPa]	Bending Strength after Thermal Shock [MPa]
Wo-CBPC	10.5	3.5
Wo-CBPC + GNPs	20	4.0
Wo-CBPC + glass fibers	138	10.5
Wo-CBPC +graphite fibers	200	13.0
Wo-CBPC + basalt fibers	151	12.5
Wo-CBPC + SiC fibers	143	19.0

CONCLUSION

The CBPC reinforced with different fibers and GNPs under thermal shock test has been presented in this paper. The composites have big structural damages in the first thermal cycle as shown in the microscopy images as well as in the results from the bending tests. XRD tests at high temperature showed that new crystalline phases appear when samples are heated at 1000C°. The thermal shock test by using samples as a furnace door has being showed effective to evaluate materials under oxidation environments and thermal stresses, typical in accidental fires, which opens up new applications for the CBPCs.

ACKNOWLEDGEMENTS

The authors wish to thank to the NIST-ATP Program through a grant to Composites and Solutions Inc. (Program Monitor Dr. Felix H. Wu) and to Colciencias from Colombia for the grant to H. A. C.

REFERENCES

[1] D. M. Roy, New Strong Cement Materials: Chemically Bonded Ceramics, *Science*, **235**, 651-58 (1987).

[2] G. Mosselmans, Monique Biesemans, R. Willem, J. Wastiels, M. Leermakers, H. Rahier, S. Brughmans and B. Van Mele, Thermal Hardening and Structure of a Phosphorus Containing Cementitious Model Material, *Journal of Thermal Analysis and Calorimetry*, **88**, 723–729 (2007).

[3] H. A. Colorado, C. Hiel and H. T. Hahn, Processing-Structure-Property Relations of Chemically Bonded Phosphate Ceramic Composites, *Bulletin of Materials Science*, **34**, Suppl. No. 1, 1-8, (2011).

[4] Arun S. Wash, Chemically bonded phosphate ceramics, *Elsevier*, 283p (2004).

[5] S. A. Dimitry, Characterization of reinforced chemically bonded ceramics, *Cement and Concrete Composites*, **13**, Issue 4, 257-263 (1991).

[6] H. A. Colorado, C. Hiel and H. T. Hahn, Chemically bonded phosphate ceramic composites under thermal shock and high temperature conditions, *Society for the Advancement of Material and Process Engineering*, Seattle, Washington (2010).

MICROSTRUCTURE AND PROPERTIES OF Al$_2$O$_3$ CERAMIC COMPOSITE TOUGHENED BY DIFFERENT GRAIN SIZES OF LiTaO$_3$

Yangai Liu∗, Zhaohui Huang, Minghao Fang
School of Materials Science and Technology, China University of Geosciences (Beijing)
Beijing 100083, P. R. China

ABSTRACT

Ceramic composite toughened by different grain sizes of LiTaO$_3$ was illustrated in this paper. Al$_2$O$_3$ and LiTaO$_3$ could coexist after sintering. The addition of weak LiTaO$_3$ with lower melted point contributed to the improving of sintering and mechanical properties of Al$_2$O$_3$. Non 180° domains formed in LiTaO$_3$ grains in the Al$_2$O$_3$ ceramic composite. The domain configuration in the composite was independent of the grain size of LiTaO$_3$. The bending strength and fracture toughness of 42 nm LiTaO$_3$/Al$_2$O$_3$ ceramic composite were 511 MPa and 6.13 MPa·m$^{1/2}$, respectively, which were higher than that of 3.0 μm LiTaO$_3$/Al$_2$O$_3$ ceramic composite.

INTRODUCTION

Many second phases such as nano-particles, fiber, whisker, and metallic particles have been added into ceramic materials as reinforcements to improve the mechanical properties[1-3]. In recent years, piezoelectric materials with unique microstructure have been introduced into structural ceramics and some new toughening mechanisms have been proposed [4-8]. In our studies, different grain sizes of LiTaO$_3$ were added into Al$_2$O$_3$ to prepare LiTaO$_3$/Al$_2$O$_3$ ceramic composites by hot pressing. The influences of LiTaO$_3$ additions on microstructure and mechanical properties of the composite were summarized in this paper.

EXPERIMENTAL

Commercially available Al$_2$O$_3$ powder (0~0.3 μm, High Tech Ceramic Institute, Beijing, China), LiTaO$_3$ powder (3.0 μm, Dongfang Tantalum Joint-stock Corporation, Ningxia, China) and nano LiTaO$_3$ powders (42nm) prepared by the sol-gel method using Li$_2$CO$_3$ and Ta$_2$O$_5$ as raw materials, citric acid as coordination agent, ethylene glycol as esterifying agent, polyethylene glycol as dispersing agent, were used as base materials. 15 vol% LiTaO$_3$ of different grain sizes and Al$_2$O$_3$ powders were weighed accurately and then mixed for 12 h with ethanol as a solvent for the ball milling. The slurry was dried for 6h and then packed into a carbon die and hot-pressed at 1300°C and 1350°C respectively in vacuum atmosphere to prepare ceramic composites.

The density of composite was determined by the Archimedes method. The crystalline phases were examined by X-ray Diffraction using Cu Kα radiation. The microstructure was observed by transmission electron microscopy. The bending strength and fracture toughness was evaluated via the three-point-bending technique and single-edge-notched-beam method.

∗Corresponding author. Tel.: +86-10-82322186; Fax: +86-10-82322186.
E-mail address: liuyang@cugb.edu.cn (Y. G. Liu)

RESULTS AND DISCUSSION

XRD results showed that no reaction phases occurred in LiTaO$_3$/Al$_2$O$_3$ composite [5]. LiTaO$_3$ and Al$_2$O$_3$ could coexist stably after sintering. The relative density of the ceramic composites is shown in Fig.1. The sintering property of Al$_2$O$_3$ ceramics was improved by the addition of LiTaO$_3$ and both the LiTaO$_3$/Al$_2$O$_3$ ceramic composites had a relative density of more than 97 % after sintering at temperatures lower than 1500 °C, which was much higher than that of pure Al$_2$O$_3$ ceramic sintered at the same conditions.

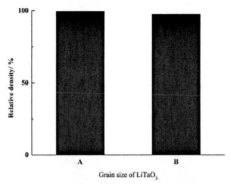

Figure 1. Relative density of LiTaO$_3$/Al$_2$O$_3$ ceramic composites with different grain sizes of LiTaO$_3$ (A: 3um; B:42nm)

Figure 2. TEM images of LTA ceramic composites

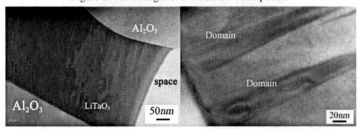

Figure 3. TEM images of domains in 42 nm LiTaO$_3$/Al$_2$O$_3$ ceramic composite

LiTaO₃ grains were distributed at grain boundaries of Al₂O₃ ceramic matrix [5]. TEM images of 42 nm LiTaO₃/Al₂O₃ ceramic composites are showed in Fig.2. The dark phase was LiTaO₃ and the grey phase was Al₂O₃ matrix. Besides 180° domains, many non-180° domains was found in both of the ceramic composites. Fig.3 showed the domains in 42 nm LiTaO₃/Al₂O₃ ceramic composite, which were similar to those in 3.0 μm LiTaO₃/Al₂O₃ ceramic composite [5, 8]. The results suggested that the domain configurations in the composites were independent of LiTaO₃ grain sizes. It was inferred that the formation of non 180° domains in LiTaO₃ grains was due to large stress caused by residual polarization of LiTaO₃ from paraelectric to piezoelectric phase after sintering.

The mechanical properties of LiTaO₃/Al₂O₃ ceramic composites with different grain sizes of LiTaO₃ addition are showed in Fig.4. **There were 4-6 specimens were tested for bending strength as well as fracture toughness for each material type.** The average bending strength and fracture toughness of 3.0 μm LiTaO₃/Al₂O₃ ceramic composite reached 492.9 MPa and 5.3 MPa·m^{1/2}, respectively [5]. The **average** bending strength and fracture toughness of 42 nm LiTaO₃/Al₂O₃ ceramic composite achieved the values of 511 MPa and 6.13 MPa·m^{1/2}, respectively. The results showed that the decrease of grain size of LiTaO₃ addition had more effect on the improvement of mechanical properties of Al₂O₃ ceramics. The toughening mechanisms of the composite were all related to the domains in LiTaO₃ grains [5, 6].

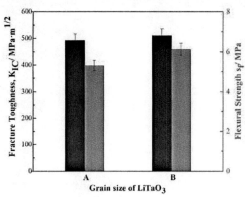

Figure 4. The bending strength and fracture toughness of LiTaO₃/Al₂O₃ ceramic composites
A:3um; B:42nm

CONCLUSION

Non 180° domains formed in both of the Al₂O₃ ceramic composites with different grain sizes of LiTaO₃. Domains in the composite were independent of LiTaO₃ grain sizes. It was inferred that the formation of non 180° domains in LiTaO₃ grains was caused by large stress. The mechanical properties of LiTaO₃/Al₂O₃ ceramic composites were improved than that of pure Al2O3 ceramics. With the decrease of grain size of LiTaO₃ addition, the mechanical properties of the composites increased. The bending strength and fracture toughness of 42 nm LiTaO₃/Al₂O₃ ceramic composite achieved the values of 511 MPa and 6.13 MPa·m^{1/2}, respectively.

ACKNOWLEDGMENTS

This study was supported by National Nature Science Foundation of China (Grant No. 51072186 and 51032007), New Star Technology Plan of Beijing(Grant No.2007A080) and the Fundamental Research Funds for the Central Universities (Grant No. 2010ZD12).

REFERENCES

[1] A. H. Heuer，Transformation Toughness in ZrO$_2$-containins Ceramic, J. Am. Ceram. Soc., **70(10)**, 689-93 (1987).

[2] X. M. Chen and B. Yang, A New Approach for Toughening of Ceramics, Mater. Lett., **33**, 237-40 (1997).

[3] B. Yang, X. M. Chen, and X. Q. Liu, Effect of BaTiO$_3$ Addition on Structures and Mechanical Properties of 3Y-TZP Ceramics, J. Euro. Ceram. Soc., **20**, 1153-57 (2000).

[4] X. M. Chen, X. Q. Liu, F. Liu, and X. B. Zhang. 3Y-TZP Ceramics Toughened by Sr$_2$Nb$_2$O$_7$ Secondary Phase, J. Euro. Ceram. Soc., **21**, 477-81 (2001).

[5] Y.G. Liu, Y. Zhou, and D.C. Jia. Domain Switching Toughening in a LiTaO$_3$ Dispersed Al$_2$O$_3$ Ceramic Composite, Scripta Mater., **47**, 63-67 (2002).

[6] Y. G. Liu, Y. Zhou, and D.C. Jia. Domain Structures and Toughening Mechanism in Al$_2$O$_3$ Matrixceramic Composites Dispersed with Piezoelectric LiTaO$_3$. Mater. Sci. Eng., A, **347**, 359-62 (2003).

[7] Y.G. Liu, Y. Zhou, and D.C. Jia. Polarization Toughening of LiTaO3 Particles Dispersed Al$_2$O$_3$ Ceramic Composite. Ceram. Int., **31**, 81-85 (2005).

[8] Y.G. Liu, Y. Zhou, and D.C. Jia. Characterization of Domain Configurations in LiTaO$_3$/Al$_2$O$_3$ Ceramic Composites. Mater. Chem. Phys., **125**, 143-47 (2011).

EFFECT OF COMPOSITION OF BORON ON THE TRIBOLOGICAL PERFORMANCE OF ALUMINA MATRIX MULTIFUNCTIONAL COMPOSITES FOR ENERGY EFFICIENT SLIDING SYSTEMS

R. Paluri
Mechanical Engineering, Texas A&M University
College Station, Texas, USA
Marine Engineering Technology, Texas A&M University
Galveston, Texas, USA

S. Ingole
Marine Engineering Technology and Marine Biology, Texas A&M University
Galveston, Texas, USA

ABSTRACT
   With upcoming stringent norms to improve energy efficiency and environmental safety, reducing frictional losses of the sliding systems are crucial. Newly developed alumina matrix multifunctional composites (AMMCs) showed promising mechanical properties and frictional performance. AMMCs have potential applications in low friction applications due the presence of low friction components in the matrix. AMMCs were synthesized using powder compaction. To optimize the composition of reinforcement (boron) and to correlate the structure and properties, it is important to understand the effect of reinforcement on the in-situ phase formation and their subsequent effect on the friction and wear properties. Therefore, this study will discuss the effect of boron on in-situ phases and friction of AMMCs. It was found that sintering temperature, composition of reinforcement have influence on the friction and wear resistance of these composites. The wear resistance of the composite was better than alumina.

INTRODUCTION
   The world reserves of primary energy and of raw materials are obviously limited. More than 90% of the present world's energy demand is supplied by the exhaustible fossil fuels (natural gas, oil, coal, and wood), which are the material basis for the chemical industry. The importance of energy is further weighted by several political and economic factors. Energy saving is of prime importance at this juncture of time. The considerable amount of energy is wasted in friction and wear losses in sliding systems. These loses are estimated to be around 4.5% of GNP. Therefore, efforts to develop new class of materials for lubrication and superior surface properties for efficient sliding systems are paramount.
   Surface properties of a material can also be improved by incorporating a desired friction modifier. Ceramic matrices with dispersed self-lubricating phases for example calcium fluoride (CaF_2), graphite and boron nitride (BN) have been used for various self-lubricating tribological applications. Jianxin *et al*. studied the effect of CaF_2 on the tribological properties of Al_2O_3/TiC ceramic composites. Cemented carbide and hardened steel were used as sliding partners. The presence of CaF_2 improved its lubrication performance and reduced the friction values. This was due to the hexagonal crystal structure of CaF_2 with easy shear at the sliding interface. The presence of dispersed CaF_2 phase caused the self lubrication effect and also acted as a solid lubricant between the ceramic surface and the sliding partner [1].

Gangopadhyay *et al.* studied the effect of graphite on friction and wear properties of alumina matrix composites. Studies were conducted on alumina-graphite composites containing 22% graphite. Steel ring was used as sliding partner. The coefficient of friction reduced from 0.5 for alumina matrix to 0.3 for the composite. The change was not noticeable due to the absence of graphite in the transfer film. A thick compacted layer formed on the surface of the composite resulted in maximum coverage of the graphite regions. Also a very less amount of graphite was observed in the transfer film which was formed on the steel sliding partner. The low wear rate of alumina-graphite composites restricted the supply of graphite to the interface. Hence, the wear of the ceramic matrix controlled the supply of solid lubricant to the interface. The formation of thick oxide layer due to large wear rate of steel further reduced the supply of graphite. Their results suggested that proper selection of combination of matrix, solid lubricant and sliding partner improves the tribological properties of ceramic matrix composites. Also, it is essential that the solid lubricant phase should not be completely covered by non-lubricating wear debris. The formation of continuous solid lubrication on the sliding surface plays a major role in further reducing the coefficient of friction [2].

Carrapichano *et al.* studied the effect of boron nitride (h-BN) on the tribological properties of silicon nitride (Si_3N_4) and its composites. In their study they used hexagonal boron nitride platelets as solid lubricant material to improve the tribological properties of the composites. It was observed that the coefficient of friction was lower for the composites compared to pure Si_3N_4 especially when the platelets were aligned parallel to the direction of sliding. The easy shear in the direction of crystallographic planes of hexagonal boron nitride particles decreased the coefficient of friction. The friction value reduced from 0.85 to 0.65 by the addition of 10% h-BN to Si_3N_4 matrix [3].

Thus it is evident that incorporation of reinforcement into matrix material to form ceramic-matrix composites has shown a considerable improvement in mechanical and tribological properties. This paper focuses on the formation of *in-situ* phases in alumina matrix composites which will contribute to the low friction and wear resistance. These multifunctional composites can be utilized for tailored tribological properties to meet the industry demand. We report here the friction behavior of alumina matrix multifunctional composites (AMMCs).

EXPERIMENTAL PROCEDURE
Composite Synthesis and Characterization

AMMCs were synthesized with varying composition of boron using powder compaction techniques. The green AMMCs were sintered at two different temperatures (1200°C and 1500°C). The details of synthesis and sample preparation can be found at in ref. [4], [5]. Phase characterization and microstructure are studied and are reported elsewhere [4], [5]. The physical properties and surface characterization of the AMMCs can be found in previous publications [6]. Sintered AMMCs are shown in Figure 1.

Figure 1. Sintered AMMC specimen

Tribological Characterization

CSM Tribometer was used to perform dry reciprocating ball-on-flat tests on AMMCs. A spherical counterpart that slides against a flat AMMC surface in a linear sliding motion was used. The ball was fixed rigidly in the holder and attached to the load arm. All tests were conducted room temperature and at a relative humidity of 50-55%. The reciprocating tests were performed using a track length of 6mm. The test was conducted for a period of 30 minutes with a normal load of 1N and a sliding speed of 1 cm/s. AISI 52100 bearing ball of 6mm diameter was used as the counterpart. Coefficient of friction was recorded using the data acquisition software. Average coefficient of friction with standard deviation was measured by utilizing the values at regular time interval during each friction test for three such tests (Figure 3). The wear volume of the spherical counterpart was determined using optical micrographic picture of wear scar.

RESULTS AND DISCUSSIONS

Authors reported previously that aluminum diboride (AlB_2), boron oxide (B_2O_3), and aluminum borate ($Al_{18}B_4O_{33}$) were formed in the matrix for all compositions of boron when AMMCs were sintered at 1200^oC. Further, it was studied that the phases of $Al_{18}B_4O_{33}$ were not observed when AMMCs sintered at 1500^oC for all compositions of boron. $Al_{18}B_4O_{33}$ has orthorhombic crystal structure. AlB_2 and B_2O_3 have hexagonal crystal structure [4], [7].

Coefficient of Friction of Pure Alumina

To study the effect of reinforcement on the friction properties of AMMCs, friction behavior of sintered pure alumina was studied. Figure 2 shows the graph of coefficient of friction with time for pure alumina. It is observed that the coefficient of friction is gradually increasing with time. The value of coefficient of friction increased from an initial value of 0.45 to 0.52 at the end of the test period. As the bulk material is uniform, the gradual increase in friction could have resulted from various factors including surface roughness, hardness, and porosity. However this value is lower than the value reported in literature (which is between 0.9 and 1) [8]. Alumina possesses low density, high hardness and strength, and good wear resistance which make it a good choice for structural applications. It also has low thermal conductivity, high electrical insulation, and chemical resistance [9]. It is often used as a reinforcement material in metals, ceramics, and polymers matrix composites and as high temperature insulation. Alumina is also used in cutting tools, grinding wheels and biomedical implants [10]. In most of these applications tribological properties of alumina are of prime importance. Its high temperature tribological applications are limited. Therefore, the *in-situ* phases those are stable at elevated temperature and provide the friction component will increase the high temperature performance of alumina.

Effect of Sintering Temperature on Coefficient of Friction of AMMCs

Figure 3 shows that coefficient of friction for AMMCs sintered at 1200°C and 1500°C. The coefficient of friction of AMMCs sintered at 1500°C was lower as compared to alumina matrix and AMMCs sintered at 1200°C. The lowest coefficient of friction for AMMCs sintered at 1200°C was 0.37 (± 0.03) for 1 wt % boron AMMCs. The highest value was for 3 wt % boron AMMCs which was 0.54 (± 0.04). For the AMMCs sintered at 1500°C the lowest was 0.19 (± 0.01) for 1 wt % boron AMMCs. Highest value was observed for 3 wt % boron AMMCs which was 0.32 (± 0.01). There is obvious difference between the friction curve of AMMCs sintered at1200°C and 1500°C. A typical friction curve for AMMCs sintered at 1200°C showed initial low friction which later increased and stabilized (Figure 4). A typical friction curves for AMMCs sintered at 1500°C showed initial high friction which later increased and stabilized (Figure 4).

Effect of Composition of Boron on Coefficient of Friction of AMMCs

The phases formed in the matrix when the AMMCs sintered at 1200°C were $Al_{18}B_4O_{33}$ and AlB_2 [4], [7]. The coefficient of friction increased with increase in composition of boron for all AMMCs sintered at 1200°C. At higher composition of boron, the presence of hard $Al_{18}B_4O_{33}$ phases might have increased which contributed to the higher friction. The lower friction compared to matrix alumina might be due the presence of AlB_2. AMMCs sintered at 1500°C, the XRD results showed the presence of AlB_2 and no dominating peaks of $Al_{18}B_4O_{33}$ [4], [7]. Within AMMCs sintered at 1500°C, the coefficient of friction increased with composition of boron. The absence of $Al_{18}B_4O_{33}$ phases might have reduced the friction value further.

It is proposed that the presence of AlB_2 phase provided low coefficient of friction in AMMCs sintered at 1200°C and 1500°C. AlB_2 has layered crystal structure which is similar to layered molybdenum disulphide (MoS_2). The molybdenum atoms are in between the layers of Sulphur atoms. Due to weak Van der Waal forces between the layers of sulfide atoms, it has low shear strength which results in low coefficient of friction [11]. Similarly, aluminum atoms are located at the centers of hexagonal prisms which are formed by the boron sheets as shown in Figure 5.

It was previously reported that for AMMC sintered at 1200°C, the porosity reduced and surface roughness (Figure 6 and 7) was also reduced [4], [7]. This fact has less effect on the friction value for AMMC sintered at 1200°C. AMMC sintered at 1500°C, the porosity increased and surface roughness (Figure 6 and 7) was also increased with composition of boron. The coefficient of friction increased with compsosition of boron for AMMCs sintered at both 1200°C and 1500°C. Higher friction for AMMCs sintered at 1500°C might also be due to the higher porosity. The presence of porosity assists the nucleation of cracks due to high stress concentration areas. As a result, cracking along grain boundaries occurs which increases with porosity. This microcracking or grain pullout results in generation of wear debris in the parh of the wear track, increasing the contact area thereby increasing the coefficient of friction [12]. This phenamenon is might be presence at smaller scale in case of AMMCs sintered at 1200°C.

Wear behavior of AMMCs

The wear volume of counter part (steel ball) is listed in table 1. The wear volume was comparatively lower for the steel balls used when slid against AMMCs sintered at 1200°C sintered at 1500°C. The lowest wear volume was observed when slid against 3 wt % boron

AMMC sintered at 1200°C. The higher wear rate for the 3 wt % Boron AMMCs sintered at 1500°C might be due to higher porosity similarly explained in case of friction.

Table I. Wear volume of the steel ball at different sintering temperatures.

Parameters	1 % Boron	2 % Boron	3 % Boron
AMMC sintered at 1200°C	14.23 x 10^{-12} m^3	4.01 x 10^{-12} m^3	2.016 x 10^{-12} m^3
AMMC sintered at 1500°C	3.62 x 10^{-12} m^3	2.575 x 10^{-12} m^3	4.42 x 10^{-12} m^3

CONCLUSIONS

The boron composition and the sintering temperature have strong influence on the friction and wear behavior of AMMCs. The composition of boron and sintering temperature can control the type of inter-metallic phase formation as reinforcement and thus the wear and friction. Hexagonal crystal structure of AlB$_2$ and its high temperature stability will make AMMCs potential candidate for high temperature sliding systems. It is important to know the effect of individual phases. It is aimed to study this aspect further.

ACKNOWLEDGEMENTS

We acknowledge Dr. Hong Liang (Professor, Dept. of Mechanical Engineering, Texas A&M University) for her valuable guidance and timely discussion. Special thanks to Mr. Kevin Win of Marine Engineering Technology of TAMUG for assistance during specimen preparation and technical help. Ms. Paluri would like to thank Mrs. and Mr. Penuel in manuscript preparation.

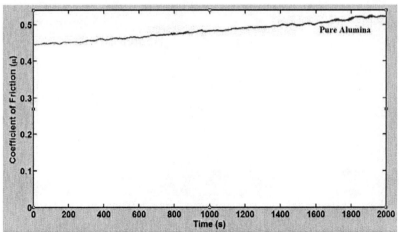

Figure 2. Coefficient of friction for pure alumina

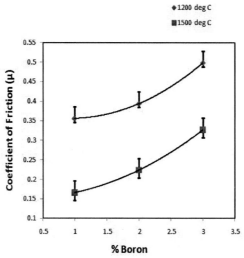

Figure 3. Average coefficient of friction of the composites

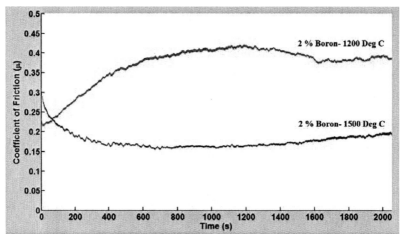

Figure 4. Coefficient of friction for a) 1 wt%, b) 2 wt%, c) 3 wt% boron AMMCs sintered at 1200°C and 1500°C.

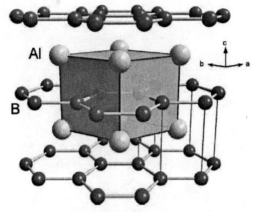

Figure 5. Schematic of layered crystal sturcture of AlB$_2$ [13].

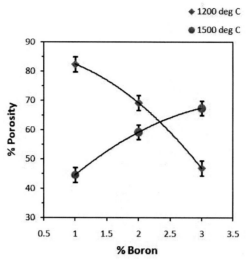

Figure 6. Porosity of composites with different wt % of boron sintered at 1200°C and 1500°C

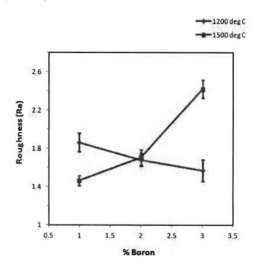

Figure 7. Surface roughness of composites with different wt % of boron for composites sintered at 1200°C and 1500°C

REFERENCE

[1]Jianxin, D., et al., *Tribological behaviors of hot-pressed Al₂O₃/TiC ceramic composites with the additions of CaF₂ solid lubricants*. Journal of the European Ceramic Society, 2006. **26**(8): p. 1317-1323.

[2]Gangopadhyay, A. and S. Jahanmir, *Friction and wear characteristics of Silicon Nitride-Graphite and Alumina-Graphite composites*. Tribology Transactions, 1991. **34**(2): p. 257-265.

[3]Carrapichano, J.M., J.R. Gomes, and R.F. Silva, *Tribological behaviour of Si₃N₄-BN ceramic materials for dry sliding applications*. Wear, 2002. **253**(9-10): p. 1070-1076.

[4]Paluri, R., *Development and characterization of novel alumina based ceramic matrix composites for energy efficiency sliding applications*, in *Mechanical Engineering* 2010, Texas A&M University: College Station. p. 106.

[5]R. Paluri and S. Ingole. *Development and Mechanical Characterization of Novel Alumina-based Composites for Reduced friction and Wear applications*. in *TMS 2011 Annual Meeting & Exhibition*. 2011. San Diego, California: TMS.

[6]R. Paluri and S. Ingole, *Surface characterization of novel alumina-based composites for energy efficient sliding systems*. Journal of Materials (JOM), 2011. **63**: p. 77-83.

[7]R. Paluri and S. Ingole, *Effect of composition of boron on the formation of in-situ phases and microstructure of alumina composites for energy efficient sliding systems*. Ceramic Transactions, 2011: p. In review.

[8]Pasaribu, H.R., J.W. Sloetjes, and D.J. Schipper, *Friction reduction by adding copper oxide into Alumina and Zirconia ceramics*. Wear. **255**(1-6): p. 699-707.

[9]Esposito, L. and A. Tucci, *Microstructural dependence of friction and wear behaviors in low purity alumina ceramics*. Wear, 1997. **205**(1-2): p. 88-96.

[10]Naglieri, V., et al., *Follow-up of zirconia crystallization on a surface modified alumina powder*. Journal of the European Ceramic Society, 2010. **30**(16): p. 3377-3387.

[11]Park, K.T., et al., *Surface Structure of Single-Crystal MoS2(0002) and Cs/MoS2(0002) by X-ray Photoelectron Diffraction*. The Journal of Physical Chemistry, 1996. **100**(25): p. 10739-10745.

[12]He, Y., et al., *Influence of porosity on friction and wear of tetragonal Zirconia polycrystal*. Journal of the American Ceramic Society, 1997. **80**(2): p. 377-380.

[13]Loa, I., et al., *Crystal structure and lattice dynamics of AlB₂ under pressure and implications for MgB₂*. Physical Review B, 2002. **66**(13): p. 134101.

AN INVESTIGATION INTO SOLID-STATE EXPANSION OF CERAMIC MATERIALS

Ariane Erickson
Montana Tech of the University of Montana
Butte, MT, U.S.A.

C. Hank Rawlins
eProcess Technologies U.S.
Butte, MT, U.S.A.

ABSTRACT
Solid-state expansion of ceramic materials can be induced by a variety of methods. Thermal and physical energy absorption, gas and liquid phase reactions, surface diffusion and bombardment, and crystallographic transformations can all produce physical expansion of ceramic materials. Physical expansion in brittle ceramics often leads to cracking and failure; however, there are cases where controlled solid state expansion improves ceramic material properties. A key interest of our research is harnessing the energy of expansion for beneficial use. One example use is expandable mortars and cements; where expansion induced by a chemical reaction is harnessed for mining and quarrying. This project is focused on understanding the thermochemical processes and crystallographic changes from ceramic expansion and propose novel uses of this phenomena for the energy and materials industries.

INTRODUCTION

Ceramic materials are known for their outstanding chemical and physical stability. When an application calls for a solid-state component to maintain shape against corrosive attack or at very high temperatures or against abrasive wear, usually a ceramic is the most cost-effective material to meet this performance target. Compared to metals and polymers, ceramics are more corrosion resistant, especially in aqueous environments, have much higher melting temperatures, and exhibit higher hardness and compressive strength.

Even with their outstanding ability to maintain solid shape and form, at certain conditions ceramic materials will undergo solid-state expansion. The strength of the covalent bonds present in ceramics prevents gross or rapid expansion under normal thermal-mechanical changes however the enlargement that does occur is measureable. Solid-state expansion is typically undesirable and the materials engineer is tasked to provide suitable materials selection to minimize the magnitude of this effect. Undesirable effects from expansion may occur in the material or component itself, such as cracking, intergranular separation, or spalling, or due to the net effect of the component expansion on its surroundings, such as swell interference, component mismatch, or delamination.

Whereas solid-state expansion has been deemed a material design aspect to minimize failure, the goal of this project is to study the mechanisms and conditions under which significant expansion transpires in solid matter and investigate applications that can be harnessed for beneficial use. Expansion of materials is a challenge for engineers in many different disciplines and an understanding could aid in assuaging material failures, as well as identifying novel uses for existing materials.

SOLID STATE EXPANSION MECHANISMS

The most familiar form of solid-state expansion is the volume increase induced by thermal changes. This effect is well quantified and commonly applied in material selection and design. In ceramic materials with strong covalent or ionic bonds the net effect of thermal expansion is less than a few percent volume increases. A wide variety of other mechanisms exist that induce expansion in solid matter, and their effect may range from tenths of a percent volume increase to several hundred percent.

Bismuth shows a volume increase of 2.5% when solidifying from the liquid state[1,2], refractory slags reacting with water result in 6-9% volume swelling which generate forces strong enough for building foundation buckling and failure[3], and austenitic stainless steel exhibits 20-23% volume increase when subjected to high radiation doses[4].

Although in many documented instances, expansion is viewed as an unwanted physical change, it can also be harnessed for beneficial use. Expandable mortars and cements have exploited expansion to create a non-explosive, silent product that produces no vibrations, yet is still capable of breaking down large rocks and structures.

The research in this paper was undertaken to identify, categorize, describe, and quantify the various mechanisms known to cause expansion in solid materials. These mechanisms can be broadly categorized into chemical reaction, phase change, and physical induction. The conditions by which the material expands, as well as, the physical and atomic changes it undergoes in order to achieve expansion were studied. Quantitative data is provided, typically based on the molar volume change from the standard state to the enlarged state, or by referencing published measured data. With quantitative data, the magnitude and effect of these mechanisms can be compared.

CHEMICAL REACTION

Solid-state expansion can be induced through the exposure of a solid substance to a non-solid, such as an aqueous or gaseous medium. Expansion occurs when the solid and medium interact; completely altering the chemical composition of the condensed phase.

A chemical reaction is dependent on thermodynamic and kinetic favorability and; therefore, is often conducted under conditions at which the reaction will progress more rapidly; such as elevated temperatures. For the purpose of this paper, all reactions were considered at ambient conditions, which are defined as 25°C and 1 atmosphere pressure. To determine the prevalent phases, HSC Chemistry 7.1 equilibrium composition diagrams were used[5].

Gas-Solid Reaction

When a material is exposed to an environment rich in a gaseous medium, it will either be inert and remain unchanged or undergo a chemical reaction. For a gas-solid reaction, the chemical reaction will change the solid composition, creating a new material. This new material can possess a different lattice shape and structure, thus changing the overall dimensions of the material.

In Table I, a list of elements reacted in an oxygen-rich environment are shown with the resulting volume change. The thermodynamic favorability is shown as the Gibbs free energy of the reaction in kilojoules. The stability temperature, up to which each reaction spontaneously converts reactants to products, is calculated in degrees Celsius. The molar volume of each substance was also calculated to determine the percent volume change.

Table I. Reaction of elements with an abundance of oxygen gas. Data adapted from *HSC Database*[5]. Molar volume data adapted from the *U.S. Geological Survey Bulletin 1452*[6].

Species	Mineral Name	Reaction Equation	ΔG (kJ)	Stability Temperature (°C)	Molar Volume (cm³)	Volume Change (%)
$Al_{(s)}$	Aluminum				10.00	
$Al_2O_{3(s)}$	Corundum	$Al_{(s)} + 0.75\, O_{2\,(g)} \rightarrow 0.5\, Al_2O_3\,_{(s)}$	-791.1	5774	12.79	27.9%
$Fe_{(s)}$	Iron				7.09	
$Fe_2O_{3(s)}$	Hematite	$Fe_{(s)} + 0.75\, O_{2\,(g)} \rightarrow 0.5\, Fe_2O_3\,_{(s)}$	-372.1	2956	15.14	113.4%
$Fe_3O_{4(s)}$	Magnetite	$Fe_{(s)} + 0.67\, O_{2\,(g)} \rightarrow 0.33\, Fe_3O_4\,_{(s)}$	-337.4	3802	14.84	109.3%
$FeO_{(s)}$	Wuestite	$Fe_{(s)} + 0.5\, O_{2\,(g)} \rightarrow FeO_{(s)}$	-250.6	4285	12.00	69.2%
$Mn_{(s)}$	Manganese				7.35	
$MnO_{(s)}$	Manganosite	$Mn_{(s)} + 0.5\, O_{2\,(g)} \rightarrow MnO_{(s)}$	-362.8	5130	13.22	79.8%
$MnO_{2\,(s)}$	Pyrolusite	$Mn_{(s)} + O_{2\,(g)} \rightarrow MnO_2\,_{(s)}$	-465.1	2536	16.61	125.9%
$Si_{(s)}$	Silicon				12.06	
$SiO_{2\,(s)}$	Silicon Dioxide Glass	$Si_{(s)} + O_{2\,(g)} \rightarrow SiO_2\,_{(s)}$	-856.4	4726	27.27	126.2%
$SiO_{2\,(s)}$	Quartz	$Si_{(s)} + O_{2\,(g)} \rightarrow SiO_2\,_{(s)}$	-856.3	4628	22.69	88.2%
$SiO_{2\,(s)}$	Cristobalite	$Si_{(s)} + O_{2\,(g)} \rightarrow SiO_2\,_{(s)}$	-853.1	4529	25.74	113.5%
$SiO_{2\,(s)}$	Tridymite	$Si_{(s)} + O_{2\,(g)} \rightarrow SiO_2\,_{(s)}$	-853.8	4622	26.53	120.1%
$SiO_{2\,(s)}$	Coesite	$Si_{(s)} + O_{2\,(g)} \rightarrow SiO_2\,_{(s)}$	-850.5	4663	20.64	71.2%
$SiO_{2\,(s)}$	Stishovite	$Si_{(s)} + O_{2\,(g)} \rightarrow SiO_2\,_{(s)}$	-802.8	4058	14.01	16.2%
$Ti_{(s)}$	Titanium				10.63	
$TiO_{2\,(s)}$	Anatase	$Ti_{(s)} + O_{2\,(g)} \rightarrow TiO_2\,_{(s)}$	-886.0	4802	20.52	93.0%
$TiO_{2\,(s)}$	Rutile	$Ti_{(s)} + O_{2\,(g)} \rightarrow TiO_2\,_{(s)}$	-889.4	5753	18.82	77.0%
$Zr_{(s)}$	Zirconium				14.02	
$ZrO_{2\,(s)}$	Baddeleyite	$Zr_{(s)} + O_{2\,(g)} \rightarrow ZrO_2\,_{(s)}$	-1042.5	6556	21.15	50.9%

All shown reactions are thermodynamically favorable, as shown by the negative Gibb's free energy. An expansion occurs for all elements when reacted with oxygen. The calculated molar volume increase ranged from a low of 16.2% in the transformation of silicon to stishovite, to 126.2% in the silicon to silicon dioxide glass.

Elements reacted with hydrogen gas were also studied. Only results in which the reaction was determined to be thermodynamically stable at room temperature were included in Table II.

Table II. Reaction of elements with an abundance of hydrogen gas. Data adapted from *HSC Database*[5]. Molar volume data adapted from the *U.S. Geological Survey Bulletin 1452*[6].

Species	Mineral Name	Reaction Equation	ΔG (kJ)	Stability Temperature (°C)	Molar Volume (cm³)	Volume Change (%)
$Ti_{(s)}$	Titanium				10.63	
$TiH_{2\,(s)}$	Titanium (II) Hydride	$Ti_{(s)} + H_{2\,(g)} \rightarrow TiH_2\,_{(s)}$	-105.1	775	13.31	25.2%
$Zr_{(s)}$	Zirconium				14.02	
$ZrH_{2\,(s)}$	Zirconium Dihydride	$Zr_{(s)} + H_{2\,(g)} \rightarrow ZrH_2\,_{(s)}$	-129.2	924	16.65	18.8%

Only reactions with elemental titanium and zirconium proved to be thermodynamically spontaneous. Titanium and zirconium exposed to an abundance of hydrogen gas expanded 25.2% and 18.8% by volume, respectively.

Reactions with oxide materials and the gaseous species carbon dioxide and sulfur dioxide were calculated and are shown in Table III and Table IV.

Table III. Reaction of oxides with an abundance of carbon dioxide gas. Data adapted from *HSC Database*[5]. Molar volume data adapted from the *U.S. Geological Survey Bulletin 1452*[6].

Species	Mineral Name	Reaction Equation	ΔG (kJ)	Stability Temperature (°C)	Molar Volume (cm³)	Volume Change (%)
$BaO_{(s)}$	Barium Oxide				25.59	
$BaCO_{3\ (s)}$	Witherite	$BaO_{(s)} + CO_{2\ (g)} \rightarrow BaCO_{3\ (s)}$	-215.5	1533	45.81	79.0%
$CaO_{(s)}$	Lime				16.76	
$CaCO_{3\ (s)}$	Calcite	$CaO_{(s)} + CO_{2\ (g)} \rightarrow CaCO_{3\ (s)}$	-130.4	886	36.93	120.3%
$CaCO_{3\ (s)}$	Aragonite	$CaO_{(s)} + CO_{2\ (g)} \rightarrow CaCO_{3\ (s)}$	-130.0	865	34.15	103.7%
$CaCO_{3\ (s)}$	Vaterite	$CaO_{(s)} + CO_{2\ (g)} \rightarrow CaCO_{3\ (s)}$	-127.4	618	37.63	124.5%
$FeO_{(s)}$	Wuesite				12.00	
$FeCO_{3\ (s)}$	Siderite	$FeO_{(s)} + CO2_{(g)} \rightarrow FeCO_{3\ (s)}$	-21.7	146	29.38	144.8%
$K_2O_{(s)}$	Dipotassium Monoxide				40.38	
$K_2CO_{3\ (s)}$	Potassium Carbonate	$K_2O_{(s)} + CO_{2\ (g)} \rightarrow K_2CO_{3\ (s)}$	-349.4	2842	60.35	49.5%
$Li_2O_{(s)}$	Dilithium Monoxide				14.76	
$Li_2CO_{3\ (s)}$	Lithium Carbonate	$Li_2O_{(s)} + CO_{2\ (g)} \rightarrow Li_2CO_{3\ (s)}$	-176.6	1606	35.02	137.3%
$MgO_{(s)}$	Periclase				11.25	
$MgCO_{3\ (s)}$	Magnesite	$MgO_{(s)} + CO_{2\ (g)} \rightarrow MgCO_{3\ (s)}$	-64.1	398	28.02	149.1%
$MnO_{(s)}$	Manganosite				13.22	
$MnCO_{3\ (s)}$	Rhodochrosite	$MnO_{(s)} + CO_{2\ (g)} \rightarrow MnCO_{3\ (s)}$	-52.7	343	31.07	135.0%
$Na_2O_{(s)}$	Disodium Monoxide				25.88	
$Na_2CO_{3\ (s)}$	Sodium Carbonate	$Na_2O_{(s)} + CO_{2\ (g)} \rightarrow Na_2CO_{3\ (s)}$	-276.3	2127	41.86	61.7%
$SrO_{(s)}$	Strontium Oxide				20.69	
$SrCO_{3\ (s)}$	Strontianite	$SrO_{(s)} + CO_{2\ (g)} \rightarrow SrCO_{3\ (s)}$	-183.2	1171	39.01	88.6%

Again all shown reactions of oxides with carbon dioxide gas are thermodynamically spontaneous. The largest expansion occurs as periclase transforms to magnesite with an increase in volume of 149.1%.

The same oxide materials as shown previously were then considered in an abundance of sulfur dioxide gas. The results of these reactions are shown in Table IV.

Table IV. Reaction of oxides with an abundance of sulfur dioxide gas. Data adapted from *HSC Database*[5]. Molar volume data adapted from the *U.S. Geological Survey Bulletin 1452*[6].

Species	Mineral Name	Reaction Equation	ΔG (kJ)	Stability Temperature (°C)	Molar Volume (cm³)	Volume Change (%)
$BaO_{(s)}$	Barium Oxide				25.59	
$BaSO_{4\ (s)}$	Barite	$BaO_{(s)} + 1.5\ SO_{2\ (g)} \rightarrow BaSO_{4\ (s)} + 0.5\ S_{(g)}$	-387.2	1530	52.10	103.6%
$CaO_{(s)}$	Lime				16.76	
$CaSO_{4\ (s)}$	Anhydrite	$CaO_{(s)} + 1.5\ SO_{2\ (g)} \rightarrow CaSO_{4\ (s)} + 0.5\ S_{(g)}$	-268.2	1039	45.94	174.0%
$FeO_{(s)}$	Wuesite				12.00	
$Fe_2(SO_4)_{3\ (s)}$	Ferric Sulfate	$FeO_{(s)} + 2.5\ SO_{2\ (g)} \rightarrow 0.5\ Fe_2(SO_4)_{3\ (s)} + S_{(g)}$	-131.4	297	65.39	444.9%
$FeSO_{4\ (s)}$	Iron (II) Sulfate	$FeO_{(s)} + 1.5\ SO_{2\ (g)} \rightarrow FeSO_{4\ (s)} + 0.5\ S_{(g)}$	-124.2	464	41.62	246.8%
$K_2O_{(s)}$	Dipotassium Monoxide				40.38	
$K_2SO_{4\ (s)}$	Arcanite	$K_2O_{(s)} + 1.5\ SO_{2\ (g)} \rightarrow K_2SO_{4\ (s)} + 0.5\ S_{(g)}$	-547.4	2323	65.50	62.2%
$Li_2O_{(s)}$	Dilithium Monoxide				14.76	
$Li_2SO_{4\ (s)}$	Lithium Sulfate	$Li_2O_{(s)} + 1.5\ SO_{2\ (g)} \rightarrow Li_2SO_{4\ (s)} + 0.5\ S_{(g)}$	-309.9	1449	49.37	234.5%
$MgO_{(s)}$	Periclase				11.25	
$MgSO_{4\ (s)}$	Magnesium Sulfate	$MgO_{(s)} + 1.5\ SO_{2\ (g)} \rightarrow MgSO_{4\ (s)} + 0.5\ S_{(g)}$	-219.2	826	45.25	302.3%
$MnO_{(s)}$	Manganosite				13.22	
$MnSO_{4\ (s)}$	Manganese Sulfate	$MnO_{(s)} + 1.5\ SO_{2\ (g)} \rightarrow MnSO_{4\ (s)} + 0.5\ S_{(g)}$	-144.2	523	43.62	229.9%
$Na_2O_{(s)}$	Disodium Monoxide				25.88	
$Na_2SO_{4\ (s)}$	Thenardite	$Na_2O_{(s)} + 1.5\ SO_{2\ (g)} \rightarrow Na_2SO_{4\ (s)} + 0.5\ S_{(g)}$	-443.6	1952	53.33	106.1%
$SrO_{(s)}$	Strontium Oxide				20.69	
$SrSO_{4\ (s)}$	Celestite	$SrO_{(s)} + 1.5\ SO_{2\ (g)} \rightarrow SrSO_{4\ (s)} + 0.5\ S_{(g)}$	-328.0	1268	46.25	123.6%

When an abundance of sulfur dioxide gas was reacted with specific oxides, the largest expansion occurred in wuestite as it transformed into ferric sulfate. A theoretical expansion of 444.9% was calculated. This was the largest determined theoretical volume expansion calculated for any of the gaseous chemical reactions studied.

Oxides reacted with hydrogen gas were also studied, but since no reaction proved to be thermodynamically stable under ambient conditions, no results were included.

Liquid-Solid Reaction

As a solid material is placed in contact with an overabundance of liquid water, a chemical reaction may occur creating a new solid phase. Table V summarizes elements and oxides that spontaneously react with pure water to create new solid compounds. Some of these phases may be soluble in water, but when precipitated they are a stable solid phase.

Table V. Reaction of elements and oxides with an abundance of water. Data adapted from *HSC Database*[5]. Molar volume data adapted from the *U.S. Geological Survey Bulletin 1452*[6].

Species	Mineral Name	Reaction Equation	ΔG (kJ)	Stability Temperature (°C)	Molar Volume (cm³)	Volume Change (%)
$Al_{(s)}$	Aluminum				10.00	
$Al(OH)3$	Gibbsite	$Al_{(s)} + 3H_2O_{(l)} \rightarrow Al(OH)_{3\ (s)} + 2\ H_{2(g)}$	-443.9	206.85	31.96	219.6%
$AlO(OH)$	Boehmite	$Al_{(s)} + 2H_2O_{(l)} \rightarrow AlO(OH)_{(s)} + 2\ H_{2(g)}$	-444.3	2836	19.54	95.4%
$AlO(OH)$	Diaspore	$Al_{(s)} + 2H_2O_{(l)} \rightarrow AlO(OH)_{(s)} + 2\ H_{2(g)}$	-449.2	2810	17.76	77.6%
$BaO_{(s)}$	Barium Oxide				25.59	
$Ba(OH)_2 \cdot 8H_2O_{(s)}$	Barium hydroxide octahydrate	$BaO_{(s)} + 9H_2O_{(l)} \rightarrow Ba(OH)_2 \cdot 8H_2O_{(s)}$	-134.6	281	144.71	465.5%
$CaO_{(s)}$	Lime				16.76	
$Ca(OH)_{2\ (s)}$	Portlandite	$CaO_{(s)} + H_2O_{(l)} \rightarrow Ca(OH)_{2\ (s)}$	-57.8	1149	33.06	97.2%
$Fe_{(s)}$	Iron				7.09	
$Fe(OH)_{2\ (s)}$	Iron (II) Hydroxide	$Fe_{(s)} + 0.5\ O_{2\ (g)} + H_2O_{(l)} \rightarrow Fe(OH)_{2\ (s)}$	-249.8	1630	26.43	272.7%
$FeO(OH)_{(s)}$	Goethite	$Fe_{(s)} + 0.75\ O_{2\ (g)} + 0.5\ H_2O_{(l)} \rightarrow FeO(OH)_{(s)}$	-371.9	2145	20.82	193.6%
$Li_2O_{(s)}$	Dilithium Monoxide				14.76	
$LiOH$	Lithium Hydroxide	$Li_2O_{(s)} + H_2O_{(l)} \rightarrow 2LiOH_{(s)}$	-79.5	2260	32.88	122.8%
$2LiOH \cdot H_2O_{(s)}$	Lithium Monohydrate	$Li_2O_{(s)} + 3H_2O_{(l)} \rightarrow 2LiOH \cdot H_2O_{(s)}$	-95.2	516	55.58	276.6%
$MgO_{(s)}$	Periclase				11.25	
$Mg(OH)_{2\ (s)}$	Brucite	$MgO_{(s)} + H_2O_{(l)} \rightarrow Mg(OH)_{2\ (s)}$	-27.2	531	24.63	119.0%
$Mn_{(s)}$	Manganese				7.35	
$Mn(OH)_{2\ (s)}$	Manganese (II) Hydroxide	$Mn_{(s)} + H_2O_{(l)} + 0.5\ O_{2\ (g)} \rightarrow Mn(OH)_{2\ (s)}$	-380.5	2093	27.29	271.0%
$MnO_{(s)}$	Manganosite				13.22	
$Mn(OH)_{2\ (s)}$	Manganese (II) Hydroxide	$MnO_{(s)} + H_2O_{(l)} \rightarrow Mn(OH)_{2\ (s)}$	-17.7	387	27.29	106.4%
$SrO_{(s)}$	Strontium Oxide				20.69	
$Sr(OH)_2_{(s)}$	Strontium Hydroxide	$SrO_{(s)} + H_2O_{(l)} \rightarrow Sr(OH)_{2\ (s)}$	-82.6	1772	33.55	62.2%

Many hydrated compounds are formed with a maximum expansion of 465.5% shown during the formation of barium hydroxide octahydrate from barium oxide.

PHASE CHANGE

Crystalline solids undergo phase change both during melting/freezing (liquid-solid) and in the solid-state as polymorphic transformations. Phase transformations are induced by pressure and temperature changes. As temperature and pressure varies the physical atomic bonds are rearranged and adjusted, which can often result in expansion of the solid material.

Phase Change (Solid-Solid)

A crystallographic change that occurs within an elemental substance that remains solid is considered an allotropic transformation. The crystal structure alteration does not change the chemical composition of the material, but can have a radical effect on the physical characteristics of the substance. When a compound, such as an oxide ceramic, undergoes the same crystal structure alteration it is considered a polymorphic transformation[7].

In general, these transformations can occur in two forms, displacive or reconstructive. During a displacive transformation, no atomic bonds are broken, but the atoms are restructured into a different pattern. Reconstructive transformations transpire when chemical bonds are broken, and new crystal structures are formed from a complete reordering of atoms[8].

In the Table VI, several elements and their respective allotropes as well as their volume change during a solid phase transformation is shown.

Table VI. Elements and their allotropes. Data adapted from the *CRC Handbook of Chemistry and Physics*[9].

Allotropes	Pearson Symbol			Temperature (°C)	Pressure (atm)	Molar Volume (cm³/mol)	Volume Change from Standard	Volume Change from Previous
αAl	c	F	4	25	1	10.00		
βAl	h	P	2	25	>20.5	8.32	-16.8%	-16.8%
C$_{Graphite}$	h	P	4	25	1	5.30		
C$_{Diamond}$	c	F	8	25	>60	3.42	-35.5%	-35.5%
C$_{hd}$	h	P	4	25	HP	3.42	-35.5%	0.0%
αFe	c	I	2	25	1	7.09		
γFe	c	F	4	> 912	1	7.30	2.9%	2.9%
σFe	c	I	2	>1394	1	7.59	7.0%	3.9%
εFe	h	P	2	25	>13	6.29	-11.3%	-17.1%
αMn	c	I	58	25	1	7.35		
βMn	c	P	20	>710	1	7.58	3.2%	3.2%
γMn	c	F	4	>1079	1	8.66	17.8%	14.2%
σMn	c	I	2	>1143	1	8.80	19.7%	1.6%
αSi	c	F	8	25	1	12.06		
βSi	t	I	4	25	>9.5	8.55	-29.1%	-29.1%
γSi	c	I	16	25	>16	11.00	-8.8%	28.7%
σSi	h	P	4	25	>16→1	11.82	-1.9%	7.5%
αTi	h	P	2	25	1	10.63		
βTi	c	I	2	>882	1	10.88	2.4%	2.4%
ωTi	h	P	3	25	HP→1	10.46	-1.6%	-3.9%
αZr	h	P	2	25	1	14.02		
βZr	c	I	2	>863	1	14.15	1.0%	1.0%
ωZr	h	P	2	25	HP→1	20.56	46.7%	45.3%

Commonly documented allotropic, or single element, changes occurs in the iron system. Under standard conditions, α-Fe is the stable phase, but as the metal is heated to 912°C, the crystal structure shifts from body-centered cubic to face-centered cubic. To accommodate the lattice alteration as α-Fe is transformed into γ-Fe, a volume increase occurs of approximately 2.95%.

Silica is a unique material that undergoes numerous polymorphic transformations that are both reconstructive and displacive in nature. Figure 1 illustrates the nature of the phase transitions in the silica system, as well as, the accompanied expansion characteristics. The highest magnitude change occurs between low and high cristobalite, which experiences a volume change of >3 %.

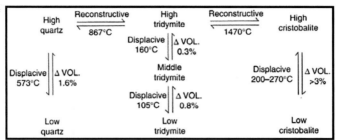

Figure 1. Phase transformations in silica. Figure taken from *Modern Ceramic Engineering*[10].

Multiple polymorphic phases in a compound are a trait that can be manipulated such as in zirconia for example. Table VII demonstrates the expansion characteristics of zirconia polymorphs.

Table VII. Phase transformations in zirconia. Data adapted from the A. Umeri study[11].

	Crystal Structure	Temperature Range (@ Ambient Pressure)		Volume Expansion
		°C	°C	
m-ZrO₂	Monoclinic	298.15	1170	
t-ZrO₂	Tetragonal	1170	2370	-4.50%
c-ZrO₂	Cubic	2370	2680	-2.30%

Zirconia expands as it cools from cubic to tetragonal, and tetragonal to monoclinic; its ambient phase. Through the use of additives, the high temperature cubic phase can be stabilized and therefore, zirconia can be useful over a wider temperature range without the concern of phase transformations. Because such a large volume increase occurs as pure zirconia is cooled from the tetragonal to monoclinic, by restricting this change, a phenomenon called transformation toughening can be induced. The restrained material will not be allowed to expand and will therefore be forced to cool with tetragonal precipitates creating a metastable phase that exhibits high toughness as well as high strength; which is profound for ceramic materials. A microphotograph of a toughened zirconia structure is shown in Figure 2.

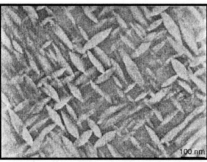

Figure 2. Crystal Structure of toughened zirconia. Photo taken from *Modern Ceramic Engineering*[10].

Phase Change (Liquid-Solid)

Most materials, when cooled from a liquid phase to a condensed solid phase will contract and increase in density; however, there are some very profound exceptions. Table VIII illustrates a few elemental species that expand upon cooling to a solid.

Table VIII. Liquid and solid phase density for elements and water. Data from the *Metals Handbook*[1]. Melting and density data for germanium, water, and zirconium adapted from the *CRC Handbook of Chemistry and Physics*[9]. Initial solid density for bismuth, antimony and gallium from *Chemistry of the elements*[2]. Solid density for titanium taken from *Physical Metallurgy Principles*[12].

Element	Melting Temperature °C	Liquid at Melting Point		First Solid Formed			Percent Volume Change (Liquid→Solid)
		Density (g/cm³)	Molar Volume (cm³/mol)	Phase	Density (g/cm³)	Molar Volume (cm³/mol)	
Aluminum	660	2.38	11.36	α-Al	2.70	10.00	-12.0%
Antimony	631	6.53	18.65	α-Sb	6.68	18.22	-2.3%
Bismuth	271	10.05	20.79	α-Bi	9.81	21.31	2.5%
Gallium	30	6.08	11.47	α-Ga	5.90	11.81	3.0%
Germanium	938	5.60	12.97	α-Ge	5.32	13.65	5.2%
Iron	1538	6.98	8.00	δ-Fe	7.36	7.59	-5.2%
Manganese	1246	5.95	9.23	δ-Mn	6.24	8.80	-4.7%
Silicon	1414	2.57	10.93	α-Si	2.33	12.06	10.3%
Titanium	1668	4.11	11.65	β-Ti	4.35	11.00	-5.5%
Water	0	1.00	18.02	Ice	0.92	19.65	9.1%
Zirconium	1855	5.80	15.73	βZr	6.45	14.15	-10.0%

The most notable of these materials is water. Because ice is less dense than water, icebergs can float on the surface of water and lakes freeze over from the top while the lake bottom remains unfrozen.

An important metal that undergoes an expansion from the liquid to solid phase is bismuth. Because of this unique property, bismuth was used in the early print type. Once cast, bismuth expands upon solidification making the letters sharper and ultimately the type more readable, giving the company that manufactured the bismuth lettering an edge and selling point[13].

PHYSICAL INDUCTION

Solids-state materials may undergo volume change from physical factors that do not involve a chemical reaction or phase change. These factors are termed physical induction. These physical mechanisms do not change the chemical composition of the materials but instead substantially change the structure. These mechanisms are grouped into three broad categories based on the type of external factor inducing the change in the solid material. Thermal expansion arises from energy being transferred to the material which expands the lattice structure. Radiation expansion occurs when high energy particles impact atoms in the lattice structure creating high disorder that results in expansion and diffusion expansion occurs when a high chemical potential gradient occurs forcing external atoms into the lattice interstices which expands the crystalline arrangement.

Thermal Expansion

Expansion occurs within the lattice structure when the atomic kinetic energy of a material is amplified due to an increase in thermal energy. The acquired thermal energy induces atomic vibrations, causing each atom to behave as if they possess a larger atomic radius. The vibrations increase the atomic spacing of atoms within the material, expanding the lattice structure. This results in the overall dimensional increase of the substance.

The degree of the volume increase in each substance is dependent on the crystal structure, as well as, the atomic chemical binding energy. A higher strength atomic bond will require a larger amount of energy to expand than a weak bond. The relative binding energy of four main bonding mechanisms is shown in Table IX.

Table IX. Relative Binding energies of four main chemical bonds. Adapted from *The Science and Engineering of Materials*[14].

Bond	Binding Energy (kcal/mol)
Ionic	150 - 370
Covalent	125 - 300
Metallic	25 - 200
Van der Waals	< 10

Because each material can exhibit a mixture of bond types, the linear coefficient of thermal expansion is used to determine the degree of expansion each material will experience for a specified temperature range. In general, ceramic and oxide materials, due to the strength of the ionic bonding, experience little thermal expansion, whereas covalently bonded polymers will experience large expansions from thermal energy increases. This trend is shown in Figure 3.

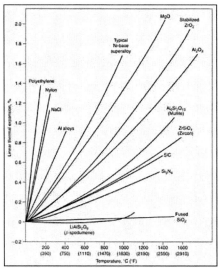

Figure 3. Comparison of linear thermal expansion of common polymers, ceramics, and metals. Graph taken from *Modern Ceramic Engineering*[10].

Figure 3 shows the ceramic materials, such as fused silica and silicon carbide, with the smallest amount of thermal expansion for each increase in temperature. This signifies that the bonding in these ceramic materials is much stronger than those in polymers or even metals. Polymeric materials, like nylon and polyethylene, show large expansions at small increases in temperature, insinuating a weak atomic bonding. Metallic substances, such as aluminum, fall between the ceramic and polymers experiencing moderate thermal expansion for each increase in energy.

The coefficient of thermal expansion is a commonly measured value used in material design. This value quantifies the volume change of a solid material in response to temperature change. As most polycrystalline solids exhibit isotropic expansion, this value can be taken as the coefficient if linear thermal expansion and used to calculate the change in length per unit length of material (i.e. strain). Table X shows values for linear thermal expansion of select elements and ceramic oxides.

Table X. Linear coefficient of thermal expansion data for select elements and oxides. Data adapted from the *CRC Handbook of Chemistry and Physics*[9]. Thermal data for silicon, and carbon adapted from *Metals Handbook*[1]. Oxide data from *Fundamentals of Ceramics*[8]. Silica, alumina, and concrete data from *Materials Science and Engineering*[7].

Elements	Coefficient of Linear Thermal Expansion ($\mu m/m$)*K	Ceramics	Coefficient of Linear Thermal Expansion ($\mu m/m$)*K
Al	23.1	Al_2O_3 (90% pure)	7
C $_{(graphite)}$	0.6 - 4.3	Al_2O_3 (99.9% pure)	7.4
Fe	11.8	Concrete	10.0 - 13.6
Mn	21.7	SiO_2 $_{(fused)}$	0.4
Si	2.62	TiO_2	8.5
Ti	8.6	ZrO_2 $_{(monoclinic)}$	7.0
Zr	5.7	ZrO_2 $_{(tetragonal)}$	12.0

Radiation Induced Expansion

Understanding the effects and mechanisms by which material properties are altered due to radiation is critical for reactor service life. Since the physical properties can drastically change due to radiation exposure, it is important to identify where the induced alteration is beneficial or detrimental.

One of the most documented physical changes in irradiated materials is the void swelling austenitic stainless steel and nickel based alloys. Carbonaceous substances, such as graphite, have also been noted to increase in volume when irradiated.

In an experiment conducted by Watkin, Gittus, and Standring[15] a total volume increase of 20-23% was noted for solution treated specimens of FV 548 or austenitic stainless steel. The test was conducted at a controlled set of temperatures, and the material was irradiated in doses of 12, 30, and 48 displacements per atom (dpa). Figure 4 is a graphical representation of this data.

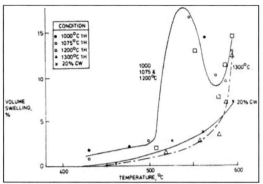

Figure 4. Temperature versus percent volume swelling for solution treated austenitic stainless steel. Graph taken from *Irradiation Effects in Crystalline Solids*[4].

The swelling of the stainless steel specimen is said to be the result of void introduction into the crystal lattice. The phenomenon of void addition was also observed in Nimonic PE163, or a high nickel alloy. A volume increase of 2-3.5% was measured for the cast alloys[4].

When a material is exposed to radiation, the crystal structure is bombarded with ionizing particles, creating disorder within a material's lattice in as little as 10^{-12} seconds. It is speculated that a neutron moving at a high speed can displace approximately 100 surrounding atoms throughout the material with one collision of a carbon atom[4]. This shows that in a short amount of time, an entire crystal structure can be rearranged and distorted. In metals, the neutron collision creates voids, often large enough to be filled by gas. The larger the created void, the more distorted the lattice structure becomes and the greater the material swells.

Graphite is a common material used as a moderator in nuclear reactors. Because of this, the physical expansion properties due to radiation exposure have been closely observed. Figure 5, is a graphical representation of the experimental volume expansion and contraction of graphite monocrystals.

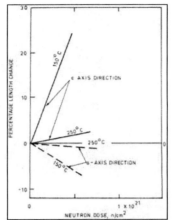

Figure 5. Increase in length of graphite due to increased neutron dose. Graph taken from *Irradiation Effects in Crystalline Solids*[4].

It should be noted that two different expansion directions were measured for the crystalline material, as depicted by the c-axis and a-axis expansion lines. This means that the material experienced anisotropic expansion.

A special characteristic of graphite is the unique ability to store the energy from radiation within the crystal structure. Because of the stored energy, point defects such as dislocations and vacancies have an easier time diffusing through the structure and conglomerating. Although this causes expansion, the increase in volume is also accompanied by stress concentrations and anisotropic expansion.

In graphite, a hexagonal structure, the increased rate of diffusion leads to preferential orientation along one plane and therefore swelling in only one direction. In ceramic materials, it is also documented that the ionized particles can also displace electrons creating charges ions within the structure. The ions attract charge and preferentially orient creating mismatch within the lattice, such as between planes or surrounding other charged ions causing anisotropic expansion.

Due to the energy storage of radiation within a material, and therefore the increased rate of diffusion, the formation of metastable phases is not uncommon. Materials that experience diffusion based phase transformations, not normally spontaneous at under low temperature conditions, can transform with ease when irradiated. These transformations can be detrimental since the physical properties of the material will be altered as well as the lattice structure. Areas of precipitate formation and dissolution that are not consistent with the bulk material can also occur due to the increased diffusion rate[4].

Although most studies show the harmful effects of radiation on materials, such as degradation, there is also promise for further research. Because irradiated materials contain an excess of energy within the lattice, under carefully controlled conditions, this can be manipulated to enhance mechanisms such as diffusion, precipitation, phase changes, increased stability, and nucleation[16,17].

Diffusion (Plasma and Ion) Induced Expansion

Diffusion is defined as the total flux of a species and can describe species ranging from molecules, to atoms, to electrons. The movement of these species can be influenced by a number of physical and chemical factors including concentration gradient, temperature, time, and composition of the matrix as well as the diffusing particles. In metals, diffusion occurs must more readily than in ionic ceramics, due to the charge differences that the particle must overcome.

Boronizing, nitriding, and carburizing are common diffusion-based material modification methods used to increase the surface hardness of many metals and some ceramics. To surface harden a material with one of these methods, a large concentration gradient of the diffused species is created often accompanied by a pressure or temperature increase to make the diffusion more rapid. The diffusive species is then forced into the surface of the substance displacing atoms within the matrix, filling vacancies, or inserting into the crystal structure interstitially. The chemical composition of the substance's surface is altered as well as the crystal structure[12].

Diffusive expansion is most commonly used for carburizing steel or nitriding. During this process, the insertion of carbon into the crystal structure of the steel often forms an expanded lattice. This phenomenon was documented in the formation of "expanded austenite" in a stainless steel matrix. To achieve the documented 5-12% lattice expansion, nitrogen contents of 10 – 20 atomic percent were necessary. The insertion of nitrogen atoms into the face-centered cubic austenitic stainless steel lattice resulted in a bulk increase of several micrometers for the material[18].

Diffusion does not only occur in metals or with carbon, nitrogen or boron atoms. Diffusive expansion can be achieved through the insertion of any ion into a lattice structure as long as the kinetic energy is favorable. An example of this is the expansion documented in gallium nitride due to calcium and argon ions. At ion doses of approximately 10^{15} cm^{-2}, an expansion of 14% and 9% was measured in the (0002) planar spacing of gallium nitride due to argon and calcium respectively[19].

CONCLUSION

At least seven different mechanisms for solid-state expansion have been identified along with the preliminary thermodynamic and crystallographic processes involved. These mechanisms induce volume increases from a few percent to several hundred percent.

The largest expansion seen for the species studied was experienced by materials that were undergoing chemical composition changes induced by a reaction. The addition of interstitial or additional atoms into the lattice stretched, distorted, and transformed the material allowing for swelling. The largest expansion of 465.52% was experienced as barium oxide interacted with liquid water to form barium hydroxide octahydrate. This expansion was followed closely by the 444.9% expansion of the gas-solid reaction transforming wuestite to ferric sulfate.

As a material underwent a phase change, especially from a high pressure phase, a large expansion was documented. The largest were seen as zircon and silicon transformed from a high pressure phase to standard state and experienced a 45.3% and 28.7% lattice expansion respectively.

Additional expansions were seen under the following conditions. As a material is bombarded with neutrons, enlargements of 20-23% were seen in the irradiated austenitic stainless steel specimens. As positively charged argon ions were diffused into the surface of gallium nitride an expansion of 14% was documented. Finally, as a liquid is cooled to a solid phase, like with pure silicon an expansion of 10.3% can be induced.

The study of these mechanisms is part of a larger study on the beneficial use of solid-state expansion. Possible applications for induced or controlled expansion include;

• Expandable bioceramic that can be induced to grow along with the patient
• Hard expansion for mudjacking in building foundation repair
• Self-healing cements to repair cracks caused by thermal stress
• MEMS triggers and switches to provide mechanical action based on temperature
• Porous ceramics that absorb contaminants swell to seal off leaks

Expansion of solid materials is a challenge for engineers in many disciplines and further understanding aids in preventing failures, as well as identifying novel uses current materials.

REFERENCES

[1]Davis, J.R., [ed.]. *Metals Handbook*. 2nd. Materials Park : ASM International Handbook Committee, (1998).

[2]Greenwood, N. N. and Earnshaw, A. *Chemistry of the elements*. 2nd. Oxford : Butterworth-Heinemann, (1997).

[3]Crawford, Carl B. and Burn, Kenneth N. Building Damage from Expansive Steel Slag Backfill. *Research Paper No. 422*, Ottawa : National Research Council of Canada,Vol. Division of Building Research (1970).

[4]Gittus, John. *Irradiation Effects in Crystalline Solids*. London : Applied Science Publishers LTD, (1978).

[5]Roine, Antti, et al., *HSC Chemistry 7*. s.l. : (c) Outotec, Research Center, 1974-2011. Version 7.14.

[6]Robie, Richard A., Hemingway, Bruce S. and Fisher, James R. *Thermodynamic Properties of Minerals and Related Substances at 298.15K and 1 Bar (10^5 Pascals) Pressure and at Higher Temperatures*. Washington, D.C : U.S. Geological Survey Bulletin 1452, (1984).

[7]Callister, William D. and Rethwisch, David G. *Materials Science and Engineering An Introduction*. 8th. Hoboken : John Wiley & Sons, Inc., (2010).

[8]Barsoum, Michel W. *Fundamentals of Ceramics*. [ed.] B. Cantor and M.J. Goringe. New York : Taylor & Francis Group, (2003).

[9]W.M. Haynes, Ph.D., [ed.]. *CRC Handbook of Chemistry and Physics*. 91st. New York : CRC Press, (2010).

[10]Richerson, David W. *Modern Ceramic Engineering*. 3rd. Boca Raton : Taylor and Francis Group, (2006).

[11]Umeri, A., Study of Zirconia's ageing for applications in dentistry. *Scienza e Technologia dei Materiali*, Vol. XXII. (2009/2010).

[12]Reed-Hill, Robert E, Abbaschian, Lara and Abbaschian, Reza. *Physical Metallurgy Principles*. 4th. Stamford : Cengage Learning, (2009).

[13]Burke, James. *Connections*. s.l. : Simon & Schuster, (2007).

[14]Askeland, Donald R. and Phule, Pradeep P. *The Science and Engineering of Materials*. 5. s.l. : Nelson, (2006).

[15]Watkin, J.S., Gittus, J.H. and Standring, J. The influence of alloy constitution on the swelling of austenitic stainless steels and nickel based alloys. *Scottsdale, AZ : Am. Nuc. Soc/ASTM/ERDA, International Conference on Radiation Effects in Breeder Reactor Structural Material*, .(1977).

[16]Billington, Douglas S. *Radiation Damage to Materials*. (c) McGraw-Hill Companies, AccessScience, (2008).

[17]Anno, J. N. *Notes on Radiation Effects on Materials*. Washington: Hemisphere Publishing Corporation, (1984).

[18]Mandl, S. and Rauschenbach, B. Concentration dependent nitrogen diffusion coefficient in expanded austenite formed by ion implantation. American Institute of Physics, *Journal of Applied Physics*, Vol. 91. (June 15, 2002)

[19] Liu, C., et al., Lattice expansion of Ca and Ar ion implanted GaN. American Institute of Physics, Vol. 71, (October 20, 1997).

PROPERTIES OF SHOCK-SYNTHESIZED ROCKSALT-ALUMINIUM NITRIDE

Kevin Keller[1], Thomas Schlothauer[1], Marcus Schwarz[2], Erica Brendler[3], Kristin Galonska[1], Gerhard Heide[1], and Edwin Kroke[2]

[1]Institute for Mineralogy, TU Bergakademie Freiberg, Brennhausgasse 14, 09599 Freiberg, Germany

[2]Institute for Inorganic Chemistry, TU Bergakademie Freiberg, Leipziger Straße 29, 09599 Freiberg, Germany

[3]Institute for Analytical Chemistry, TU Bergakademie Freiberg, Leipziger Straße 29, 09599 Freiberg, Germany

ABSTRACT

Successful syntheses of the rocksalt-type of aluminium nitride were carried out with shock wave recovery experiments. The sixfold, octahedral coordination of Al atoms in rs-AlN is proved with ^{27}Al MAS NMR showing an isotropic chemical shift of 2 ppm. Nanosized rs-AlN is oxidised in air at temperatures above 530°C. At 1100°C in vacuum the metastable rs-AlN transforms back into the wurtzitic structure. Preliminary experiments indicate a good chemical resistance of rs-AlN against acids and bases.

INTRODUCTION

With its outstanding thermal and mechanical properties aluminium nitride is a promising substrate material for high-power electronic applications[1]. At pressures from 14 to 23GPa the wurtzitic aluminium nitride (w-AlN) undergoes a phase transition to rocksalt structure (rs-AlN) as reported for static high-pressure experiments[2-4]. Sintered bodies of w-AlN/rs-AlN showed high hardness ($\leq$ 4500HV), high electric resistance and a thermal conductivity of up to 600 W/m K[5]. Rs-AlN was recently also successfully synthesised with shock wave dynamic high-pressure experiments using the flyer-plate method[6]. The fine greyish powder (obtained in amounts of up to 2 g per experiment), which can be gathered from the recovery containers, contains up to 50% of the high-pressure AlN phase. Thus, the shock wave method allows to examine chemical and thermal properties of the high-pressure phase rs-AlN. Structural characterisation by ^{27}Al MAS NMR and XRD has been carried out. The thermal stability in air and under vacuum, as well as chemical resistance against strong acids and bases have been examined for the first time.

EXPERIMENTAL

Nanopowder of w-AlN with an average grain size of 20 nm was used as starting material. The rs-AlN is synthesised with shock wave recovery experiments using the flyer-plate method. A metal plate is accelerated by a high-explosive charge, strikes the sample container and therefore generates high pressure and high temperature conditions in the sample powder, which causes the phase transition into rs-AlN. A detailed description of the experiments is given in[6].

Phase analysis is carried out with powder X-ray diffraction (XRD) with Cu-radiation and fixed slit (3000TT diffractometer from Seifert). Quantification was done using the Rietveld full-pattern fitting method. Structural characterisation was carried out with magic angle spinning NMR on a Bruker Avance 400 MHz WB spectrometer working at 9.4 Tesla. Samples were rotated at 25 kHz in a 2.5 mm ZrO$_2$ rotor. The resonance frequency for ^{27}Al was 104.29 MHz. Chemical shifts were recorded relative to 1 M AlCl$_3$ (0 ppm). For the 2D Triple Quantum MAS NMR (MQMAS) measurements 40 t$_1$-increments with 1s experiment repetition time were aquired.

Thermal Analysis of the nanocrystalline w-AlN powder (Plasmachem) and a shock-synthesised sample with 25% rs-AlN, as well as a submicron w-AlN (ABCR/ H.C. Starck) with a grain size of 0,8-1,8 μm, were carried out. Oxidation resistance of the different powders were investigated with TG-DTA (Setaram 92-16.18) up to 1100°C with a heating rate of 10 K/min. Thermogravimetric analyses in vacuum and mass spectrometry with direct coupling (STA 409 CD/3/403/5/G, Netzsch) up to 1300°C with 10 K/min heating rate were applied. The mass-charge-quotient (m/z) was detected with 100ms per step for m/z from 1-100. Ex-situ phase analysis (XRD) was done after cooling. Chemical stability against water, HCl, H_3PO_4, H_2SO_4, HNO_3, nitrohydrochloric acid and caustic soda was tested for one hour with 500 mg of the sample and 4 ml of the testing substance. Samples were centrifuged three times, dried at 80°C and a phase analysis with XRD was performed.

RESULTS

Synthesis
        Phase Analysis of the shock-synthesised samples shows a mixture of mostly aluminium nitride (w-AlN/rs-AlN) and minor amounts of corundum and γ-AlON (see figure 1). Depending on synthesis conditions the yield of rs-AlN ranges from 0-50%. The high amount of up to 15% of oxygen-bearing phases results from the high oxygen content (12-14%) of the commercial w-AlN starting powder. A minor content of ammonium chloride is likewise accounted to the contamination of w-AlN powder.

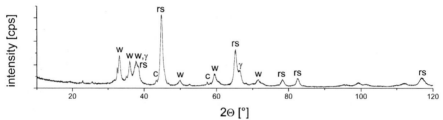

Figure 1. Diffraction pattern of shock-synthesised nano aluminium nitride with 38% w-AlN (w), 48% rs-AlN (rs), 3% corundum (c), 10% γ-AlON (γ) and <1% NH_4Cl.

Structural Characterisation
        The lattice constant a of rs-AlN was determined to be 4.046 ± 0.003Å, which correlates with the data from other authors[2,4]. Based on the used Rietveld model a grain size of 10-25 nm for rs-AlN and 15-30 nm for w-AlN was calculated. An exact grain size determination is hindered by the nature of nanomaterial and therefore the simulation of line broadening (grain size, microstrain, surface effects) is not trivial. Furthermore because of the multi component phase mixture and broad diffraction peaks, the overlapping complicates a reasonable peak profile fitting. The cone end of peaks are relatively narrow, but at the bottom a large broadening can be noticed. It is assumed that the core of the nano particles are more or less strain free, while in the surfaces a lot of microstrain is accumulated. A bimodal grain size/microstrain model for w-AlN giving a better fitting, but no quantitative data for grain sizes and strains. Complementary investigations (especially with HR-TEM) are necessary for a better understanding.
        The NMR spectra of samples consist of a symmetrical peak at 113 ppm, which can be attributed to the well known AlN_4-environment of w-AlN (see figure 2). A wide shoulder at about 60-70 ppm can be assigned to poorly ordered AlO_4-units or a mixed tetrahedral AlON-environment of the γ-AlON phase. At about 0 ppm an overlapping peak can be observed, which is comprised of two peaks at about 12 ppm and 2 ppm. The signal with the positive chemical shift is referred to the AlO_6-group of corundum and γ-AlON. The small negative shifted peak is assigned to the AlN_6-polyeder of the rs-AlN

phase. The intensities and relative peak heights of the AlN₄ and AlN₆ signals correlate roughly with the XRD phase analysis of aluminium nitride (wurtzitic/ rocksalt).

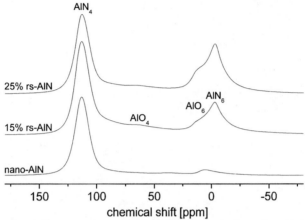

Figure 2. ^{27}Al MAS NMR spectra of nano-AlN and shock-synthesised samples with 15% rs-AlN and 25% rs-AlN.

Because the AlN₆/AlO₆-coordination overlay each other, MQMAS measurements were carried out to separate the chemical shift and the quadrupole interaction of the ^{27}Al-nucleus (see figure 3). The quadrupole interaction of ^{27}Al influences the chemical shifts and causes a line broadening in the 1D-spectrum. On the isotropic (F1) axis of the 2D-plot almost symmetrical peaks with the isotropic chemical shifts can be seen (see figure 3a). The splitting of the AlO₆/AlN₆-peak is shown in figure 3b. The isotropic shifts are 20 ppm for AlO₆ and 2 ppm for AlN₆. The AlO₆-group indicates a quadrupole-splitting in F2, in contrast to the AlN₆- and AlN₄-peaks. The excitation of the AlN₄-environment is more effective under the chosen experimental conditions, thus the intensity for AlN₄ is higher than for the AlN₆, though the phase content is less. The isotropic chemical shift for AlN₄ is 118 ppm. The peak for AlO₄ is too weak and therefore cannot be detected here. The crystallinity of the shocked samples is obviously poor. As the AlN peaks show no quadrupole-splitting, but an evenly distributed peak broadening due to less long-range order. This insight is confirmed by X-ray analysis, depicting a small amorphous portion at 2Θ = 10-30°.

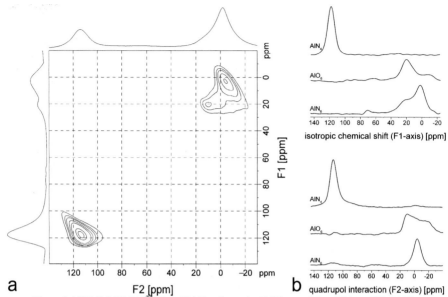

Figure 3. MQMAS NMR (^{27}Al) at 25 kHz of sample AN33 containing 48% rs-AlN. (a) The quadrupole interaction is drawn in the F2-dimension and isotropic chemical shift on F1-axis. At the F2-axis the 1D-NMR spectrum is plotted. (b) Sections through peak maxima of AlN$_4$-, AlO$_6$- and AlN$_6$-polyeder in F1- and F2-axis.

Thermal Stability

When heating the shock-synthesised sample with 25% rs-AlN in air, a small decreases of the sample mass with a total weight loss of 2.3% can be observed (see figure 4). Because of the hygroscopicity of nano-AlN, the powder gradually reacts to amorphous Al(OH)$_3$ when it is exposed to air. Besides, water is physically adsorbed at the surface. At heating to 500°C these chemically and physically bonded water is stepwise evaporised to form AlOOH and at higher temperatures Al$_2$O$_3$. Two larger mass increases of 8.5% between 530-940°C and of 5.4% between 940-1200°C can be detected, which are accompanied with a positive heat flow, indicating the stepwise oxidation of the components. In the first step starting at about 530°C corundum and γ-AlON is formed from AlN. Also the high-pressure phase rs-AlN reacts in this way and shows no preferable oxidation resistance compared to w-AlN. At higher temperatures up to 1200°C all components react to the stable corundum phase (α-Al$_2$O$_3$).

When heating the sample to 1300°C in vacuum a total weight loss of 2.5% can be detected. The major mass decrease of 1.8% occurs between 50-380°C and another 0.5% up to 900°C. The phase analysis after heating in vacuum demonstrates that up to 1100°C no notable compositional changes occur. Between 1100-1300°C the rs-AlN reconverts to some extent to w-AlN. Coincidental the content of corundum and γ-AlON increases.

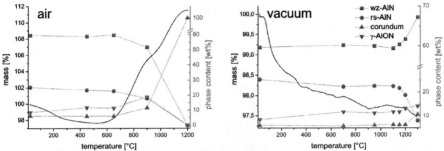

Figure 4. TG of sample AN22 (60% w-AlN, 25% rs-AlN, 6% corundum and 9% γ-AlON) heated at 10K/min in air and vacuum. Grey curves denote results of the ex-situ XRD analysis.

In the MS spectra (see figure 5) of the nanosized w-AlN starting powder heated in vacuum the highest ion currents are detected for the m/z-ratios 14 (N$_2$ - not shown), 16 (O$_2$ and NH$_2$), 17 (OH and NH$_3$), 18 (H$_2$O), 28 (N$_2$, CO, AlH) and 32 (O$_2$ – not shown). A massive weight loss and sharp peak for m/z 15-19 at about 100°C indicates the release of superficially bonded water, hydroxide-groups and NH$_x$-fragments. Up to 500°C further water, hydroxides and oxygen, but also N$_2$ (m/z 14/28) and NH (m/z 15) degasses. Nitrogen (m/z 28) and Al-H-compounds (m/z 28-30 – not shown) are released in the whole temperature range. In a wide temperature-region with the maximum at about 400°C Cl (m/z 35/37) and HCl (m/z 36/38 – not shown) is released from the samples. For the nanocrystalline w-AlN the m/z 27 signal (Al) show peaks below 100°C, at about 250°C, 420°C, 550°C and the tallest at 870°C. It is supposed that different Al-components (e.g. amorphous AlCl$_3$ and Aluminium hydroxides) were formed during the synthesis of the nanocrystalline w-AlN starting material leading to multiple decomposition reactions. Until about 800°C a variety of other fragments are degassing showing the high contamination and reactivity of the nanopowder. At temperatures higher than 900°C no considerable weight loss can be detected. The overall weight loss for the nano-AlN powder is about 6.6%. In comparison, the submicron w-AlN powder did not show such a immense degassing caused by contamination and water absorption and is comparatively clean. The weight loss here is only 0.6%.

All degassing until 800°C, including hygroscopic bonded water, described for the starting w-AlN material can be relocated and assigned in the TG-MS signal of the synthesised samples, but with a comparatively lower intensity. The shock-synthesised samples (by way of example AN22) also absorb water and oxygen at the particle surfaces and react to hydroxides, but the release at m/z 16-18 is comparatively higher at lower temperatures (<100°C) than at temperatures >200°C. The overall weight loss of sample is only about 2.5% indicating that a large fraction of the volatile components already removed during high-pressure synthesis. Also Al (m/z 27) reveal a similar behaviour like the starting powder, but more gas is released already at <100°C and at 250°C, 400°C and 550°C than at 900°C, indicating that the portion of volatile fragments is higher than that of more strongly bound. Chlorine (m/z 35/37) and HCl (m/z 36/38 – not shown) degasses at lower temperatures in a wide region with a maximum at 300°C showing basically the decomposition of ammonium chloride. Almost no nitrogen (m/z 14/28) is unleashed over the whole temperature-range.

When the rocksalt structure collapses at temperatures between 1100-1300°C, the amounts of Al$_2$O$_3$ and γ-AlON increase. It is supposed that the rs-AlN phase contains some residual oxygen, so that rs-AlN transforms to w-AlN and coincidental new oxygen-bearing phases are generated. The total mass loss at temperatures of 1100-1300°C is only 0.25% and comparatively low. No degassing takes place in this temperature range, which is significantly different from that of the starting powder. Whether the rocksalt structure of AlN can incorporate some oxygen, in which amount and which effect this has on the stability of the structure has to been examined with further experiments. The assumption that the rs-AlN phase contains oxygen is confirmed by the fact that the oxygen-content calculated from the crystalline phases is lower than the total oxygen-content determined with elementary analysis.

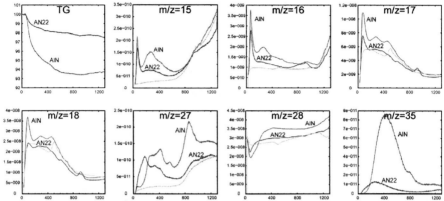

Figure 5. TG plot and MS spectra of nano-AlN starting powder and shock-synthesised sample AN22 (60% w-AlN, 25% rs-AlN, 6% corundum and 9% γ-AlON). The light grey curve is the baseline.

Chemical Stability

The w-AlN of the sample product partly reacts with water at room temperature (also in diluted acids and bases) to insoluble γ-AlO(OH) (boehmite) causing a mass increase (see table 1). For the reaction with HCl a small amount of ammonium chloride, respectively ammonium sulphate and nitrate for the reaction with H_2SO_4 and HNO_3/nitrohydrochloric acid is formed. The w-AlN is more stable towards concentrated H_3PO_4 than other acids, caused by the formation of a phosphate-based protective layer on the AlN particles[7]. The measured mass increase can again be explained for most part by the formation of boehmite (theoretical increase of 46% per mole AlN). The calculated mass increase from phase analysis for the reaction with H_2SO_4 and H_3PO_4 shows much smaller values, indicating that additional non-crystalline phases were formed upon etching. Corundum is extremely stable against all the used acids, whereas the γ-AlON dissolves slightly. The high-pressure phase rs-AlN is stable against water, bases and all used diluted and concentrated acids and shows almost no solubility. Nevertheless, further long term experiments have to been carried out.

Table 1. Mass increase after solubility-test with water, caustic soda and different acids.

Chemical	Concentration	Mass change Δm [%]		Comment
		Gravimetric	Phase analysis	
Distilled water	-	+17	+16	
NaOH	5%	+20	+18	
NaOH	20%	+20	+17	
HCl	1 molar	+19	+16	<1% NH$_4$Cl formed
H$_3$PO$_4$	>85%	+35	+6	turquoise sample powder, non-crystalline phases formed?
H$_2$SO$_4$	concentrated	+52	+8	no boehmite formed, 3% (NH$_4$)$_2$SO4
HNO$_3$	concentrated	+34	+22	1% NH$_4$NO$_3$
Nitrohydrochloric acid	3HCl:1HNO$_3$	+19	+17	

CONCLUSION

Aluminium nitride with rocksalt structure was characterised with ^{27}Al Solid State NMR showing a signal of the symmetrical AlN$_6$-peak with an isotropic chemical shift of about 2 ppm overlapped by the signal of the AlO$_6$-group. The rs-AlN reacts in air to corundum beginning at about 530°C and show no special oxidation resistance in comparison to w-AlN. In vacuum the rs-AlN is stable up to 1100°C and reconverts to w-AlN at higher temperatures. The rise of the corundum and γ-AlON content during this phase transformation is giving evidence that oxygen could have an important influence on phase transition and stability. The rs-AlN is chemical stable against water, diluted and concentrated acids and bases.

REFERENCES

[1]O. Ambacher, Growth and applications of Group III-nitrides, *Journal of Physics D: Applied Physics*, **31**, 2653-2710 (1998)

[2]H. Vollstädt, E. Ito, M. Akaishi, S. Akimoto, and O. Fukunaga, High Pressure Synthesis of Rocksalt Type of AlN, *Proceedings of the Japan Academy, Ser. B*, **66**, 7-9 (1990).

[3]M. Ueno, A. Onodera, O. Shimomura, and K. Takemura, X-ray observation of the structural phase transition of aluminum nitride under high pressure, *Physical Review B*, **45**, 10123-6 (1992).

[4]Q. Xia, H. Xia, and A.L. Ruoff, Pressure-induced rocksalt phase of aluminum nitride: A metastable structure at ambient condition, *Journal of Applied Physics*, **73**, 8198-200 (1993).

[5]H. Vollstädt, and H. Recht, Verfahren zur Herstellung von kubischem Aluminiumnitrid [DD000000292903A5], *Patent* (1991).

[6]K. Keller, T. Schlothauer, M. Schwarz, G. Heide, and E. Kroke, Shock wave synthesis of aluminium nitride with rocksalt structure, *High Pressure Research*, **31**, (2011), in press.

[7]M. Oliveira, S. Olhero, J. Rocha, and J.M.F. Ferreira, Controlling hydrolysis and dispersion of AlN powders in aqueous media, *Journal of Colloid and Interface Science*, **261**, 456-63 (2003).

Modeling

ENVIRONMENTAL BARRIER COATING (EBC) DURABILITY MODELING; AN
OVERVIEW AND PRELIMINARY ANALYSIS

A. Abdul-Aziz, R. T. Bhatt, J. E. Grady and D. Zhu
NASA Glenn Research Center
Cleveland, Ohio 44135

ABSTRACT

A study outlining a fracture mechanics based model that is being developed to investigate crack
growth and spallation of environmental barrier coating (EBC) under thermal cycling conditions is
presented. A description of the current plan and a model to estimate thermal residual stresses in the
coating and preliminary fracture mechanics concepts for studying crack growth in the coating are
also discussed. A road map for modeling life and durability of the EBC and the results of FEA
model(s) developed for predicting thermal residual stresses and the cracking behavior of the coating
are generated and described. Further initial assessment and preliminary results showed that
developing a comprehensive EBC life prediction model incorporating EBC cracking, degradation
and spalling mechanism under stress and temperature gradients typically seen in turbine
components is difficult. This is basically due to mismatch in thermal expansion difference between
sub-layers of EBC as well as between EBC and substrate, diffusion of moisture and oxygen though
the coating, and densification of the coating during operating conditions as well as due to foreign
object damage, the EBC can also crack and spall from the substrate causing oxidation and recession
and reducing the design life of the EBC coated substrate.

INTRODUCTION

The need for increasing the cycle efficiency and reducing NOx emission and noise in future
gas turbine engines has promoted development of fiber reinforced ceramic matrix composites
(FRCMC), specifically SiC fiber reinforced SiC matrix composites (SiC/SiC) for the structural hot
section components [1]. These materials not only are lighter, but also can operate at about 200 $^\circ$C
hotter than current metallic materials. In dry air, these materials form a protective silica layer on the
surface and thus are stable at temperatures up to 1300^0C for long-term applications. However in
combustion environments containing moisture, these materials show surface recession due to
decomposition of protective silica and active oxidation of the substrate, and the rate of recession
also increases with increasing gas velocity [2-6]. To reduce moisture diffusion and surface recession
issues, a variety of environmental barrier coatings (EBC) have been developed. The upper-use
temperatures of these EBCs vary between 1200 to 1650 ^{0}C depending on the composition [7-
11].These coatings also recess with time at temperatures > 1400^0C, but rate of recession is an order
of magnitude or two lower than that of the uncoated substrate at the same temperature. For
successful use of CMCs in turbine components, these coatings need to be prime reliant which
means that life of the coated CMCs depends on coating and not the uncoated substrate life. In other
words if the coating spalls off, the CMC substrate life is drastically reduced .This design concept is
directly in contrast with the design concept of thermal barrier coated metal parts in current engines
where design life of TBC coated metal is based on the life of uncoated metal under engine operating
conditions. Currently, failure mechanisms of EBC under steady state and transient conditions are
not fully understood or models to predict life of EBC do not exist. Therefore, determining the
factors influencing EBC stability, mechanisms of EBC degradation and failure modes, and
developing models to predict durability and life of an EBC alone and CMC after the EBC failure are
critical for introducing EBC coated CMC components in engines.

The overall goals of this study are several: (1) To generate thermo-mechanical and crack growth properties for as-coated model EBC system; (2) To model thermal residual stress in the as-processed EBC; (3) To model failure behavior of EBC under 3-point flexural tests; (4) To develop model to predict conditions under which cracks in the coating are deflected and penetrated through-the-thickness; (5) To monitor crack initiation and crack growth and spallation of the coating with thermal cyclic conditions with and without imposed mechanical stress via NDE and electro-optical methods; and (6) To develop and validate fracture mechanic based FEA models to predict life of the coating. However, in this study, a road map for modeling life and durability of the EBC and the results of FEA model(s) developed for predicting thermal residual stresses in the coating and the cracking behavior of the coating are discussed.

EBC PROCESSING METHODS

Generally an EBC system is made up out of two or more layers of coating, in which each layer serves a specific purpose. The composition, total thickness and the processing methods for EBC can vary depending on the component type and intended life requirements of the coating. Currently EBC system is deposited by four methods: (a) Air plasma spray (APS); (b) Electron beam assisted physical vapor deposition (EB-PVD); (c) sputtering; and (d) slurry deposition. The microstructure, elastic and fracture properties of the coating deposited by above methods can vary significantly. A typical microstructure of an APS coated EBC on SiC/SiC substrate shown in figure 1 indicates crack pattern on the top coat, and pore size and micro-crack distribution in sub-layers of the EBC system.

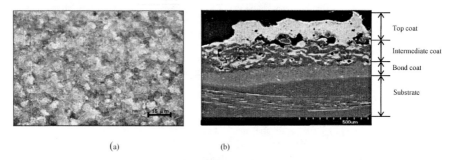

(a)   (b)

Figure 1. Typical microstructure of plasma sprayed EBC on SiC/SiC composites:
a) Top view (optical micrograph), and (b) cross-sectional view with SEM.

Since the thermal and mechanical properties of the coatings and the substrate are different and since the sub-layers are applied at different processing temperatures, thermal residual stresses are generated in the coating. Under engine operating conditions, oxygen and moisture permeates through the coating, and the coating thickness changes due to densification. Therefore microstructure, phase composition, thermal residual stresses, thermal and mechanical properties of the coating continuously changes with exposure time. Depending on the magnitude and nature of residual stresses, the coating may crack after deposition or after exposing the coated substrate to turbine operating conditions.

EBC FAILURE MODES

EBC can spall-off from the substrate during thermo-mechanical cycling or due to foreign object damage and failure modes can vary depending on the exposure conditions. Under thermal cycling condition without stress two failure modes have been observed [12-13]:

(1) Formation of through-the-thickness cracks spanning from the top coat to bond/intermediate coat interface or to bond coat/substrate interface followed by formation and growth horizontal crack along the respective interfaces, linkage of the cracks and eventual spallation of the coating. Also, coating interface reactions and internal pore formation further accelerate coating spallation. Schematics of the generalized failure mechanisms are shown in figure 2.

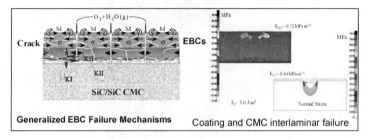

Generalized EBC Failure Mechanisms   Coating and CMC interlaminar failure

Figure 2. Failure mechanisms in EBC

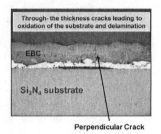

Perpendicular Crack

Figure 3. Optical photographs of the cross-sections of coated silicon nitride thermally cycled at $1382^{0}C$ for 100hrs in a moisture containing environment

Stress contour from a finite element based analysis shows the inter-laminar failure of the coating and CMC stress distribution due to thermal and mechanical loading, figure 2. Observation of the above mechanism in an EBC coated Si_3N_4 subjected to thermal cycling test in a moisture environment at $1382^{0}C$ for 100hrs is shown figure 3.

(2) Diffusion of oxygen and moisture through the coating to the substrate/bond coat interface followed by oxidation of substrate, formation of pore, linkage of the pore and eventual spallation of the coating. This mechanism is observed when the thermal expansion of the substrate

closely matched with that of the coating. An example of this mechanism is shown in figure 4 which shows pore formation at the substrate/coating interface leading to coating delamination.

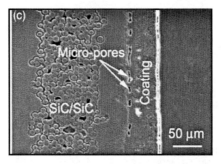

Figure 4. SEM photograph of a cross-section of mullite/Gd$_2$SiO$_5$ coated MI–SiC/SiC composite coupons after 200 thermal cycles at 1350°C in a moisture environment.

EBC MODEL

As indicated above, the EBC coated CMC substrate when exposed to flowing combustion environment containing moisture several things happens to the coating simultaneously: EBC recesses, moisture and oxygen from the flowing environment diffuse through the coating, atomic diffusion occurs between the sub-layers of the coating, coating sub-layers densifies and thermal residual stresses within the coating change. As a result, the microstructure and physical and mechanical properties of the coating continuously evolves with exposure time at a fixed temperature or during temperature/stress cycles. Due to thermal expansion mismatch between sub-layers of the EBC and between the EBC and the substrate as well as continuously evolving microstructure of the EBC, spallation of the EBC occurs by two different mechanisms as stated above. Therefore development of a life prediction model requires the constitutive properties of the as-deposited EBC layers and variation of these properties as the microstructure of the coating evolves during exposure with and without stress, failure modes of the EBC, crack initiation, propagation, and spallation times, recession rate of the coating, magnitude of the thermal residual stress in the as-processed EBC and during exposure, diffusion rate oxygen and moisture within the coating as shown in bubble diagram of figure 5. For predicting life of the EBC coated CMC turbine components under engine operating conditions, a comprehensive model incorporating all the modeling elements shown in the bubble diagram is needed.

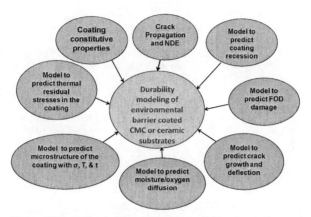

Figure 5. Durability modeling of environmental barrier coating.

Because of the complexity of modeling life of EBC, a simplified fracture mechanics model will be developed initially without incorporating diffusional phenomena, coating recession and sintering effects. This model when fully developed should predict thermal residual stresses, crack initiation, propagation, and spallation (delamination) behavior of EBC.

A model to predict thermal residual stress in the coating has been developed and results have been documented and discussed in reference [14]. Currently, the crack propagation behavior in an EBC coated CMC with single edge notch subjected to external stress and thermal cycling is being studied by computed tomography (CT) to further perform the combined NDE/FEA modeling.

ANALYTICAL MODELING

Two options are being considered to model the EBC durability; (1) **physics based approach** that focus on the layout described in figure 5 and shall include supportive experimental initiatives to complement and verify the analytical data. It is expected to record the baseline microstructure and crystallinity of EBC and thickness variation of individual EBC layers and expose the EBC coated CMC in air and moisture environments at temperatures up to 1400^0C for up to 500 hours and thermal cycle it without stress in a moisture environment. Crack propagation phenomenon will be addressed via inducing an artificial crack or an indentation in the EBC. Crack propagation is to be conducted under stress and thermal cycling conditions.

The ultimate goal is to establish a stress-strain failure baseline using experimentally driven materials data and examine the impingement of the crack behavior at the interface of the EBC/CMC stack to enable deriving a correlation leading to establishing when the crack is arrested or advanced by either penetrating the interface or deflecting into the interface. The analytical model will attempt to account for environmental and operational conditions such as oxidation, interface roughness, creep, sintering, failure assessments under thermal cycling, and failure assessments under thermo-mechanical fatigue cycling, impact modeling of foreign object damage (FOD). Modeling is to include both isothermal and thermal gradient loading conditions. Life prediction will follow and it will be based upon acquiring optimum data and results. Such data will be confined to obtaining information relevant to the constitutive properties of the substrate and the coating, influence of time, temperature, stress and moisture on the coating, coating recession information, diffusion rates of species from the substrate to the coating and within the coating, measured values of the residual stresses in the coating, coating deposition conditions and coating failure criteria and damage information. Finite element will be the base numerical method used for these studies along with supportive commercial software such as MSC/MARC [14], MSC/Patran [15] and Abaqus [16].

(2) **NDE-FEA** approach is to be performed along the proposed physics based methodology to correlate imaging data obtained from sources such as computer tomography (CT), X-rays etc…and apply reverse engineering principals by using the finite element computational method to model tested specimen under thermal and mechanical loading conditions. Segmented images can be exported as stereographic (STL) or Initial Graphics Exchange Specification (IGES) files for computer aided design (CAD) geometry and finite element analysis. This methodology will allow using a broad range of image visualization, processing and segmentation tools available to incorporate and utilize state of the art facilities at NASA GRC to execute NDE related tests and assessments to evaluate CMC/EBC coated specimens and permit a combined NDE-computational verification.

Unique meshing capabilities for creating 3D image based models of supreme accuracy and sophistication is very feasible [17]. The modeling plan is to consist of; (1) conducting computer tomography scans using baseline-unloaded specimen; (2) test load the specimen under three point or four bend conditions at various loads and re-scan the loaded specimen and assess damage and structural changes; (3) thermal cycle tested specimen at three time intervals of 10, 50 and 100 hours and re-scan at different time intervals; (4) perform image processing, segmentation and 3-D volume rendering of tested specimen. Crack measurements and other micro-deformities if any for each test case will be also performed for crack dimensioning purpose ; (5) The final segment of this modeling approach will include analyzing the as tested specimen geometry including all structural deformities and defects for each test case using the NDE-CT scan information. A follow up diagram will incorporate replicating the outlined plan using a specimen with an artificially induced crack on the top surface (at the top coating) not to exceed 20 mm.

However, it should be noted that due a certain design limitations the CT scanning will only embrace a specified region of the test specimen with a designated area size and location. This is due to the fact that the highest and desirable CT scan resolution of ~ 5 μm can only be achieved if a small area is targeted. As a result, this will hinder the ability to analyze the entire specimen and will only enable modeling the area of interest. Nevertheless, suitable assumptions and application of appropriate boundary conditions should compensate for these limitations.

PRELIMINARY RESULTS

A general finite element model was developed for estimating the residual stresses in coated substrates based on the known processing conditions of the coating, the specimen geometry of the coated substrate, and the thermo-mechanical properties of the coated layers and the substrate. The analytical modeling assumed that the substrate is maintained at 1200^{0}C during deposition of the plasma spray coating and then cooled to room temperature. The analysis performed did not include effects of conditions such as creep and fracture propagation due to simplicity and unavailability of other supportive data. However, plans to incorporate these effects and other relevant ones are planned to be addressed in the proposed analytical modeling presented in this article. Figure 7 shows the predicted residual stresses for the EBC system shown in Table I. The material property data for the coating constituents and the substrate and details of the FEA model used for predicting thermal residual stresses are from reference [18]. Part (a) of figure 7 represents the predicted in plane and through-the-thickness thermal residual stresses in the coating as a function of the normalized distance.

Table I. Analytical cases and coating systems considered.

System	Coating Thickness
Top Coat- Barium Strontium aluminum silicate (BSAS)	75 μm
Intermediate coat Mullite + (BSAS)	75 μm
Bond coat-silicon	75 μm
SiC-SiC substrate	3 mm

Included in the results for comparison is the fracture strength range of standalone EBC layers (the hatched band). It is clear that the in-plane or X-Y stresses are much higher than the through-the-thickness stresses; in fact the through-the-thickness stresses are negligible. The intermediate coat experienced higher stress than the bond or the topcoats mainly due to a greater difference in the coefficient of thermal expansion (CTE) mismatch between the intermediate coat and the substrate. Figure 7 (b) shows the predicted in plane and through-the-thickness thermal residual stresses in the substrate. In general, the in-plane stresses at the substrate/coating interface is mostly compressive and small, but with increasing distance away from the interface the stresses decreased gradually and then reached a value close to zero near the mid section of the substrate. This is seems to be applicable to the in-plane stresses along the Y-axis. See Figure 6 for direction axis and coating-substrate geometry layout.

The coating and substrate thickness is designated by the symbols l_c and l_s and they represent the incremental thickness of the coating and the substrate respectively, while L_c and L_s are the total thickness of each entity. The location at the substrate/coating interface is represented by a ratio of zero for l_c/L_c, while a ratio of unity represents maximum coating thickness. Similar convention is used concerning the ratio of the substrate thickness arrangement i.e. l_s/L_s corresponds to maximum substrate thickness. In figure 7(a) and 7(b) the stresses along the Z-axis and X-Y axes are shown as a function of the normalized distance of the thickness (Z-axis-direction), for both the substrate and the coating. The normalized distance is defined as the ratio of the length increment divided by the total thickness.

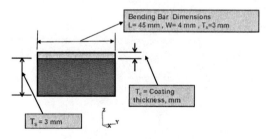

Figure 6. Two dimensional representation of the coating-substrate layout showing through thickness and in-plane directional axis.

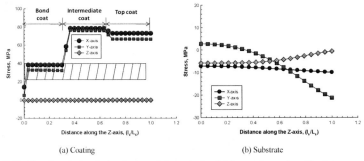

(a) Coating   (b) Substrate

Figure 7. Predicted variation in in-plane and through-the-thickness thermal residual stresses with normalized Z distance for coated SiC/SiC composite.

Also noticed is that the predicted stresses in the EBC are much greater than the fracture strength of the coating layers which suggests that the EBC coating on the SiC/SiC composite is possibly micro/macro cracked in the as-coated condition. Therefore actual residual stresses in the coating may be much smaller than the predicted stress, and more details concerning these results and analyses can be found in reference [18].

CONCLUSIONS

This study described two combined analytical/experimental options to model the EBC durability. Option one is expressed in a bubble chart illustrating the technical modeling approach and listings of all the corresponding elements that have to be addressed in order to asses and model the durability issues of the EBC. The second option is portrayed in an NDE-FEA modeling approach using imaging data obtained from sources such as computer tomography (CT) and advanced X-ray systems. Reverse engineering principals via the finite element computational method is enlightened and the modeling procedure to analyze tested specimen under thermal and mechanical loading conditions is also described.

Results obtained from the finite element analysis relating to the residual stress effects showed that z-direction (through thickness) stresses are small and negligible, but maximum in-plane stresses can be significant depending on the composition of the EBC constituent layer and the distance from the substrate.

Future work will include recording the baseline microstructure and crystallinity of EBC and thickness variation of individual EBC layers and expose the EBC coated CMC in air and moisture environments at temperatures up to $1400^{0}C$ for up to 500 hours and thermal cycle it without stress in a moisture environment. Crack propagation phenomenon is to be investigated via inducing an artificial crack or an indentation in the EBC and analyses will be conducted under stress and thermal cycling conditions.

REFERENCES

[1] D. Brewer, "HSR/EPM Combustion Materials Development Program" *Mat. Sci. and Eng.*, **A261**, p284-291, (1999).

[2] E.A. Gulbransen and S.A. Jansson, "The High-Temperature Oxidation, Reduction, and Volatilization Reactions of Silicon ad Silicon Carbide," *Oxidation of Metals*, **4**[3] p181-201, (1972)

[3] P.J. Jorgensen, M.E. Wadsworth and I.B. Cutler, "Oxidation of Silicon Carbide," *J. Am. Ceram. Soc.*, **42** (12): p613-616 (1959).

[4] J.L. Smialek, R.C. Robinson, E.J. Opila, D.S. Fox, and N.S. Jacobson, "SiC and Si_3N_4 Recession Due to SiO_2 Scale Volatility under Combustor Conditions," *Adv. Composite Mater*, **8** [1], 33-45 (1999).

[5] K. L. More, P. F. Tortorelli, and L. R. Walker, "Effects of High Water Vapor Pressures on the Oxidation of SiC-Based Fiber-Reinforced Composites," *Mater. Sci. Forum*, **369–372**, 385–94 (2001).

[6] K. L. More, P. F. Tortorelli, and L. R. Walker, "High-Temperature Stability of SiC-Based Composites in High-Water-Vapor-Pressure Environments," *J. Am. Ceram. Soc.*, **86** [8] 1272–81 (2003)

[7] K.N. Lee, "Current status of environmental barrier coatings for Si-based ceramics," *Surface and Coatings Technology*, 133-134, **1**, 1-7 (2000)

[8] K.N. Lee, D.S. Fox, R.C. Robinson, and N.P Bansal, "Environmental Barrier Coatings for Silicon-Based Ceramics. High Temperature Ceramic Matrix Composites," High Temperature Ceramic Matrix Composites, Edited by W. Krenkel, R. Naslain, H. Schneider, Wiley-Vch, Weinheim, Germany, 224-229 (2001).

[9] K.N. Lee, D.S. Fox, N.P Bansal, "Rare Earth Environmental Barrier Coatings for SiC/SiC Composites and Si_3N_4 Ceramics," *J. Eur. Ceram. Soc.*, **25** [10] 1705-1715 (2005).

[10] D.M. Zhu, N.P. Bansal and R.A. Miller, "Thermal Conductivity and Stability of HfO2-Y2O3 and $La_2Zr_2O_7$ Evaluated for 1650°C ," *Advances in Ceramic Matrix Composites IX*, N.P. Bansal, J.P. Singh, W.M. Kriven and H. Schnneider (eds.), The American Ceramic Society, Westerville, Ohio, **153**, 331-343, (2003).

[11] D.M. Zhu, R.A. Miller and D.S. Fox, "Thermal and Environmental Barrier Coating Development for Advanced Propulsion Engine Systems," NASA TM-2008-215040, January (2008).

[12] D.M. Zhu, and R.A Miller, "Thermal Conductivity and Elastic Modulus Evolution of Thermal Barrier Coatings under High Heat Flux Conditions" *J. Therm. Spray Tech.*, **9**, [2], 175-180 (2000).

[13] S. Ramasamy , S.N. Tewari, K.N. Lee, R.T. Bhatt, D. S. Fox, "Mullite–gadolinium silicate environmental barrier coatings for melt infiltrated SiC/SiC composites," Surface & Coatings Technology 205 (2011) 3578–3581.

[14] MSC/PATRAN Graphics and Finite Element Package. The MacNeal-Schwendler Corporation, 2007, Costa Mesa, CA.

[15] Marc General Purpose Finite Element Analysis Program, the MacNeal-Schwendler Corporation, 2007, Costa Mesa, CA.

[16]) Abaqus Finite Element Code, Abaqus Inc., 534 Forest Avenue, Palo Alto, CA 94301.

[17] *ScanIP*, Simpleware Ltd, Innovation Centre, University of Exeter, Rennes Drive, EX4 4RN, UK.

[18] Ali Abdul-Aziz and Ramakrishna T. Bhatt, "Modeling Of Thermal Residual Stress In Environmental Barrier Coated Fiber Reinforced Ceramic Matrix Composites"; *Journal of Composite Materials, September 21, 2011; 0021998311414950, first published on September 21, 2011*

Author Index